Errata

Advances in Atomic, Molecular, and Optical Physics
Volume 46

THE REFERENCES FOR THE FIGURES LISTED BELOW WERE
INADVERTENTLY OMITTED:

Figure 2, page 248   (From Kessel *et al.*, 1978.)

Figure 3, page 249   (From Dworetsky *et al.*, 1967, where details are given.)

Figure 4, page 250   (From Tolk *et al.*, 1976.)

Advances in
ATOMIC, MOLECULAR, AND OPTICAL PHYSICS

VOLUME 46

Editors

BENJAMIN BEDERSON
New York University
New York, New York

HERBERT WALTHER
Max-Plank-Institut für Quantenoptik
Garching bei München
Germany

Editorial Board

P. R. BERMAN
University of Michigan
Ann Arbor, Michigan

M. GAVRILA
F. O. M. Institut voor Atoom-en Molecuulfysica
Amsterdam, The Netherlands

M. INOKUTI
Argonne National Laboratory
Argonne, Illinois

W. D. PHILIPS
National Institute for Standards and Technology
Gaithersburg, Maryland

Founding Editor

SIR DAVID BATES

Supplements

1. *Atoms in Intense Laser Fields*, Mihai Gavrila, Ed.
2. *Cavity Quantum Electrodynamics*, Paul R. Berman, Ed.
3. *Cross Section Data*, Mitio Inokuti, Ed.

ADVANCES IN

ATOMIC, MOLECULAR, AND OPTICAL PHYSICS

Edited by

Benjamin Bederson
DEPARTMENT OF PHYSICS
NEW YORK UNIVERSITY
NEW YORK, NEW YORK

Herbert Walther
UNIVERSITY OF MUNICH AND
MAX-PLANK INSTITUT FÜR QUANTENOPTIK
MUNICH, GERMANY

Volume 46

ACADEMIC PRESS
A Harcourt Science and Technology Company

San Diego   San Francisco   New York
Boston   London   Sydney   Tokyo

This book is printed on acid-free paper.

Copyright © 2001 by ACADEMIC PRESS

All Rights Reserved.
No part of this publication may be reproduced or transmitted in any form or by any means, electronic or mechanical, including photocopy, recording, or any information storage and retrieval system, without permission in writing from the Publisher.

The appearance of the code at the bottom of the first page of a chapter in this book indicates the Publisher's consent that copies of the chapter may be made for personal or internal use of specific clients. This consent is given on the condition, however, that the copier pay the stated per copy fee through the Copyright Clearance Center, Inc. (222 Rosewood Drive, Danvers, Massachusetts 01923), for copying beyond that permitted by Sections 107 or 108 of the U.S. Copyright Law. This consent does not extend to other kinds of copying, such as copying for general distribution, for advertising or promotional purposes, for creating new collective works, or for resale. Copy fees for pre-2000 chapters are as shown on the title pages. If no fee code appears on the title page, the copy fee is the same as for current chapters.
1049-250X/01 $35.00

Explicit permission from Academic Press is not required to reproduce a maximum of two figures or tables from an Academic Press chapter in another scientific or research publication provided that the material has not been credited to another source and that full credit to the Academic Press chapter is given.

Academic Press
A Harcourt Science and Technology Company
525 B Street, Suite 1900, San Diego, California 92101-4495, USA
http://www.academicpress.com

Academic Press
Harcourt Place, 32 Jamestown Road, London NW1 7BY, UK
http://www.academicpress.com

International Standard Serial Number: 1049-250X

International Standard Book Number: 0-12-003846-3

PRINTED IN THE UNITED STATES OF AMERICA
01 02 03 04 05 06 SB 9 8 7 6 5 4 3 2 1

Contents

CONTRIBUTORS vii

Femtosecond Quantum Control
Tobias Brixner, Niels H. Damrauer, and Gustav Gerber

I. Introduction 1
II. Experimental Techniques 6
III. One-Parameter Quantum Control in the Gas Phase 12
IV. Adaptive Femtosecond Pulse Shaping 23
V. Many-Parameter Quantum Control in the Gas Phase 29
VI. Many-Parameter Quantum Control in the Liquid Phase 40
VII. Summary and Outlook 46
VIII. Acknowledgments 48
IX. References 48

Coherent Manipulation of Atoms and Molecules by Sequential Laser Pulses
N. V. Vitanov, M. Fleischhauer, B. W. Shore, and K. Bergmann

I. Introduction 57
II. Principles of Coherent Excitation 67
III. Three-State STIRAP: Theory 82
IV. Three-State STIRAP: Experiments 103
V. STIRAP-Like Population Transfer in Multistate Chains 117
VI. Adiabatic Momentum Transfer 126
VII. Branched-Chain Excitation 135
VIII. Population Transfer via a Continuum of Intermediate States 143
IX. Extensions and Applications of STIRAP 156
X. Propagation Phenomena 162
XI. Applications of STIRAP in Quantum Optics and
    Quantum Information 173
XII. Summary and Outlook 179
XIII. Acknowledgments 180
XIV. References 180

Slow, Ultraslow, Stored, and Frozen Light
*Andrey B. Matsko, Olga Kocharovskaya, Yuri Rostovtsev,
George R. Welch, Alexander S. Zibrov, and Marlan O. Scully*

I. Introduction 191
II. Group Velocity: Kinematics 193
III. Slow Light 196

v

IV. Ultraslow Light 209
V. Storing and Retrieving Quantum Information 214
VI. The Ultimate Slow Light: Frozen Light 219
VII. Applications 225
VIII. Conclusion 236
IX. References 237

Longitudinal Interferometry with Atomic Beams

S. Gupta, D. A. Kokorowski, R. A. Rubenstein, and W. W. Smith

I. Introduction 243
II. Interference in Ion–Atom Collisions 245
III. Differentially Detuned Separated Oscillatory Fields: DSOF 252
IV. Amplitude Modulation and Rephasing 255
V. Detecting Longitudinal Momentum Coherences 258
VI. Discussion of the Semiclassical Approximation 259
VII. Apparatus and Experimental Techniques 260
VIII. Density Matrix Deconvolution 263
IX. Search for Longitudinal Coherences in an Unmodified
    Atomic Beam 268
X. Conclusion 272
XI. Acknowledgments 273
XII. References 273

SUBJECT INDEX 277

CONTENTS OF VOLUMES IN THIS SERIES 289

Contributors

Numbers in parentheses indicate the pages on which the authors' contributors begin.

K. BERGMANN (55), Fachbereich Physik, Universität Kaiserslautern, 67653 Kaiserslautern, Germany

TOBIAS BRIXNER (1), Physikalisches Institut, Universität Würzburg, Am Hubland, D-97074 Würzburg, Germany

NIELS H. DAMRAUER (1), Physikalisches Institut, Universität Würzburg, Am Hubland, D-97074 Würzburg, Germany

M. FLEISCHHAUER (55), Fachbereich Physik, Universität Kaiserslautern, 67653 Kaiserslautern, Germany

GUSTAV GERBER (1), Physikalisches Institut, Universität Würzburg, Am Hubland, D-97074 Würzburg, Germany

S. GUPTA (243), Massachusetts Institute of Technology, Cambridge, MA 02139

OLGA KOCHAROVSKAYA (191), Department of Physics and Institute for Quantum Studies, Texas A&M University, College Station, TX 77843, and Institute of Applied Physics, RAS, 603600 Nizhny Novgorod, Russia

D. A. KOKOROWSKI (243), Massachusetts Institute of Technology, Cambridge, MA 02139

ANDREY B. MATSKO (191), Department of Physics and Institute for Quantum Studies, Texas A&M University, College Station, TX 77843

YURI ROSTOVTSEV (191), Department of Physics and Institute for Quantum Studies, Texas A&M University, College Station, TX 77843

R. A. RUBENSTEIN (243), Massachusetts Institute of Technology, Cambridge, MA 02139

MARLAN O. SCULLY (191), Department of Physics and Institute for Quantum Studies, Texas A&M University, College Station, TX 77843, and Max-Planck-Institut für Quantenoptik, D-85748 Garching, Germany

B. W. SHORE (55), Lawrence Livermore National Laboratory, Livermore, CA 94550

W. W. SMITH (243), Physics Department, University of Connecticut, Storrs, CT 06269-3046

N. V. VITANOV (55), Helsinki Institute of Physics, PL 9, 00014 University of Helsinki, Finland

GEORGE R. WELCH (191), Department of Physics and Institute for Quantum Studies, Texas A&M University, College Station, TX 77843

ALEXANDER S. ZIBROV (191), Department of Physics and Institute for Quantum Studies, Texas A&M University, College Station, TX 77843, and P. N. Lebedev Institute of Physics, RAS, Moscow, 117 924, Russia

Advances in
ATOMIC, MOLECULAR, AND OPTICAL PHYSICS

VOLUME 46

FEMTOSECOND QUANTUM CONTROL

TOBIAS BRIXNER, NIELS H. DAMRAUER, and GUSTAV GERBER

Physikalisches Institut, Universität Würzburg, Am Hubland, D-97074 Würzburg, Germany

I. Introduction 1
II. Experimental Techniques. 6
   A. Overview 6
   B. Setup. 8
III. One-Parameter Quantum Control in the Gas Phase 12
   A. Introduction. 12
   B. Control by Pulse Energy. 13
   C. Control by Pulse Duration. 16
   D. Control by Pump-Probe Delay 17
   E. Control by Phase-Sensitive Excitation 19
   F. Control by Linear Chirp 21
IV. Adaptive Femtosecond Pulse Shaping. 23
   A. The Technology 23
   B. Experimental Examples 26
V. Many-Parameter Quantum Control in the Gas Phase 29
   A. Introduction. 29
   B. Testing the Technology with $Fe(CO)_5$ 30
   C. Selective Dissociation and Complexity Analysis with $CpFe(CO)_2Cl$ 34
   D. Accessing the "Problem of Inversion" with Atomic Calcium 37
VI. Many-Parameter Quantum Control in the Liquid Phase. 40
   A. Introduction. 40
   B. Experiment 41
   C. Results and Discussion. 43
VII. Summary and Outlook 46
VIII. Acknowledgments 48
IX. References. 48

I. Introduction

Controlling the outcome of chemical reactions—i.e., guiding the formation of desired products at the expense of starting materials and unwanted products—is a central theme in modern chemistry. At the very least, it is the basic application of a vast knowledge of chemical bonding and reactivity gained in the laboratory and in the study of molecular quantum mechanics and statistical mechanics. The traditional methods of control can be divided into those that exploit the thermodynamic properties of reactants and products and those that exploit the kinetic properties of the reactions themselves. In the first case, reactants are designed—for example, by exploiting chemical functionality—so that desired products are thermodynamically favored. One can then change macroscopic variables such as temperature, pressure, and concentration to exert control. Kinetic control concerns reactions in which substantial barriers exist to desired product channels (typically on ground-state adiabatic reaction surfaces), thus making rates of formation of wanted products too slow or competitive with side-product formation. Traditionally, chemists have dealt with such barriers—and thus with the problem of kinetic control—by designing appropriate catalysts.

In view of both thermodynamic and kinetic control, light can be thought of as having the potential to be a very powerful tool. For example, one can imagine using excited states to change the thermodynamic starting place of a reactant species or to overcome (or circumvent) barriers inhibiting desired reactions. Nonetheless, one has only to open a common organic or inorganic chemistry textbook to realize the basic shortcoming of photochemistry in achieving a prominent role in the methodology of synthetic control. There are, of course, a number of well-understood and used photochemical routes to product formation—especially those dealing with unsaturated organic precursors (Kan, 1966); however, these numbers pale in comparison to the volumes of literature associated with more traditional synthetic methods. The responsibility for this shortcoming lies in the complexity of the excited-state electronic structure of many molecular reactants and the speed at which energy is statistically distributed within the molecule and to its surroundings following photoexcitation. Using typical laboratory photochemical variables such as intensity and wavelength, product yields are difficult to predict and even more so to control.

The discovery and development of laser technology brought new promise to the field of photochemistry as a whole. It was initially believed that intense coherent light could be used for selective bond activation or cleavage reactions (Ambartsumian and Letokhov, 1977; Letokhov, 1977; Bloembergen and Yablonovitch, 1978; Zewail, 1980; Jortner et al., 1981). The basic idea was that the wavelength of the light could be tuned to be in resonance with the vibrational energy of a particular chemical bond and then quanta deposited until reaching the appropriate barrier height or the dissociation limit. However, with the exception of a few very special cases, the promise of this methodology was never realized because, for the vast majority of molecular systems, the localization of energy within a single vibrational mode is rapidly destroyed by intramolecular energy redistribution (Bloembergen and Zewail, 1984; Elsaesser and Kaiser, 1991; Nesbitt and Field, 1996; Gruebele and Bigwood, 1998; Wong and Gruebele, 1999). That is, the influence of the complex couplings among the different molecular modes was not properly accounted for.

While the problem of exerting control over the outcome of photochemical reactions is a difficult one, theoreticians and experimentalists have made substantial progress in the past 15 years. The emerging discipline can be termed "quantum optical control," and it makes full utilization of the relevant properties of the molecular Hamiltonian. By exploiting the broad range of quantum interference effects, control is possible in a large number of physical systems. The special case of optical control of molecular dynamics—which mainly concerns us here—was reviewed comprehensively in the book by Rice and Zhao (2000). Additional review articles were provided by Warren et al. (1993), Gordon and Rice (1997), Zare (1998), and Rabitz et al. (2000).

In the initial development of quantum control, three main methodologies emerged, which can be broadly divided into the categories of frequency-domain control, time-domain control, and adiabatic control. At first sight, they appeared to be quite different, but now it is clear that they merely emphasize different aspects of a general methodology (Rice and Zhao, 2000): adjusting the spectral–temporal characteristics of the applied coherent light field so that a given initial wavefunction is transformed into the desired target wavefunction at some later time.

The "spectral-domain technique" was proposed by Brumer and Shapiro and relies on quantum-mechanical interferences (constructive and destructive) between wavefunctions leading to competing product channels (Brumer and Shapiro, 1986; Shapiro et al., 1988). Exerting control is possible by adjusting the relative phase of two incident continuous-wave (CW) laser beams which both couple initial and final states, but via different pathways. First experimental demonstrations of this idea were carried out on atoms (Chen et al., 1990) and small molecules (Park et al., 1991). In a recent work, for example, the competition between autoionization and predissociation product states of gas-phase DI molecules was controlled (Zhu et al., 1995).

The second methodology—the time-domain technique—was initially proposed by Tannor, Kosloff, and Rice (Tannor and Rice, 1985; Tannor et al., 1986). The first required condition is the initiation of coherent nuclear motion in a molecular species by interaction with a short laser pulse. This wavepacket then visits different nuclear configurations of the molecule. The opportunity for control arises because a second pulse can be fired when the wavepacket is visiting a nuclear configuration which is preferentially associated with a given reaction channel on a second potential energy surface. Experimental demonstration of the Tannor–Kosloff–Rice scheme was achieved in our laboratories while studying the wavepacket dynamics of the diatomic molecule Na_2 (Baumert et al., 1991; Baumert and Gerber, 1994). It was shown that precise control over the relative timing between ultrashort pump and probe pulses allowed one to preferentially form Na_2^+ at the expense of Na^+ and vice versa. Experiments where laser-induced fluorescence was used to monitor the controlled evolution of the molecular wavepacket were carried out by Potter et al. (1992) and by Herek et al. (1994).

The third (adiabatic) control technique was realized by Bergmann and co-workers (Gaubatz et al., 1988, 1990; Bergmann et al., 1998) and is known as stimulated Raman adiabatic passage (STIRAP). In the initial formulation, two suitably timed laser interactions are used for complete population transfer in three-state (Λ) systems. The "pump" laser couples initial and intermediate state, while the "Stokes" laser links intermediate and final state. The electric fields are large enough to generate many cycles of Rabi oscillations. Analysis in a "dressed-state" picture of the system shows that efficient transfer from initial to final state can be achieved with a "counterintuitive" pulse sequence, where the Stokes laser precedes (but overlaps) the pump laser. In this way, the transient population of the intermediate state remains always zero, and possible (dissipative) population losses from that state (by radiative decay, for example) are avoided.

In the above examples, one variable (i.e., the phase difference between two CW lasers or the time delay between two laser pulses) is changed and the effect noted in experimental observables. Therefore, this type of experiment belongs to a class that we refer to as single-parameter control. In this method, the single variable is called the "control knob," as it represents the experimenter's "handle" on the modulations that are induced. When changes do take place, they are governed by what is termed the "control surface." This does not encompass all possible modulations, only those that are explored within the range of the control knob that is turned. The basis of one-parameter quantum control lies in simple manipulations of the input laser field $E(t)$ to produce a change in the experimental outcome. For systems where the control surface is predictable, this can be effective. However, an important question arises as to how control can be achieved in molecular species where one-parameter changes are not useful or where the control surface is not obvious. This is the case for complex molecular systems as well as for any process that simply demands nontrivial excited-state evolution.

In approaching this problem, one can imagine that what is needed are complex electric field shapes $E(t)$ beyond those available with typical one-parameter changes. The first investigation in that context was published by Huang *et al.* (1983). The authors established sufficient (but not necessary) conditions (mainly that the system have a discrete spectrum in absence of the control field), under which it is always possible to design an electric field that actively transforms a given initial quantum-mechanical wavefunction into any given target wavefunction at some later time. Unfortunately, this and other (Ramakrishna *et al.*, 1995) proofs of existence of quantum control do not reveal how the required electric field can be found. However, this type of problem can be addressed by optimal control theory, where in the context of quantum control the following question is asked: With a given target distribution of photoproducts and the quantum-mechanical equations of motion, what is the shape of the electric field required to guide the temporal evolution of the system appropriately? This application of optimal control theory was first reported by Shi *et al.* (1988), Peirce *et al.* (1988), and Kosloff *et al.* (1989). In most cases, the calculated electric fields exhibit a very complex structure, requiring many "control knobs" for parametrization. In the early stages of this research direction, it was not clear how such electric fields could be realized experimentally. However, new technological developments in the area of ultrafast optics began to make this task possible. Femtosecond pulse shapers, developed primarily for telecommunication purposes, allowed convenient but sophisticated manipulation of the spectral–temporal characteristics of ultrashort laser pulses (Heritage et al., 1985; Weiner et al., 1990, 1992; Wefers and Nelson, 1995a,b; Weiner, 2000). Despite this technological advance, there remained difficulties with combining optimal control theory with laser pulse shapers. First, for complex molecular systems, theory is not able to predict the optimal field because molecular excited-state potential energy surfaces are, in general, not accurately known. Second, even if theory could predict such fields, it is not clear that they could be reproduced accurately enough under laboratory conditions. Finally, the search space is simply too large to be scanned completely. Initial experiments struck a compromise by employing linear chirp as a first-order approximation to the required time-varying spectral contents of ultrashort laser pulses (Bardeen et al., 1995, 1997, 1998a,b; Kohler et al., 1995; Assion et al., 1996b; Cerullo et al., 1996). Frequency-swept infrared pulses were also applied in "ladder-climbing" processes (Broers et al., 1992; Balling et al., 1994). Although these experiments exploit the "shape" of the incident electric field, manipulation of linear chirp is still only a single-parameter control scheme, and the desire to use the full flexibility of the new pulse shaping technology remained.

A general way to approach the problem of finding the optimal field in the experiment was described by Judson and Rabitz (1992). They suggested that the experimental output be included in the optimization process. In this way, the molecules subjected to control are called upon to guide the search for an optimal field within a learning loop. The experimentally determined outcome of the light/matter interaction is evaluated by a search algorithm which then modifies the laser field $E(t)$. With the proper algorithm, automated cycling of this loop provides an effective means of finding optimal fields under constraints of the molecular Hamiltonian and the experimental conditions. No prior knowledge of potential energy surfaces or any other detail of the chemical reaction process is needed in that case. Applications of this scheme to molecular systems were first demonstrated by Bardeen et al. (1997) to control the excited-state population of a laser dye in solution and by our group to control the product distribution of gas-phase photodissociation reactions (Assion et al., 1998c; Bergt et al., 1999; Brixner et al., 2001a).Weinacht et al. (1999b) have used such a scheme to control the excitation of different vibrational modes in a molecular liquid, and Hornung et al. (2000b) have used shaped pulses in a learning loop to control vibrational dynamics in a four-wave mixing experiment.

Optimal quantum control is not limited to molecular systems. Experimental examples using femtosecond laser pulse shapers and learning loops include automated femtosecond pulse compression (Yelin et al., 1997; Baumert et al., 1997b; Efimov et al., 1998, 2000; Brixner et al., 1999; Zeek et al., 1999, 2000; Zeidler et al., 2000), optimized generation of arbitrary laser pulse shapes (Meshulach et al., 1998; Brixner et al., 2000), control of two-photon transitions in atoms (Meshulach and Silberberg, 1998; Hornung et al., 2000a), shaping of Rydberg wavepackets (Weinacht et al., 1999a), optimization of high-harmonic generation (HHG) (Bartels et al., 2000), and control of ultrafast semiconductor nonlinearities (Kunde et al., 2000). Shaped electric fields have also been suggested to be of use in the context of laser cooling (Tannor et al., 1999; Tannor and Bartana, 1999).

With automated ("learning") multiparameter control in mind, one might still ask to which degree a quantum system is controllable. Apart from the very general considerations of Huang et al. (1983), there have been investigations about the degree of control attainable with optimal control theory (Peirce et al., 1988; Zhao and Rice, 1991; Demiralp and Rabitz, 1993). One important conclusion is that, in general, there exists an infinity of solutions to a well-posed problem of quantum control. These solutions can be ordered in quality according to the degree of control obtained, and therefore, many local optima may exist in the control problem (Demiralp and Rabitz, 1993). Employing suitable "global search" algorithms (such as evolutionary algorithms), it is possible to find an optimum under the given conditions. From a practical point of view (for example, in controlling photoproduct distributions), many of these local optima are very close to the global optimum, so that reaching one of them is often sufficient.

The preceding paragraphs have briefly outlined the historical development of quantum control. What follows in this chapter is a review of efforts made in our laboratory at achieving control over the outcome of light/matter interactions. In Section II we present an overview of the experimental techniques employed in our control experiments. In Section III we present examples of one-parameter schemes where gas-phase photoprocesses are controlled by varying laser pulse energy, pulse duration, pump-probe delay, excitation phase, and linear chirp. In Section IV we present the technique which generalizes these control methods by combining active multiparameter pulse shaping with a global search algorithm in a learning loop. This methodology is referred to as "adaptive femtosecond pulse shaping." Applications of this many-parameter control scheme to different systems in the gas phase are discussed in Section V, whereas liquid-phase many-parameter optimization is shown in Section VI. Finally, we conclude our review in Section VII. Many experiments described herein are among the first of their kind and represent what we believe to be important steps toward quantum control of complex molecular systems.

II. Experimental Techniques

A. OVERVIEW

The work on femtosecond quantum control has been ongoing for nearly 10 years in our laboratories. Therefore, a number of techniques have been used, and the

FEMTOSECOND QUANTUM CONTROL   7

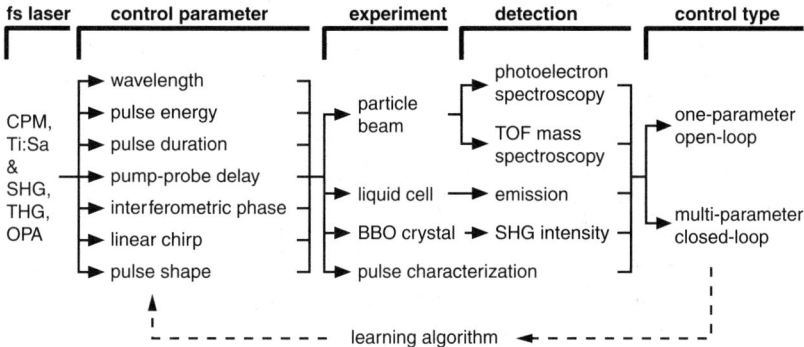

FIG. 1. Schematic of our experimental setup. The discussion of our quantum control experiments follows the five categories labeled in the top part of the figure. Femtosecond (fs) laser pulses are delivered by a colliding-pulse mode-locked (CPM) laser or by a titanium:sapphire (Ti:Sa) oscillator and amplifier. Frequency conversion is possible by second-harmonic generation (SHG), third-harmonic generation (THG), and optical parametric amplification (OPA). Different control parameters are employed to modify the laser pulses before they enter the actual experimental apparatus. Detection signals are then processed depending on the type of control experiment, either in an open-loop configuration or in a closed-loop configuration where a learning algorithm optimizes the control parameter settings iteratively.

technology has evolved. To organize our experimental efforts it is useful to provide a summarizing schematic of our control methodology. This is shown in Fig. 1. Some details about the experimental setup are given in Section II.B.

The starting point on the left of the diagram is the femtosecond laser system which is the common theme in each of our experimental efforts at control. It is of course very well known that ultrafast laser pulses can be used to study fast kinetic events in physics, chemistry, and biology. Beyond time resolution, however, ultrafast laser pulses contain an important and useful physical property, which is their inherently large spectral bandwidth. The utility lies in the ability of such sources to generate well-defined vibrational wavepackets following molecular excitation (which is often electronic in nature). This is possible when multiple vibrational eigenstates of a given potential energy surface (PES) are coherently excited with a fixed-phase relationship. Because of this requirement, ultrashort laser pulses with intrinsically large bandwidths are ideally suited. In many if not all of the control experiments described herein, it is such wavepacket motion which is manipulated and guided to particular photophysical outcomes.

The second category in Fig. 1 is labeled the "control parameter" (also called "control variable") and corresponds to the physical technique manipulated by the experimenter to exert control. Because of the common starting place of femtosecond laser pulses, the control variable always manipulates these excitation pulses. The methods used vary in complexity, starting from rather simple ideas such as manipulating pulse energy or duration to more sophisticated methods which include pump-probe delay control and laser pulse shaping. Experiments performed with each member of the "control variable" category are discussed sequentially in this chapter.

The third and fourth categories are intimately coupled. They are the experiment performed with the varied laser pulse and the detection used to evaluate the outcome. The origins of control in our laboratory involved gas-phase experiments where the modified laser was focused into an interaction region with a molecular or atomic beam prepared inside a high-vacuum chamber. Detection included photoelectron spectroscopy and time-of-flight (TOF) mass spectrometry. More recently we have also considered liquid-phase experiments involving laser excitation of charge-transfer chromophores with shaped pulses and detection of molecular emission. Also included in the experiment category are pulse characterization techniques. In general, these are used to analyze the laser pulses used in quantum control experiments; however, as will be seen in Section IV.B, they are also employed for their own sake within an entirely "optical" realization of adaptive femtosecond pulse shaping.

The final category of Fig. 1 is the "control type," and within this we have explored two separate ideas. The first is called open-loop control, and the experiments performed generally rely on a single-parameter control variable such as pulse energy or pump-probe delay time. Detection of the experimental outcome—when the control variable is changed—allows the experimenter to determine the extent of control that is possible under the given experimental conditions. The second method is called closed-loop control and represents one of the most promising recent developments in quantum control, as has already been motivated in Section I. In this automated technique, the experimental outcome is processed by a learning algorithm which can then suggest changes to the control variable. Cycling through this loop provides a means of optimizing the laser field for a given control task specified by the experimenter. We have used closed-loop control schemes of this type to explore control variables with extremely large parameter spaces. Such experiments are simply impossible with open-loop control methods. In the first sections of this chapter, we describe experimental examples where open-loop one-parameter control was achieved, while in the later sections we describe our efforts using closed-loop techniques and multiparameter control variables.

B. SETUP

Our current femtosecond laser system is shown in Fig. 2. It consists of a home-built Ar^+-pumped Ti:sapphire oscillator which delivers 50-fs, 3-nJ pulses at a center wavelength of 800 nm and a repetion rate of 80 MHz. These pulses seed a commercial chirped-pulse amplification (CPA) system (Quantronix) which has been modified by changes to the grating stretcher/compressor setup. This system

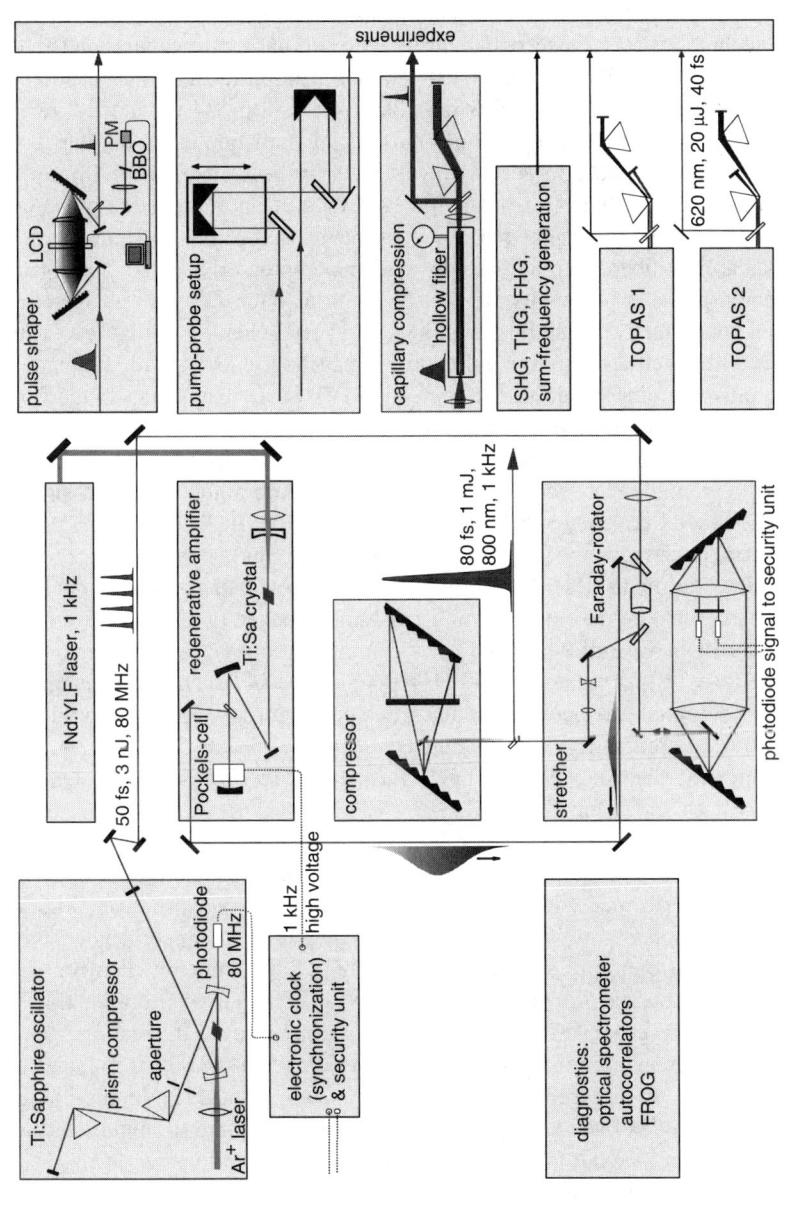

FIG. 2. Current femtosecond laser system. Specific details can be found in the text.

outputs 80-fs, 1-mJ pulses at a center wavelength of 800 nm with a repetition rate of 1 kHz determined by the Nd:YLF laser which pumps the regenerative amplifier. These pulses can be frequency-converted by a number of nonlinear optical techniques such as second-, third-, or fourth-harmonic generation (SHG, THG, FHG), and traveling-wave optical parametric amplification of superfluorescence (TOPAS). With TOPAS, it is possible to cover a broad wavelength range in the infrared (from 1200 up to 2600 nm). Using sum-frequency-mixing techniques it is also possible to access wavelengths in the visible (from 480 up to 800 nm). Increasing the laser bandwidth and thereby shortening the pulse duration to below 20 fs at 800 nm are possible by self-phase modulation in a gas-filled hollow-fiber setup followed by a prism compressor. Single-color or two-color pump-probe experiments are realized with a Mach-Zehnder interferometer and a computer-controlled translation stage. Very general modulations of the femtosecond spectral–temporal pulse profile can be achieved with frequency-domain pulse shaping (as described in Section IV). Laser diagnostics range from continuous monitoring of the pulse energy, pointing stability, and optical spectrum to complete pulse characterization methods such as intensity autocorrelation (in scanning and single-shot configuration), interferometric autocorrelation, single-shot SHG-based frequency-resolved optical gating (FROG) (Trebino et al., 1997), and spectral interferometry (Lepetit et al., 1995). It should be noted that some of the earlier experiments described herein were carried out using an additional laser system based on a home-built colliding-pulse-modelocked (CPM) ring dye laser. Equipped with two excimer-pumped bow-tie amplification stages, this laser system produced 80-fs, 40-μJ pulses at a 100-Hz repetition rate and a center wavelength of 620 nm. Additional wavelengths for pump and probe purposes could be produced by selecting and reamplifying different frequency components from a white-light continuum generated in a methanol cell. Further details of this laser system are described by Baumert and Gerber (1995).

Gas-phase quantum control experiments have been performed in a number of different atomic/molecular beam machines. Most of the earlier experiments on Na_2 and Na_3 were performed with a linear time-of-flight (TOF) spectrometer (Fig. 3). This apparatus consists of two differentially pumped high-vacuum chambers with a nominal background pressure of 2×10^{-7} torr in the interaction chamber. The molecules are prepared almost completely (>90%) in the lowest vibrational level of their electronic ground states. For Na_2 this can be achieved by heating a solid sample in an oven maintained at a temperature between 500 and 600°C followed by adiabatic expansion of gaseous species through a 200-μm nozzle. Na_3 production is enhanced at higher oven temperatures or by coexpanding the sodium molecular beam with argon at 4 bar (seeded-beam technique). In the interaction region, a (focused) femtosecond laser beam then produces parent ions, ionic fragments, and photoelectrons which are analyzed in the two TOF setups. Measuring the time interval between the laser interaction and the photoproduct detection provides

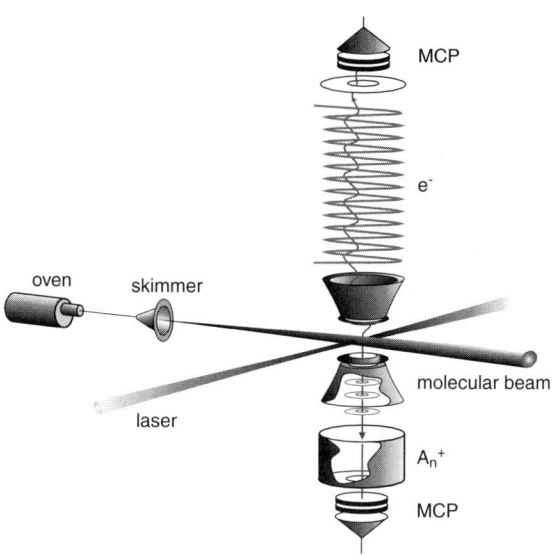

FIG. 3. Time-of-flight (TOF) spectrometer. In this schematic, laser beam and molecular beam are crossed in the interaction region where photoproducts are generated. Electron (e^-) kinetic energies are analyzed with a "magnetic bottle" spectrometer (upper part). Cations (A_n^+) are separated according to their mass in a linear TOF spectrometer (lower part) or in reflectron geometry (not shown). Detection for both electron and ion spectrometers is achieved using micochannel plates (MCP) or microsphere plates (MSP).

fragment mass spectra and photoelectron kinetic energy distributions. In cases where high photoelectron energy resolution and detection efficiency are required, we can use a "magnetic bottle" spectrometer. In this setup, photoelectrons out of a 2π solid angle are guided onto the detector by a suitably designed magnetic field. In cases where high mass resolution of ionic photoproducts is required, we use a reflectron TOF setup. In this apparatus, the molecules [such as $Fe(CO)_5$ or $CpFe(CO)_2Cl$ (vide infra)] are prepared in a simple effusive beam at room temperature or after heating to approximately 100°C. With the reflectron geometry, we achieve a mass resolution of 1900 at 196 amu, which is sufficient to analyze single-mass-unit losses and isotope distributions in these larger molecules. In all the machines, ions and photoelectrons are detected by microchannel plates (MCPs) or microsphere plates (MSPs). The detected signals can be processed in a number of ways, including boxcar integration followed by analog-to-digital conversion or direct transient analysis in a digital oscilloscope or with a fast transient recorder.

The setup for our control experiments in the liquid phase is described as follows (Brixner et al., 2001b). The laser beam is directed into a flowing sample ($\sim$2 mm) of the investigated molecule [$Ru(dpb)_3$](PF_6)$_2$ (where dpb = 4,4'-diphenyl-2,2'-bipyridine) in a methanol solution. Emitted radiation (nominally phosphorescence) is collected at an approximate right angle, spectrally filtered at 620 nm with a monochromator to remove scattered pump light, detected with a photomultiplier tube, and integrated on a microsecond time scale. During optimizations, two different signals are monitored and used in the optimization goals. The first is the molecular emission, while the second is second-harmonic generation (SHG) in BBO of the same laser beam used to excite the sample.

III. One-Parameter Quantum Control in the Gas Phase

A. INTRODUCTION

In this section we describe a number of one-parameter quantum control experiments on molecules in the gas phase. An open-loop setup is sufficient since the number and range of "control settings" are small. In these studies, the accessible range of settings is explored and the optimum result on that particular "control surface" can be determined.

Historically, one of the first control parameters considered in active quantum control was the excitation wavelength. As already alluded to in Section I, it was believed that choosing the light frequency "in resonance" with the vibrational frequency of the bond to be broken would lead to resonant amplification of the vibrational amplitude and finally to selective dissociation (Ambartsumian and Letokhov, 1977; Letokhov, 1977; Bloembergen and Yablonovitch, 1978; Zewail, 1980; Jortner *et al.*, 1981). However, this "mode-selective" photochemistry works in general only for small molecules, since the vibrational motion is coupled in a complicated way to many atoms within the molecule, and hence, what has been termed intramolecular vibrational redistribution (IVR) leads to a loss of the selective excitation and to an overall "heating" of the molecule (Bloembergen and Zewail, 1984; Elsaesser and Kaiser, 1991; Nesbitt and Field, 1996; Gruebele and Bigwood, 1998; Wong and Gruebele, 1999).

Apart from this approach of purely vibrational excitation, femtosecond pulses can be used to induce vibrational wavepacket motion on electronically excited states. Here the choice of excitation frequencies also plays an important role. Depending on the center wavelength of the broad laser spectrum, wavepackets can be prepared on different electronic states involving different vibrational eigenstates. These Franck–Condon transitions serve as starting place for wavepacket motion. Some of these starting places might then be better suited than others to achieving a given control target. As an example from our own laboratory, we observed different photoproduct distributions after single-pulse 800-nm compared to 400-nm excitation in the organometallic molecule $Fe(CO)_5$ (Bañares *et al.*, 1998).

However, we would like to go beyond this initial excitation step and influence the wavepacket motion subsequent to the first Franck–Condon transition. We do this in the following experiments on one-parameter control, which illustrate the basic mechanisms of the coherent manipulation of quantum systems.

B. CONTROL BY PULSE ENERGY

The most easily accessible experimental control parameter is the integrated laser pulse energy. By simple attenuation, the laser intensity in the interaction region with gas-phase species can be varied, leading to observed changes in the dynamic behavior of the system. As an example, we were able to transfer population to different electronic states of Na_2, depending on the laser intensity (Baumert *et al.*, 1992). The electronic state population was monitored by analyzing the induced vibrational wavepacket dynamics. In the case of high intensities, ground-state vibrational wavepackets were induced by impulsive stimulated Raman scattering (also known as stimulated emission pumping). This was the first experimental demonstration of a theoretical proposal by Hartke *et al.* (1989).

Exploiting electronic coherences in Rabi-type processes is described here in more detail for the triatomic molecule Na_3 (Baumert *et al.*, 1993, 1997a). The Na_3 excitation scheme using laser pulses near 618 nm is shown in Fig. 4. One-photon absorption excites the electronic B state. Additional absorption of one photon leads to the ionic ground state Na_3^+, which can be detected by the mass spectrometer. If three photons are absorbed by the neutral ground state, dissociative ionization can take place and the ionic fragments Na_2^+ and Na^+ are observed. Due to geometric differences among the photon-coupled states, configuration-dependent

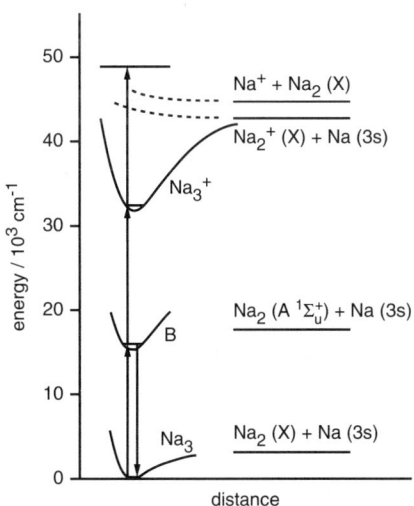

FIG. 4. Potential-energy diagram of Na_3 at 618-nm excitation wavelength.

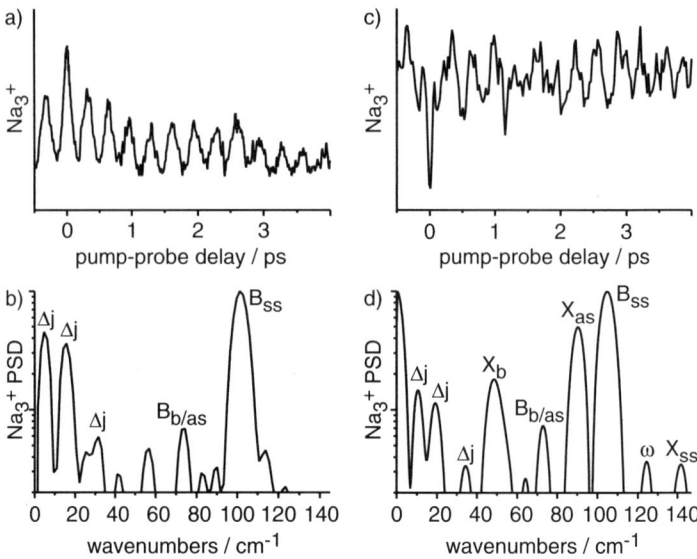

FIG. 5. Pump-probe transients (a, c) of Na_3^+ and power spectral density (PSD) after Fourier transformation (b, d). The pump and the probe pulses are centered at 618 nm and employ low (a, b) and high (c, d) laser intensities.

Franck–Condon transition probabilities are expected, leading to the generation of vibrational wavepacket motion. We can use this motion to assign population as well as to determine the degree of control that is achieved during attenuation.

The transient Na_3^+ signal was recorded (Figs. 5a and 5c) and Fourier-analyzed (Figs. 5b and 5d) for two different pump laser intensities. The results show that wavepacket dynamics in the case of low pump laser intensities (Figs. 5a and 5b) are dominated by a contribution at 105 cm^{-1}, which is assigned to the symmetric stretch oscillation (B_{ss}) of the molecule in the excited B state. A smaller component at 72 cm^{-1} is attributed to the degenerate bending (B_b) and asymmetric stretch (B_{as}) normal modes in the same electronic state. The contributions below 40 cm^{-1} are assigned to pseudo-rotations of the B state. In the case of high pump laser intensities (Figs. 5c and 5d), the transient signal is markedly different. Most notably, additional vibrational contributions at 50, 90, and 140 cm^{-1} arise. These can be assigned to the bending mode X_b, the asymmetric stretch mode X_{as}, and the symmetric stretch mode X_{ss}, respectively. In this case, however, these modes are active on the electronic ground state of Na_3. The intense pump laser pulse has therefore coherently transferred population back to the electronic ground state, initiating ground-state wavepacket motion which is subsequently probed by two-photon ionization. Note that in this experiment, the second (probe) laser pulse is used only to detect the vibrational wavepacket motion. Analysis of the vibrational frequencies then yields the information on which electronic surface the population had been prepared by the pump laser pulse. The control process itself is due solely to the different pump pulse intensities. The same information about vibrational dynamics was also obtained with the zero-electron-kinetic-energy (ZEKE) technique (Baumert et al., 1993). This was the first experiment where photoelectrons were detected following femtosecond excitation.

Adjusting the laser pulse intensity therefore makes it possible to control molecular electronic population transfers in Rabi-type processes (impulsive stimulated Raman scattering). If even higher laser intensities are employed, the potential energy surfaces themselves are manipulated by the electric field (Frohnmeyer et al., 1999). These light-induced dressed-state potentials can lead to electronic transitions which are normally forbidden in the low-field limit. This happens because the light field perturbs the shape of the involved potential energy surfaces in such a way that Franck–Condon windows are opened at different nuclear configurations. Very high intensities can furthermore lead to the observation of field ionization followed by Coulomb explosion (Assion et al., 1998a, 2001), with the possibility of accessing and controlling another manifold of exit channels.

In our second example (Bañares et al., 1998; Bergt et al., 1999) we address not only electronic population transfers but also branching ratios of different photofragmentation processes in the $Fe(CO)_5$ species. Upon irradiation with femtosecond laser pulses, $Fe(CO)_5$ can be ionized in a multiphoton process, or it can undergo ligand-loss photochemistry. Having concentrated on the ratio of direct ionization [producing the undissociated parent ion $Fe(CO)_5^+$] versus complete fragmentation and ionization (producing the bare Fe^+), we determined the $Fe(CO)_5^+/Fe^+$ ratio for a number of 800-nm laser pulse intensities produced by simple attenuation of 80-fs pulses (Fig. 6, solid circles). For increasing pulse energies (and therefore increasing intensities), the $Fe(CO)_5^+/Fe^+$ ratio decreases. Although the mechanism of this behavior is debated, it has been suggested that with higher intensities it is easier to reach higher lying electronically excited energy surfaces, possibly in the ionic $Fe(CO)_5^+$ continua, or, as another possibility, to induce multielectron excitations. In both cases, this can potentially lead to fragmentation of the molecule, and therefore to a lower $Fe(CO)_5^+/Fe^+$ ratio. If, on the other hand, the pulse energy (and therefore the laser intensity) is lower, the probability of reaching such energetic states is reduced. Since the lowest ionization threshold in the $Fe(CO)_n^+$ fragment series appears for $Fe(CO)_5^+$, lower laser intensities favor the $Fe(CO)_5^+/Fe^+$ ratio.

The previous examples have shown that it is possible to influence photoproduct distributions by different laser pulse intensities. However, one must be careful using the term "intensity," since different types of intensity variation may lead to completely opposite behavior of the molecular system. Using different pulse energies as control parameter, the main mechanisms are based on coherences between different electronic states, as especially exploited with Rabi cycling or the coupling of field-dressed potentials. On the other hand, if intensity variation

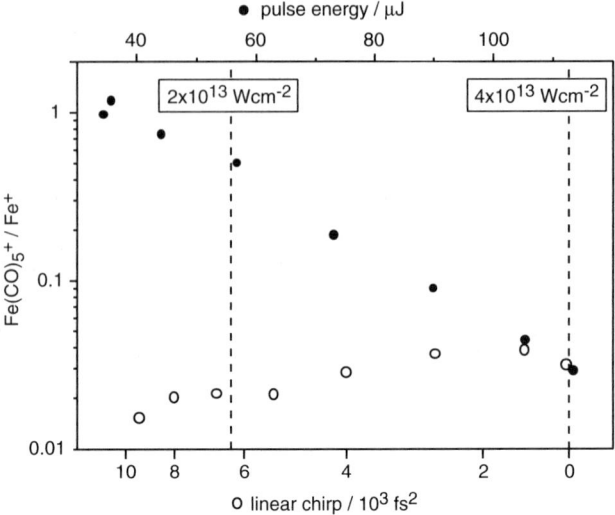

FIG. 6. Intensity dependence of pure ionization versus complete fragmentation in Fe(CO)$_5$. Decreasing the temporal peak intensity of the laser pulse (from 4×10^{13} W/cm^2 on the right side of the figure to below 2×10^{13} W/cm^2 on the left side) is achieved either by attenuating the laser pulse energy from 113 to 34 μJ at a constant pulse duration of 80 fs (solid circles), or by increasing the pulse duration at a constant pulse energy of 113 μJ (open circles).

is achieved by changing the pulse duration, vibrational coherence plays a more important role, and the vibrational wavepacket dynamics on one electronic state can be changed by different pulse lengths as described in the following section.

C. CONTROL BY PULSE DURATION

In this section, different femtosecond laser pulse intensities are achieved by variation of the pulse duration at constant energy. For example, if we use a pulse shaper to apply quadratic phases to our excitation pulses, their pulse duration increases. We have examined the change in the Fe(CO)$_5^+$/Fe$^+$ ratio under these conditions (Bañares et al., 1998; Bergt et al., 1999), and the results are plotted again in Fig. 6 (open circles). Note that the horizontal peak intensity axis is common to both open and solid circles. In this experiment, increasing intensity leads to an increase of the Fe(CO)$_5^+$/Fe$^+$ ratio. This behavior is then opposite to the "intensity" dependence of the previous section. Therefore, it is not sufficient to argue that static multiphoton absorption probabilities to different electronic states increase in a simple way with laser "intensity." Instead, the temporal wavepacket evolution on one (or several) excited-state potential energy surfaces of Fe(CO)$_5$ must be taken into account.

In a simplified argument, the beginning of the laser pulse initiates wavepacket motion on some excited state, leading to dissociation of the carbonyl ligands.

If the pulse duration is sufficiently long, light intensity remains at the end of the dissociation process for transferring wavepacket population into the ionic continuum. In this circumstance, Fe^+ is the main photoproduct which is detected. If, on the other hand, the laser pulse is temporally short, there is no light present at the end of the fragmentation to produce Fe^+. Instead, the detectable ions are produced in the short duration of the bandwidth-limited 80-fs pulse, leading mainly to the $Fe(CO)_5^+$ ion. Smaller fragments which are detected are believed to arise from ionization of neutral fragments produced by ultrafast carbonyl ligand loss. Our analysis suggests that up to four carbonyl ligands can be lost in the first 100 fs, with the fifth being removed on a picosecond time scale (Bañares *et al.*, 1997). From this, it can be stated that the pulse duration has a profound effect on photofragment distributions resulting from the formation and evolution of vibrational wavepacket motion. Therefore, pulse duration serves as a simple one-parameter control variable. More precise "timing" of the relevant excitation processes can be achieved if the laser pulse is not simply stretched in time but if two separate pulses are used in a pump-probe configuration, as explained in the next section.

D. CONTROL BY PUMP-PROBE DELAY

As already stated in Section I, photochemical product yields may be actively controlled in the time domain using the temporal separation of individual ultrashort laser pulses as the one-parameter control variable. The technique is based on the idea of Tannor, Kosloff, and Rice (Tannor and Rice, 1985; Tannor *et al.*, 1986) and relies on the manipulation of excited-state population. Because of the experiments we have considered, our discussion is limited to include only the manipulation of vibrational wavepackets prepared as a result of coherent electronic excitation of a molecule. In the simplest experiment, two laser pulses are used. The first pulse (known as the pump pulse) is used to prepare a well-defined wavepacket by coherent superposition of vibrational eigenstates. Wavepacket propagation on this surface should be unreactive toward the products being controlled and cause only a predictable time-dependent change in the nuclear and electronic structure of the molecule. The second pulse (known as the probe or dump pulse) serves to transfer population to a reactive potential energy surface. The active control variable is the time at which this laser pulse is fired relative to the initial wavepacket motion itself (i.e., the time-dependent change in the nuclear configuration). Active control is achieved by selectively moving population to regions of the reactive surface in the proximity of desired reaction channels.

The first experimental demonstration of the Tannor–Kosloff–Rice scheme for active quantum control was made in our group (Baumert *et al.*, 1991; Baumert and Gerber, 1994) using three-photon ionization in competition with molecular fragmentation of gas-phase Na_2. The experiment has been discussed previously in detail. However, in the context of one-parameter control, it is conceptually simpler to describe subsequent control experiments on Na_2 (Baumert and Gerber,

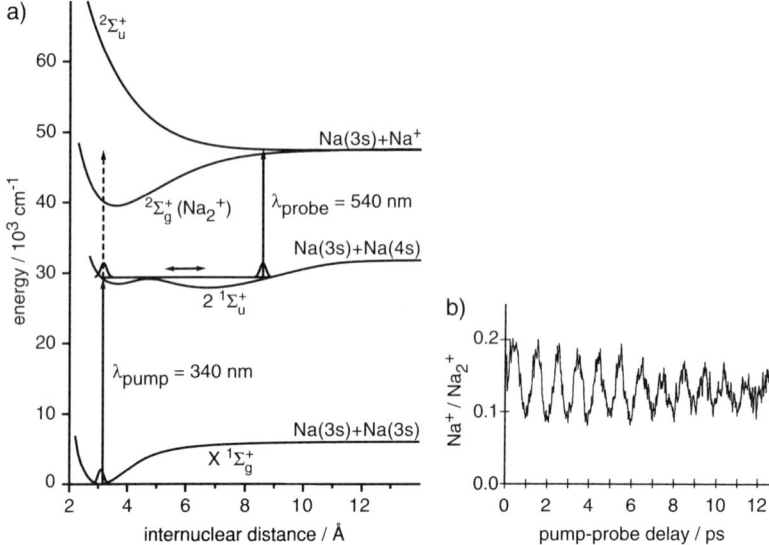

FIG. 7. Potential-energy diagram of Na_2 (a) and ratio of Na^+/Na_2^+ pump-probe transients (b) at 340-nm pump and 540-nm probe wavelength.

1994; Assion et al., 1996a; Baumert et al., 1997a) which rely on two-photon ionization and fragmentation processes. The potential energy surfaces relevant for the experiment are shown in Fig. 7a. Coherent excitation starting from the electronic and vibrational ground state using a 100-fs pulse of 340-nm light results in the formation of a vibrational wavepacket on the $2^1\Sigma_u^+$ surface. This can then propagate to its outer turning point at a large internuclear distance where absorption of a second ultrashort pulse of 540-nm light accesses the repulsive ionic state leading to $Na(3s)$ and Na^+ fragments. If, on the other hand, the 540-nm pulse is fired at the inner turning point, the bound ionic state is preferentially accessed and the Na_2^+ is detected. The results of this experiment are shown in Fig. 7b as a ratio of Na^+ versus Na_2^+ as a function of the pump-probe delay time. What is observed is an oscillatory motion corresponding to propagation of the $2^1\Sigma_u^+$ wavepacket between its turning points and a strong modulation of the product ratio. Thus, by varying the time delay between pump and probe laser pulses, one exerts control over whether the Na—Na bond is broken or kept intact.

Pump-probe delay can also be applied as a control variable in the larger $Fe(CO)_5$ molecule (Bergt et al., 1999). Again, the pump pulse prepares wavepacket motion on electronically excited potential energy surfaces, and the probe pulse interrogates this motion by multiphoton excitation into different fragment ionic continua. If the timing between pump and probe pulse is properly chosen, the ionic continuum of the desired photoproduct may be preferentially accessed. In Fig. 8, pump-probe

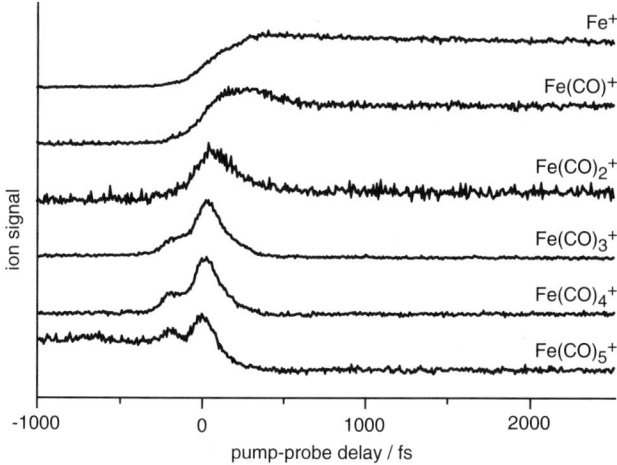

FIG. 8. Pump-probe transients of Fe(CO)$_n^+$ at 400-nm pump and 800-nm probe wavelength.

transients of the species Fe(CO)$_n^+$, $n = 0, \ldots, 5$, are shown (Bañares et al., 1997, 1998). It can be seen that different dissociation time constants are associated with different fragments, and thus the ratios between different photoproducts vary as a function of pump-probe delay. For example, the Fe(CO)$_5^+$/Fe$^+$ ratio is high for short pump-probe times, while decreasing for longer delays. These findings will be compared with many-parameter optimization in Section V.B.

E. CONTROL BY PHASE-SENSITIVE EXCITATION

We now discuss an extension of the pump-probe technique to include phase-sensitive excitation. In this experiment, a two-pulse sequence is used where each pulse generates an identical vibrational wavepacket on the same electronically excited state. If, in addition to the delay time, the optical phase between the two laser pulses is controlled, the two wavepackets interfere. For this reason the technique is also known as wavepacket interferometry (Scherer et al., 1991; Christian et al., 1993; Engel and Metiu, 1994; Blanchet et al., 1995). It exploits these interference terms in the population of excited states which arise if the measurement of the excited-state population does not determine whether the absorbed photon was taken from the first or the second laser pulse. Depending on the phase and the timing between the two pulses, the interferences strongly modulate different excited-state populations. It should therefore be possible to use this technique to exert control over photoinduced electronic populations. Wavepacket interferometry can be seen as an analogy to Young's double-slit experiment, where interference takes place if it is not determined through which slit the photon has traveled. It can

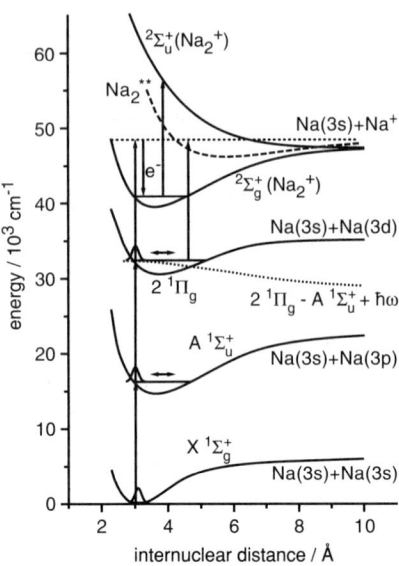

FIG. 9. Potential energy diagram of Na$_2$ at 618-nm excitation wavelength. Vibrational wavepacket propagation is initiated by coherent electronic excitation. The difference potential between the A $^1\Sigma_u^+$ and the 2 $^1\Pi_g$ state is represented by the dashed line.

thus be regarded as a femtosecond variant of the Brumer-Shapiro method initially suggested (Brumer and Shapiro, 1986) and implemented (Chen *et al.*, 1990; Park *et al.*, 1991) with continuous-wave laser sources whose relative phase is adjusted.

As an example (Baumert *et al.*, 1997a), we again consider the sodium dimer system Na$_2$. Using an excitation wavelength of 618 nm (Fig. 9), two-photon wavepacket interferometry is applied to monitor the Na$_2^+$ signal as a function of pump-probe delay. The signal obtained reflects interferences within the A$^1\Sigma_u^+$ and the 2 $^1\Pi_g$ state populations. Filtering the fast interferometric signal oscillations with the laser frequency (or multiples of it), followed by Fourier transformation, allows us to distinguish between different interference contributions (Blanchet *et al.*, 1995) and to obtain additional spectroscopic information. In this case we observe wavepacket motion in the A$^1\Sigma_u^+$ state and in the 2 $^1\Pi_g$ state (Fig. 10a). In contrast, with conventional noninterferometric pump-probe experiments (Fig. 10b), only wavepacket motion in the A$^1\Sigma_u^+$ state is accessed. Choosing the appropriate phasing therefore makes it possible to probe additional excited states. Seen from the perspective of quantum control, population transfer can be achieved between potential energy surfaces which are not accessible by phase-insensitive excitation schemes. As possible generalization of phase-sensitive excitation, femtosecond pulse shaping is desribed in Section IV.

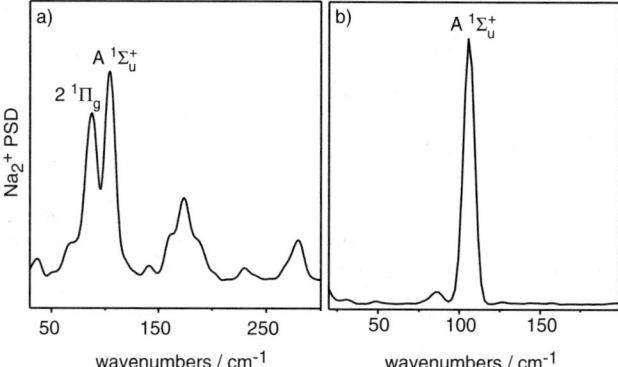

FIG. 10. Power spectral density (PSD) of the pump-probe transient Na_2^+ at 618 nm pump and probe wavelength, with (a) and without (b) interferometric timing.

F. CONTROL BY LINEAR CHIRP

The optical phase can be used as control parameter in single-pulse experiments in a sense different from that in the previous section. Here, a frequency-dependent phase structure rather than a constant additional phase is applied to a single excitation pulse. This leads to the generation of what has been referred to as "tailored" laser pulses where the frequency components of the broadband laser pulse vary throughout its temporal envelope. The idea that such tailored pulses can be used to exert control is intimately tied to the issues of wavepacket propagation discussed in the previous sections. However, the methodology goes a step further and exploits secondary interactions that may take place between photoexcited molecules and additional spectral components of a light field. These interactions take place throughout the molecule's temporal evolution. With the proper choice of the time-dependent frequency structure of the light field, it is possible to enhance or suppress certain product channels and thus to exert influence over the outcome of the light/matter interactions.

We have employed linear chirp to control multiphoton ionization in Na_2 (Assion et al., 1996b). Other examples where linearly chirped pulses have been used concern the control of excited-state population and wavepacket dynamics, as discussed by Bardeen et al. (1995) and Kohler et al. (1995). This type of chirping can be achieved with a pulse shaper, but at the time the experiments described here were done, a simple prism compressor with varying interprism separation fulfilled the same purpose. Shown in Fig. 11a are photoelectron spectra collected for Na_2 following three-photon ionization with 618-nm light using three different light fields: two with equal but opposite linear chirps (± 3500 fs^2) and one with a transform-limited pulse. An important observation that we address below is that the photoelectron yield is largest when using positively chirped laser pulses.

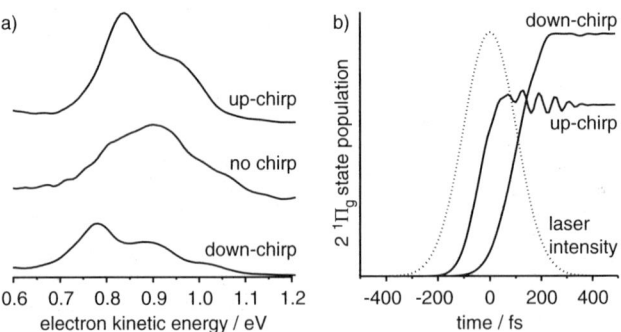

FIG. 11. Experimental photoelectron kinetic energy distributions for different pulse shapes after femtosecond excitation at 618-nm wavelength (a) and calculated $2\,^1\Pi_g$ state population as a function of time (b). The dashed line represents the laser intensity profile.

To interpret the experiments, quantum-mechanical calculations were undertaken for the Na_2 system coupled to the three different light fields in the weak-field limit using the fast Fourier transform split-operator method. Because of the potential energy surfaces involved in photoelectron production using 618-nm light (see Fig. 9), these calculations included explicit treatment of the $X^1\Sigma_g^+$, $A^1\Sigma_u^+$ and $2\,^1\Pi_g$ neutral states as well as coupling of the $2\,^1\Pi_g$ state to the continuum of the lowest ionic state $^2\Sigma_g^+$ (Na_2^+). The calculated photoelectron spectra (not shown here) are in good agreement with the measurements (Assion et al., 1996b). In addition to simulating spectra, the model was then used to calculate the time- and chirp-dependent generation of population in the $2\,^1\Pi_g$ state. Because this population is the direct precursor to the ionic state, this calculation is crucial for understanding the time and chirp dependence of photoelectron generation. The result is shown in Fig. 11b, along with the intensity profile common to these pulses.

The experiment and theory just described expose two different observations that chirp plays an important role as a control variable. The first is the chirp-dependent population yield of the $2\,^1\Pi_g$ state as seen in Fig. 11b. The second is the chirp-dependent photoelectron yield as seen in Fig. 11a. Both observations are discussed successively.

The calculations shown in Fig. 11b indicate that down-chirped pulses produce higher populations in the neutral $2\,^1\Pi_g$ than their up-chirped counterparts. This is intriguing because the chirp-dependent trend is opposite to that observed for total photoelectron yield. However, the observation can be readily understood when wavepacket evolution on the intermediate $A^1\Sigma_u^+$ state is taken into account. As the Na—Na bond stretches in its evolution toward the outer turning point, the energy separation between the $A^1\Sigma_u^+$ and $2\,^1\Pi_g$ states decreases. This is most easily seen by plotting the difference potential $2\,^1\Pi_g - A^1\Sigma_u^+$ (shown as a dashed line in Fig. 9.) When the proper down-chirped pulse is used, a resonance condition between these two states follows the temporal evolution of the pulse, leading to higher populations in the final neutral state.

With the calculation described in Fig. 11b in hand, we can address the original observation that photoelectron yield is chirp-dependent and favors an up-chirped pulse. Figure 11b indicates that when an up-chirped pulse is used, population is transferred to the final preionization state earlier in the evolution of the electric field. This occurs because the Franck–Condon maximum for the $A^1\Sigma_u^+ \leftarrow X^1\Sigma_g^+$ transition is shifted to the red of the laser bandwidth and in an up-chirped pulse, red frequencies precede the blue ones. The final ionization step from the populated $2\,^1\Pi_g$ state is of course ultimately responsible for the measured photoelectron yield. The nature of the chirp determines when population is created with respect to the remaining pulse energy necessary for this last step. This is easily seen by comparing the time-dependent rise in the $2\,^1\Pi_g$ population (for both chirp cases) to the temporal envelope of the laser pulse shown as the dotted line in Fig. 11b. For the positively chirped pulse, considerable laser intensity remains for the final ionization step. In this way, larger photoelectron yields are observed in comparison to the negatively chirped pulse despite the smaller total $2\,^1\Pi_g$ population.

The final point that we make concerns the opposing chirp dependence of the two observations discussed. These experiments have shown that when Na_2 is photoexcited with 618-nm light, linear chirp is an effective single control variable which allows one to make a competitive choice between creation of $2\,^1\Pi_g$ excited-state population and photoelectron production.

It is clear from these experiments that the phase structure of ultrashort laser pulses can be used as an optimization parameter in quantum control experiments. We might expect that for many systems, especially those which undergo complex wavepacket evolution, more complex phase profiles will be necessary to exert control. Experiments designed to explore this are described in the following sections.

IV. Adaptive Femtosecond Pulse Shaping

A. THE TECHNOLOGY

The femtosecond pulse shaper (Heritage *et al.*, 1985; Weiner *et al.*, 1990, 1992; Wefers and Nelson, 1995a,b Weiner, 2000) is the most flexible pulse-manipulation tool that we have employed in our laboratory. Details of our setup have been published previously (Brixner *et al.*, 1999). Briefly, however, the device (Fig. 12) consists of a zero-dispersion compressor in a 4f geometry, which is used to spatially disperse and recollimate the femtosecond laser pulse spectrum. Insertion of a liquid-crystal display (LCD) in the Fourier plane of the compressor provides a mechanism for convenient manipulation of the individual wavelength components. By applying voltages, the refractive indices at 128 separate pixels across the laser spectrum can be changed, and upon transmission of the laser beam

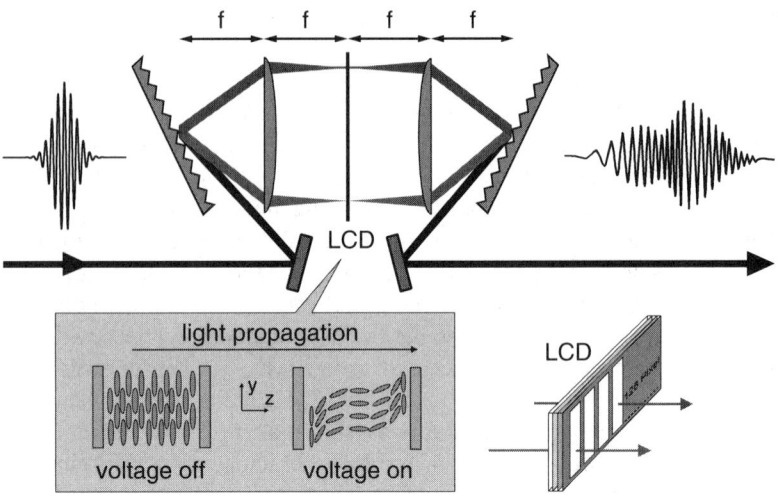

FIG. 12. Frequency-domain femtosecond pulse shaper. A zero-dispersion compressor in $4f$ configuration contains a liquid-crystal display (LCD) spatial light modulator in its Fourier plane. By adjusting the voltages of the individual LCD pixels, the liquid crystal molecules reorient themselves partially along the direction of the electric field. This leads to a change in refractive index and therefore to a phase modulation which can be independently controlled for the different wavelength components.

through the LCD, a frequency-dependent phase is acquired due to the individual pixel voltage values. In this way, an immensely large number of different spectrally phase-modulated femtosecond laser pulses can be produced. In all experiments described below, the spectral amplitudes have not been changed, and therefore the integrated pulse energy remains constant for different pulse shapes. On the other hand, by virtue of the Fourier transform properties, spectral phase changes result in phase- and amplitude-modulated laser pulse profiles as a function of time. As an extension to this technique, we recently developed computer-controlled femtosecond polarization pulse shaping, in which intensity, momentary frequency, and light polarization can be varied as functions of time within a single laser pulse (Brixner and Gerber, 2001). This opens up an immense range of applications, especially in quantum control, because the three-dimensional properties of quantum-mechanical wave functions can thus be accessed.

In a generalized control experiment, the optimal pulse is not known in advance. Since the variational space of possible pulse shapes is so huge, scanning the complete parameter space is impossible. To overcome this problem, we use a learning loop to find optimized electric fields, taking into account the experimental outcome (Judson and Rabitz, 1992). The global search method we use for this purpose is an evolutionary algorithm (Goldberg, 1993; Schwefel, 1995). Our implementation has been described in detail elsewhere (Baumert *et al.*, 1997b), but we briefly outline the technique here. Evolutionary algorithms have working schemes

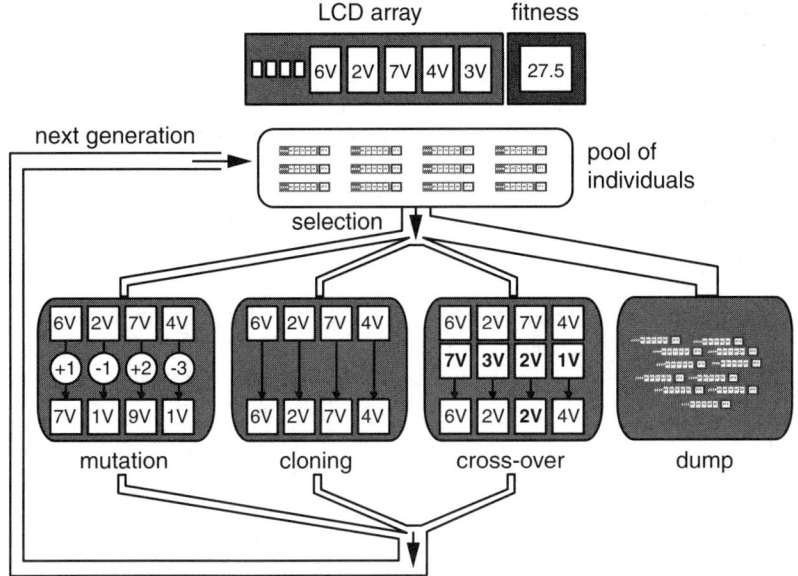

FIG. 13. Evolutionary algorithm. This basic schematic shows one cycle of the optimization procedure. Each individual within the population pool represents a specific laser pulse shape (i.e., specific LCD pixel voltage values). Depending on the fitness, an individual is either selected for reproduction or it is rejected. Reproduction operations include mutation, cloning, and crossover of the LCD voltage arrays. The children thus produced constitute the next generation, which is then evaluated in the same way. Iterating through this loop many times finally leads to optimized laser pulse shapes.

inspired by biological evolution (Fig. 13). At the beginning of the optimization, a pool of randomly initialized "individuals" (i.e., different laser pulse shapes having a "genetic code" composed of a linear array of LCD pixel voltages) is experimentally tested for control performance. The success of a pulse is judged with respect to an optimization task determined by the experimenter at the outset. Since more than one experimental observable may be optimized, a "fitness function" is employed which takes into account all required signals from the experimental output. The better the outcome of the quantum control experiment, the higher the fitness assigned to the individual laser pulse shape. After all individuals from the pool are evaluated, some percentage of the fittest laser pulses is selected for reproduction. We then use "cloning," "mutation," and "crossover" operations to produce a new pool of laser pulses. Since only the best individuals of the previous generation serve as parents of the new pool, a mechanism is provided wherein offspring can inherit good "genetic properties." It is often the case that—by virtue of crossover and mutation—some of these children are "better adapted" to the environment than their predecessors; i.e., they are better suited to achieving a given quantum control task. The new generation is then tested again, followed by selection and reproduction. By cycling this evolutionary loop a number of times, the fitness of the best individuals increases from generation to generation until an optimum is found. The most important property of this type of optimization is the generality of the search. In principle, it works for any type of system, and the user does not need to provide any type of model for the system's response. The user who has an optimization objective simply has to choose the appropriate fitness function for purposes of evaluating the individual pulses.

B. EXPERIMENTAL EXAMPLES

In our first implementation of adaptive femtosecond pulse shaping, we considered the process of second-harmonic generation (SHG) of 800-nm light in a thin BBO crystal. Since SHG is a nonlinear process, it can be used as a measure of the pulse duration. As such, if SHG is maximized in the learning loop (the imposed fitness function in this case), the end result should be the shortest possible pulse available with the given laser bandwidth (Yelin *et al.*, 1997; Baumert *et al.*, 1997b; Brixner *et al.*, 1999).

In our first example (Baumert *et al.*, 1997b), we started with titanium: sapphire oscillator laser pulses temporally broadened by propagation through a 10-cm rod of SF10 glass. The material-induced group-velocity dispersion (GVD) provides a predominant second-order spectral phase which increases the pulse duration from 80 fs to 1.2 ps. These pulses were then passed through the pulse shaper with the fitness goal programmed to maximize SHG. After approximately 150 generations within the evolutionary loop, the algorithm converged. Intensity autocorrelation traces are shown in Fig. 14 for the recompressed laser pulse

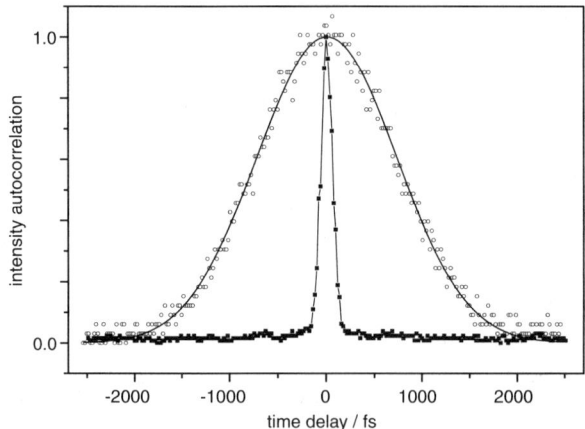

FIG. 14. Intensity autocorrelations of titanium:sapphire oscillator laser pulses temporally broadened by material dispersion (open circles) and recompressed by the adaptive pulse-shaping method (solid squares). The determined pulse durations are 1.2 ps and 80 fs, respectively.

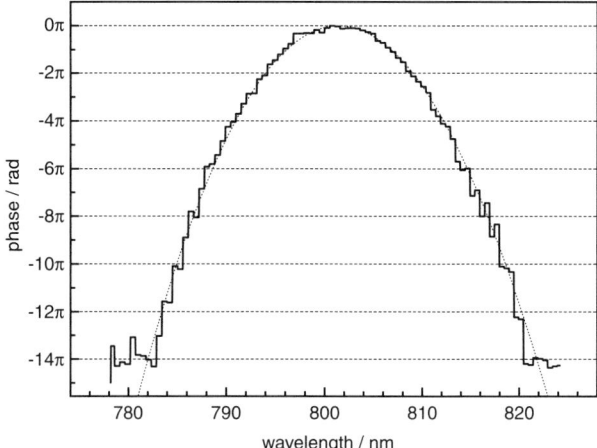

FIG. 15. Spectral phase applied with the LCD in the pulse compression experiment of Fig. 14. The phase found by the evolutionary algorithm in a 128-parameter optimization (solid line) almost perfectly resembles a quadratic phase of -2.59×10^4 fs^2 (dotted line). Phase jumps of 2π have been removed manually for this illustration.

(solid squares) and the GVD-broadened input pulse (open circles). These data show that the bandwidth limit of our oscillator can be achieved and therefore that the algorithm successfully compensates for phases introduced by the dispersive rod. As expected, the optimum phase found by the algorithm is a parabola of second order (Fig. 15). It should be noted that the optimization was run with 128 independent parameters, i.e., the individual LCD pixel voltages. For this experiment, it would have been sufficient to optimize a reduced set of parameters such as the first few coefficients in a Taylor expansion of the spectral phase. However, we are ultimately interested in optimizing highly complex quantum control problems (where a Taylor expansion may not be enough). Since the optimum in this 128-parameter search was found without previous knowledge of the input phase, the basic setup and the learning loop fulfilled their purpose well.

In our second example (Brixner et al., 1999), we considered the more technologically relevant problem of generating bandwidth-limited amplified femtosecond laser pulses from pulses which have acquired higher-order phase structures. In chirped-pulse amplification (CPA) systems, it is generally desired that phase contributions introduced by the stretcher, the amplifier, and other devices be removed by the compressor before entering the experimental apparatus. Accomplishing this with high precision requires elaborate alignment procedures. Since higher-order phase structures are very hard to remove, it is very demanding to provide bandwidth-limited laser pulses on a day-to-day basis. We therefore used the adaptive pulse shaping technology to optimize the pulse phases at the output of our CPA laser system. Again, we used maximization of SHG yield as our

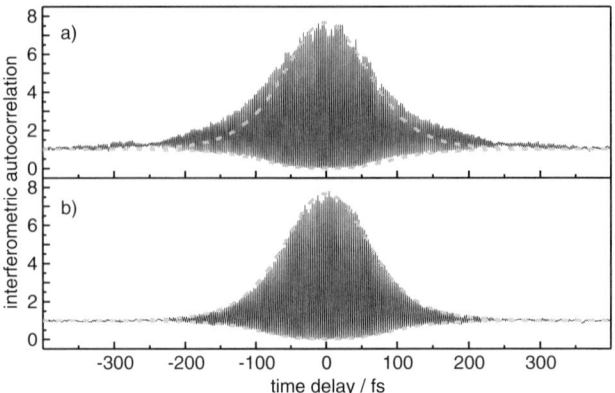

FIG. 16. Interferometric autocorrelations of titanium:sapphire amplified laser pulses before (a) and after (b) automated compression. The envelope of part (b) is included in part (a) for comparison.

optimization goal. Working with 128 independent parameters, the algorithm was able to remove all higher-order phase terms remaining in the output pulses after the conventional compressor. The interferometric autocorrelation of the input pulse (Fig. 16a) exhibits intensity contributions in the wings of the autocorrelation trace, which is a marker for uncompensated phase terms. On the other hand, the interferometric autocorrelation of the optimized laser pulse (Fig. 16b) fits well under a sech^2 envelope.

It can be recognized from the previous two experiments that the learning loop setup is able to provide bandwidth-limited oscillator or amplified femtosecond laser pulses without requiring initial phase characterization. However, in some experiments such as one-parameter quantum control with variation of linear chirp, laser pulses with specific phase profiles are needed. Very accurate chirping can be achieved by adaptive pulse shaping where complete pulse characterization is used as feedback signal. A suitable characterization technique is TADPOLE (Fittinghoff *et al.*, 1996), which is a combination of spectral interferometry (Lepetit *et al.*, 1995) and FROG (Trebino *et al.*, 1997). In this context, we have used the result from a TADPOLE measurement in a feedback loop (Brixner *et al.*, 2000). Figure 17 shows an example where a quadratic phase of 6000 fs^2 (solid line) was the target. The experimentally determined phase after application of the LCD voltages according to the approximate pulse-shaper calibration is marked with open squares. There is already qualitative agreement with the target, but a slight deviation is also seen. This deviation can be due to a number of factors such as imperfect pulse-shaper alignment, material dispersion in optical elements, or lens aberrations. After optimization (solid circles), the target phase is reached perfectly in the regions of nonvanishing intensity. Analogous to the generation of bandwidth-limited laser

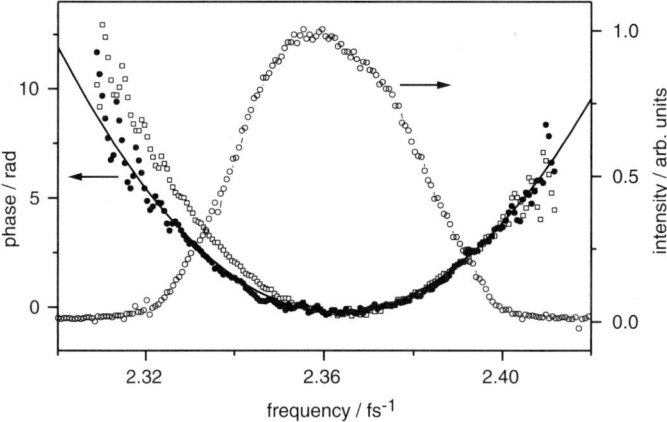

FIG. 17. Measured spectrum (open circles) and phase before (open squares) and after (solid circles) optimization, with a target phase of 6000 fs² (solid line).

pulses, the closed-loop setup eliminates the need for high-precision alignment or calibration, even when complicated laser pulses are the desired objective.

V. Many-Parameter Quantum Control in the Gas Phase

A. INTRODUCTION

Each of the examples of quantum control discussed in Section III employed a single control parameter. We now use the technology just described, which is capable of optimizing a much larger number of control variables. As has been discussed, spectral phase shaping leads to temporal phase and amplitude modulation. The adaptive many-parameter approach therefore automatically includes the one-parameter control schemes of pulse duration, pump-probe delay, interferometric excitation, and linear chirp. In the following experiments, we have used this phase-only shaping technique, but if additionally the spectral amplitude is modulated, it is in principle possible to include also the control schemes of wavelength and energy variation.

In general, if one of these "simple" control techniques already constitutes the optimum, the closed-loop learning setup should arrive at the same conclusion with the laser field shaped accordingly. If, however, more complex wavepacket guiding is needed, the automated optimization should find it as well. An intuitive picture of this process is that the wavepacket motion on the complex molecular potential energy surfaces is controlled by "offering" those wavelengths at each specific point in the temporal evolution, which are needed to transfer the wavepacket onto some other surface which is better suited to its further evolution. Many of these transient Franck–Condon windows might be envisioned throughout the control process so that an optimal steering of the wavepacket finally leads to a specific product channel. Additionally, it should be noted that with repeated Rabi cycling, states can be reached which are not accessible with a single-photon transition. This occurs when within a single Rabi cycle a photon from the blue edge of the broadband laser spectrum is absorbed while emission is stimulated by a photon from the red edge. This leaves the molecule in a higher ground-state vibrational level. Depending on where the last Rabi cycle of this sequence is stopped, the population is left in an energetically high-lying vibrational level of the ground or the excited state. This provides a mechanism for "simulating" the absorption of a photon whose wavelength may not even be included in the original laser spectrum.

Since the issue of Brumer–Shapiro two-pathway coherent control versus "optimal" quantum control is discussed frequently, we include a brief comment here. The basic idea of Brumer and Shapiro (1986) was to couple initial states with energetically degenerate final states by two simultaneously applied light fields. Interference between the two excitation pathways—resulting from the two different light fields—can then be used to modulate the transition probabilities to the individual (degenerate) final states. The best-known special case of this technique employs an excitation scheme where the two pathways correspond to absorption of one photon of frequency ω versus absorption of three photons of frequency $\omega/3$. However, this is not a necessary requisite for the two-pathway interference technique. It is also possible to use a generalized scheme where the different pathways (there might be even more than two) employ absorption of two photons (for example) of different energy in the following way (Bjorkholm and Liao, 1974; Meshulach and Silberberg, 1998, 1999). Suppose that the energy separation between initial and final states is 2ω; then in pathway 1, two photons with frequencies $\omega + \Delta\omega_1$ and $\omega - \Delta\omega_1$ could be absorbed, whereas in pathway 2, two photons with frequencies $\omega + \Delta\omega_2$ and $\omega - \Delta\omega_2$ are used. Again, by controlling the phases of these laser fields, interference terms arise and can be exploited for coherent control of final state populations. However, since the light frequencies employed need not be far apart (i.e., the method works for small $\Delta\omega_i$), the bandwidth of a femtosecond laser can be used to provide the necessary spectral components. A pulse shaper can then be employed to modify the phase relations among the different frequency components, with the effect that the Brumer–Shapiro coherent control concept is also included in the adaptive quantum control scheme with specifically shaped femtosecond laser pulses.

B. TESTING THE TECHNOLOGY WITH $Fe(CO)_5$

In the first example we employed adaptive femtosecond pulse shaping to control the photodissociation and ionization of the organometallic complex $Fe(CO)_5$ (Assion *et al.*, 1998b,c; Bergt *et al.*, 1999). The reason to start with this molecule is that we had detailed knowledge of its ionization and fragmentation dynamics studied in single-pulse and pump-probe experiments (Bañares *et al.*, 1997, 1998). These experiments showed that the molecule could be controlled with single-parameter schemes. The results of one-parameter and automated many-parameter control can therefore be compared to check the performance of the adaptive optimization scheme.

In the first series of experiments we concentrate on the two photodissociation channels explored in Sections III.B–III.D and we investigate direct ionization leading to $Fe(CO)_5^+$ versus complete fragmentation and ionization leading to Fe^+. In the learning algorithm, therefore, the fitness function was set with the goal of maximizing and minimizing the ratio ($Fe(CO)_5^+/Fe^+$) between these two channels. If, as in this case, signal ratios are to be optimized, one often has to take into account the absolute signal height. The calculated fitness value can become very high if the value in the denominator of the fitness function is very small. Since we do not want signals below the noise level to contribute substantially to the optimization procedure, we introduced a threshold value in the denominator which is used instead of the measured ion signal whenever the ion signal falls below that level (Bergt *et al.*, 1999). Using suitable threshold values, the optimization procedure was capable of controlling the $Fe(CO)_5^+/Fe^+$ ratio between 2.2 and 0.2. This result is seen in the mass spectra corresponding to the maximization (solid blocks) and the minimization condition (open blocks) shown in Fig. 18. The distribution of the remaining fragments $Fe(CO)_n^+$, $n = 1, 2, 3, 4$, was not included in the fitness function and is therefore not expected to behave in a predictable way.

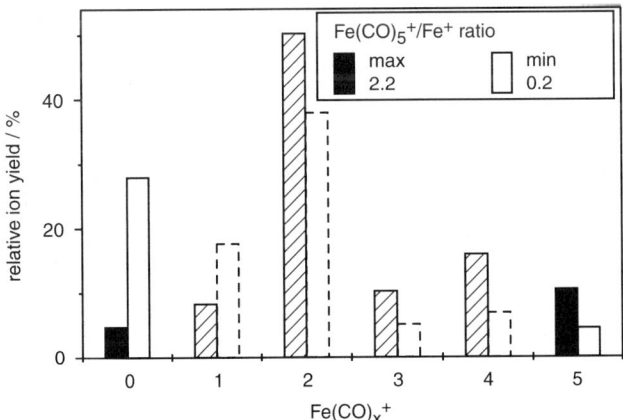

FIG. 18. Evaluation of $Fe(CO)_5$ photoproduct mass spectra. Relative ion yields are plotted for $Fe(CO)_5^+/Fe^+$ maximization (solid and shaded blocks) as well as minimization (open and dashed blocks). The $Fe(CO)_5^+/Fe^+$ ratio is switched between 2.2 and 0.2, respectively.

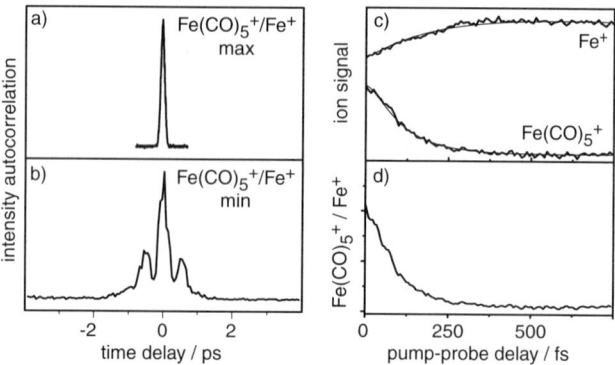

FIG. 19. Comparison of many-parameter and one-parameter control in $Fe(CO)_5$. Intensity autocorrelations are shown for automated many-parameter $Fe(CO)_5^+/Fe^+$ maximization (a) and minimization (b). Conventional pump-probe transients of $Fe(CO)_5^+$ and Fe^+ (c) are used to calculate the $Fe(CO)_5^+/Fe^+$ ratio in the one-parameter experiment as a function of pump-probe delay (d).

It is now very interesting to compare the optimum laser pulse shapes found by the algorithm with the results from the one-parameter control schemes of Section III.D (Bergt et al., 1999). Intensity autocorrelations for the maximization and the minimization experiment are shown in Figs. 19a and 19b, respectively. In the case of $Fe(CO)_5^+/Fe^+$ maximization, an ultrashort, bandwidth-limited laser pulse is achieved. On the other hand, the minimization experiment reveals a three-peak structure, indicating a temporal double pulse with a 500-fs interpulse delay. Figure 19c replots the conventional $Fe(CO)_5^+$ and Fe^+ transients of Fig. 8. Dividing the $Fe(CO)_5^+$ transient by that of Fe^+ (Fig. 19d), it can be seen that the highest $Fe(CO)_5^+/Fe^+$ ratio is achieved with overlapping pump and probe laser pulses, consistent with the bandwidth-limited pulse from the automated control experiment. For increasing pump-probe delay, the $Fe(CO)_5^+/Fe^+$ ratio drops until, at approximately 400–500 fs, the minimum is reached. This is again the result that is found by the evolutionary algorithm. The actual product ratios should not be compared quantitatively, however, since the temporal intensities in the conventional pump-probe experiment and in the automated control experiment were different. (It was seen in Section III.B that pulse energy variation can affect product distributions.) The evolutionary algorithm of course always searches for optimum results within the given experimental conditions such as pulse energy, laser bandwidth, or molecular beam properties. Nonetheless, the general correspondence between one- and many-parameter control is a very promising confirmation of the automated quantum control setup. We can conclude from our comparison that the pump laser pulse (or the first part of the optimum double pulse) induces some vibrational wavepacket dynamics on excited electronic surfaces that in the end lead to dissociation of the $Fe(CO)_5$ molecule. If the probe laser pulse (or the second part of the double pulse) arrives after the dissociation is finished, the ionization step produces bare Fe^+. If, on the other hand, the probe laser pulse arrives at a time closer to the pump pulse, the neutral dissociation processes are not complete and ionization produces more $Fe(CO)_5^+$. Successful control of this molecule constitutes the first demonstration that information about induced photoprocesses, namely, the dissociation time, can be extracted from the results of an automated many-parameter optimization experiment.

Depending on the potential energy surfaces involved in the photoprocesses, it might be advantageous to use shaped laser pulses with different center wavelengths to control the photochemical outcome of certain molecules. In this context, we conducted a series of experiments on $Fe(CO)_5$ with shaped pulses of 400-nm wavelength (Bergt et al., 1999). Due to the limited transmission of the liquid crystal molecules inside the pulse shaper, however, it is not possible to phase-shape 400-nm laser pulses directly. We therefore employed a frequency-doubling crystal (and dichroic mirrors to remove the fundamental) between the 800-nm pulse shaper and the molecular beam. In this configuration the temporal phase and amplitude information of the shaped pulses is not lost in the SHG process, and the 400-nm laser pulses thus generated can be used in automated quantum control experiments. In addition, we worked with a more versatile fitness function capable of addressing the question of absolute signal heights. This is valuable for future applications, where it is important to know if product ratios (i.e., the "purity" of the photoproducts) as well as total yields (i.e., the generation efficiency) can be controlled. The generic fitness function for this purpose was $f(x, y) = ax/y + bx + cy$, where x and y are the signal heights of the wanted and unwanted photoproduct, respectively, and a, b, and c are parameters which determine the importance of ratio optimization with respect to absolute signal optimization. This concept is also used in optimal control theory, where "penalty functions" with different weighting factors have been introduced to deal with multiple optimization goals (Peirce et al., 1988).

The results in the case of $x = Fe(CO)_5^+$, $y = Fe^+$ [i.e., $Fe(CO)_5^+/Fe^+$ maximization] are shown in Fig. 20 for three different parameter sets a, b, c. If most of the weight is put on the optimization of the $Fe(CO)_5^+/Fe^+$ ratio (left column), the highest value of the ratio within this series is achieved. The absolute signal height, however, is rather small. If more and more weight is put on the absolute product yield (middle and right columns), the total signal can be increased substantially. This, however, occurs at cost of the maximum achievable product ratio. These results show that it is indeed possible to control absolute product yields in relation to relative product yields. The experimenter simply has to specify the relative importance of these optimization goals using a suitable fitness function. It should also be noted that the results in the case of $x = Fe^+$, $y = Fe(CO)_5^+$ [i.e., $Fe(CO)_5^+/Fe^+$ minimization, not shown here] are qualitatively similar. In summary, $Fe(CO)_5^+/Fe^+$ maximization and minimization experiments show that it is

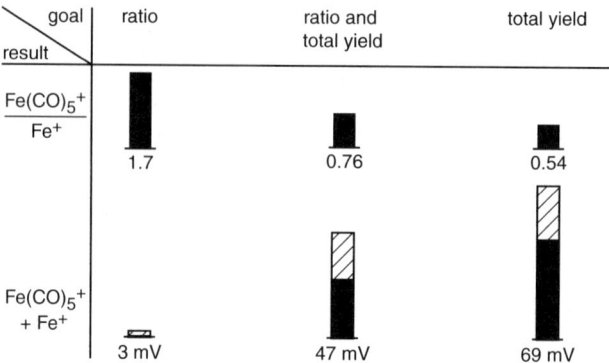

FIG. 20. Summary of results for many-parameter control in Fe(CO)$_5$ using different fitness functions. The upper line of solid blocks in this table represents the achieved Fe(CO)$_5^+$/Fe$^+$ ratios, whereas the lower line indicates the absolute individual product yields of Fe(CO)$_5^+$ (shaded) and Fe$^+$ (solid). The optimization goal was maximization of the Fe(CO)$_5^+$/Fe$^+$ ratio along with maximization of the total product yield (Fe(CO)$_5^+$ + Fe$^+$). In the left column, most weight has been put on a maximization of the Fe(CO)$_5^+$/Fe$^+$ ratio. In the middle column, both ratio and total yield have been given some weight, and in the right column, most weight was on the maximization of the total yield.

possible to control the Fe(CO)$_5^+$/Fe$^+$ ratio between "high" and "low." In addition, it is simultaneously possible to control the total product yield.

C. SELECTIVE DISSOCIATION AND COMPLEXITY ANALYSIS WITH CpFe(CO)$_2$Cl

The previous examples have successfully demonstrated multiparameter control of fragmentation versus ionization in Fe(CO)$_5$. We now consider whether control can be achieved in more complex molecules containing a variety of chemical bonding configurations. We used the metal complex CpFe(CO)$_2$Cl (where Cp = C$_5$H$_5$) as a prototype (Assion et al., 1998c). Of the many possible dissociation channels, we selected two for optimization of a branching ratio. In the first channel, single carbonyl loss and ionization leads to CpFeCOCl$^+$, while in the second channel, more complete fragmentation leads to FeCl$^+$. The two optimization goals explored were maximization and minimization of the CpFeCOCl$^+$/FeCl$^+$ product ratio. If such a ratio can be controlled, it is a significant achievement, because for the first time it would be possible to control the generation of photoproducts which arise from the breakage of different types of bonds. In other words, it is not just a matter of controlling fragmentation versus ionization. After optimization with the evolutionary algorithm, the photoproduct distribution indeed changed in the way requested by the fitness function. Figure 21 shows that the CpFeCOCl$^+$/FeCl$^+$ ratio can be switched between about 5:1 (left side, maximization) and 1:1 (right side,

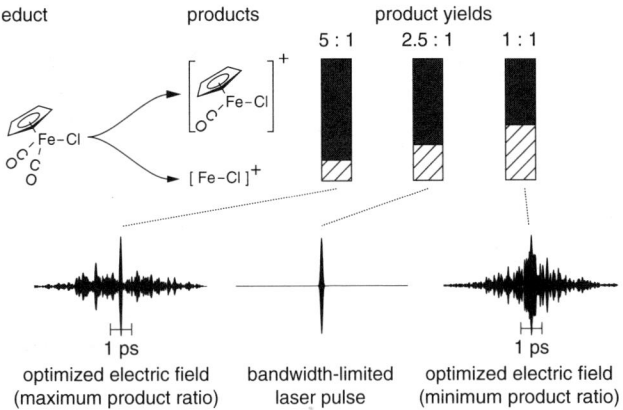

FIG. 21. Many-parameter quantum control in CpFe(CO)$_2$Cl. The top part displays relative yields of the two investigated photoproduct channels leading to CpFeCOCl$^+$ (solid blocks) and FeCl$^+$ (shaded blocks). The bottom part shows the temporal electric fields leading to a maximum (left) and minimum (right) CpFeCOCl$^+$/FeCl$^+$ ratio, as well as the electric field of a bandwidth-limited laser pulse (middle) leading to an intermediate branching ratio. The electric fields are calculated by Fourier-transforming the measured laser spectrum and the spectral phase applied by the pulse shaper.

minimization). The corresponding temporal electric laser fields are also shown in this figure and display much more complex properties than in the example of Fe(CO)$_5$. It can also be seen from Fig. 21 (middle part) that it is not possible to achieve the optimum ratios with a bandwidth-limited laser pulse. Arbitrarily phase-shaped long laser pulses (the starting place for either optimization) are not sufficient either. Successful control of the CpFeCOCl$^+$/FeCl$^+$ ratio is therefore not a trivial matter of adjusting the laser pulse "intensity." This can also be inferred from the fact that the temporal peak intensity was reduced (as compared to the bandwidth-limited pulse) in both the maximization and the minimization experiment cases, whereas in one case this "intensity" reduction led to an increase, and in the other case to a decrease, of the desired product ratio. Of course it was not just the peak intensity which changed; rather, the detailed temporal structures of the optimized pulse shapes are responsible for the results of these experiments.

In addition to "making" specific photoproducts, it is interesting to "understand" why and how it is possible to achieve control. In the previous example, control appears to require extremely complex electric fields for both optimization goals. One question is therefore: Can the degree of complexity in quantum control problems be accessed experimentally? To investigate this, we ran several optimizations with artificially reduced search parameter spaces (Bergt *et al.*, 2000; Brixner *et al.*, 2001a). Specifically, we compared 128-parameter optimization with 64- and 32-parameter optimization where sequences of two

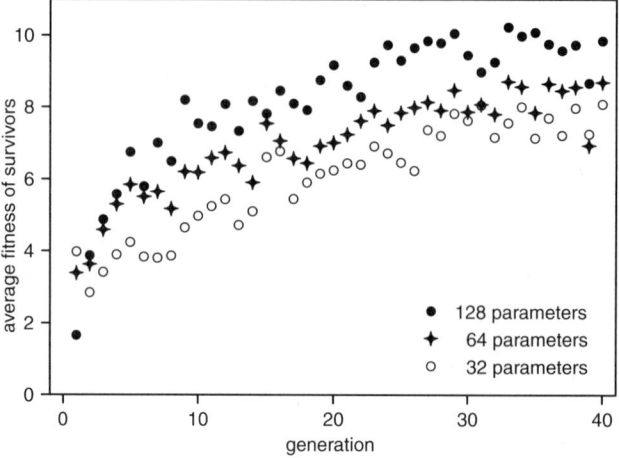

FIG. 22. Reduction of parameter space in $CpFeCOCl^+/FeCl^+$ maximization. The three experiments are run with 128, 64, or 32 independent optimization parameters such that sequences of zero, two, or four neighboring pixels, respectively, are coupled together.

or four neighboring LCD pixels, respectively, were coupled together. This coupling reduces the spectral resolution of the phase-shaping apparatus, thereby reducing the complexity of the available electric fields. The evolution of the $CpFeCOCl^+/FeCl^+$ ratio during maximization is shown in Fig. 22 for the three search parameter sets. The highest product ratio is achieved with 128 independent optimization parameters, and this decreases with each successive reduction in the parameter space. This indicates that specific features of the electric field profiles of Fig. 21 are meaningful and necessary to reach optimum control performance. As a more refined general concept of complexity reduction, it was recently suggested (Geremia et al., 2000) that the learning algorithm itself find out which control variables are needed. By introducing suitable "penalty" terms, it should be possible to reduce the complexity of the electric field (such that interpretation is not obscured by irrelevant features) without substantially diminishing the control performance.

The complexity of a control problem can be made visible using what we call correlation experiments (Bergt et al., 2000; Brixner et al., 2001a). Simultaneous to the measurement of photoproduct yields, we record the amount of SHG light produced by each individual laser pulse shape throughout the $CpFeCOCl^+/FeCl^+$ maximization. The "correlation diagram" in Fig. 23 reveals a broad distribution and therefore no clear correlation between the optimization goal and the SHG efficiency. This is most clearly seen by considering one specific SHG intensity and then observing

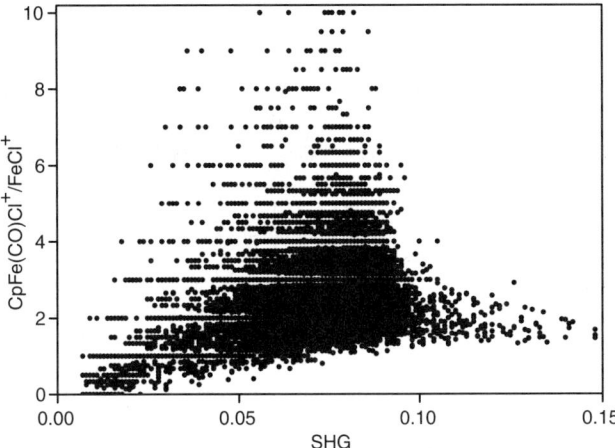

FIG. 23. Correlation diagram of $CpFeCOCl^+/FeCl^+$ product ratio versus SHG efficiency. The two observables are plotted for all individuals during the three $CpFeCOCl^+/FeCl^+$ maximizations of Fig. 22.

the large range of different $CpFeCOCl^+/FeCl^+$ values achieved, each of which depends on the exact pulse shape. Therefore, the mechanism of control active in this molecule cannot be ascribed to multiphoton absorption probabilities alone. We believe that generalized concepts of correlation diagrams can be exploited to shed light on the fundamental processes which are involved in quantum control. One could think of investigating correlations between different molecular fragments, SHG, other nonlinear signals, and also the structure of the electric field itself.

D. ACCESSING THE "PROBLEM OF INVERSION" WITH ATOMIC CALCIUM

In the previous example, complexity analysis gave us hints about control mechanisms in complex systems. What is really desired, however, is to gain detailed insight into the photoprocesses through analysis of the structure of the optimized laser field itself. This is the so-called problem of inversion (see, for example, Rabitz and Zhu, 2000). We believe that we have made steps toward achieving this during investigations of the single and double ionization in atomic calcium. The process of double ionization, in general, is not yet understood in detail, and the relevant mechanisms are still debated. One of the open questions is whether the two electrons are emitted sequentially or simultaneously. Correlated electron emission has been recently reported in the case of helium (Weber *et al.,* 2000). The technique of adaptive pulse shaping can be used to address such questions because a large number of potential excitation pathways can be explored.

38 T. Brixner et al.

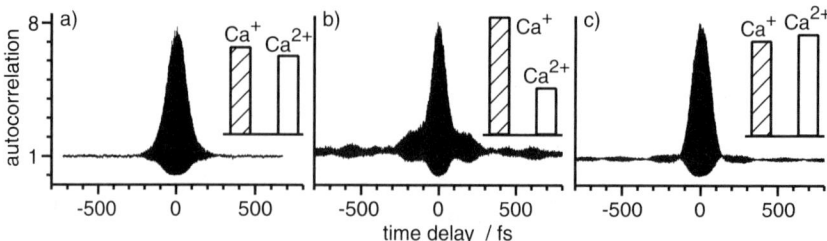

FIG. 24. Optimization of calcium single and double ionization. The Ca^+ yields (shaded blocks) and the Ca^{2+} yields (open blocks) are plotted along with interferometric autocorrelations for unmodulated laser pulses (a), laser pulses optimized for Ca^+ production (b), and laser pulses optimized for Ca^{2+} production.

In this experiment (Strehle and Gerber, 2001), we began by focusing unmodulated femtosecond laser pulses of 80 μJ energy into an atomic calcium beam. The resulting Ca^+ and Ca^{2+} yields are shown in Fig. 24a along with an interferometric autocorrelation of the bandwidth-limited laser pulse. The evolutionary algorithm was then used to maximize the amount of Ca^+ production, with a surprising result (Fig. 24b). Single ionization of atomic calcium requires at least four 800-nm photons. Because of this nonlinearity, one could expect that an optimally short laser pulse with associated maximized peak intensity would be generated. However, this is not the case, as can be seen from the "wings" in the autocorrelation trace. With specific pulse structure, it is possible to obtain more Ca^+ ions than with the bandwidth-limited laser pulse (Fig. 24a). For a possible explanation of this behavior, consider the Wigner trace (Paye, 1992) of the optimized laser pulse (Fig. 25a) and the energy level diagram of calcium (Fig. 25b). The laser pulse shown consists mainly of two contributions: a linearly up-chirped part between about -500 and $+300$ fs, in which the momentary light frequency increases from position A to position B; and an unchirped part around 0 fs, in which a broad range of frequencies is available simultaneously. In the beginning of the linearly chirped part (Fig. 25a, position A), the energy of four photons (Fig. 25b, solid arrows) is just sufficient to reach the ionization threshold I_p. However, in the subsequent temporal evolution, the momentary intensity of the laser pulse increases. This leads to a ponderomotive shift U_p (Liao and Bjorkholm, 1975), which increases the ionization potential to higher energies $I_p + U_p$ (Fig. 25b, dashed-dotted line). The density of states in the singly ionized continuum is largest in the direct proximity of the ionization threshold. An optimized transfer of population can therefore be achieved with linear up-chirp because the increasing energy of the four photons absorbed (Fig. 25b, dashed arrows) always equals the increasing energy difference between ground state and shifted ionization potential. The second contribution in the optimum pulse shape represents a broad range

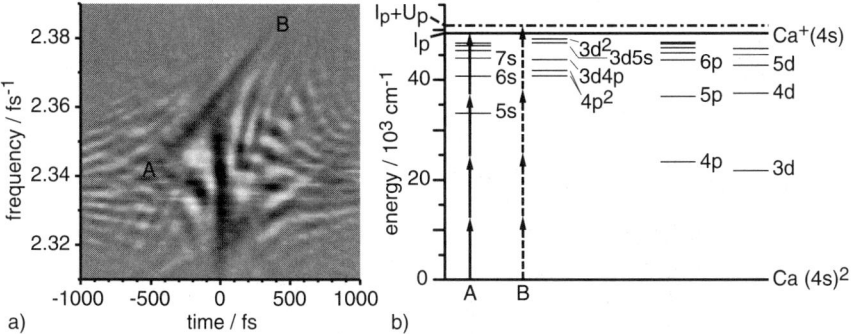

FIG. 25. Explanation of the maximized Ca$^+$ production. The optimized laser pulse shape is plotted as Wigner transform (a). The gray color seen in the outer regions of this plot indicates values of zero, darker shading indicates positive values, and lighter shading indicates negative values. The suggested ionization mechanisms are illustrated in the energy level diagram of atomic calcium (b). The weak-field ionization potential is labeled with I_p. It can be reached by four-photon absorption of intermediate photon energies (A, solid arrows). However, the ponderomotive shift U_p leads to an increase of the ionization threshold up to $I_p + U_p$ (dashed-dotted line) so that photons of higher energy are required (B, dashed arrows).

of frequencies from the lower-energy part of the laser spectrum. The associated photon energies are not sufficient to achieve four-photon ionization. In the case of the required five-photon absorption, however, final states are reached that are energetically high up in the ionic continuum. The ponderomotive shift in that case does not have a big effect on the available density of states. The unchirped contribution with associated high intensity (as seen in the distribution of simultaneously arriving frequencies around time zero) is then better suited for optimum transfer.

In an additional experiment, the evolutionary algorithm was used to maximize the amount of the doubly ionized Ca^{2+}. In the end of the optimization, the Ca^{2+} signal is higher (Fig. 24c) than in the previous two cases. Again, the optimal laser pulse is not bandwidth limited, as can be inferred from the contributions in the wings of the interferometric autocorrelation trace. In total, 12 800-nm photons are needed to reach the doubly ionized continuum from the neutral ground state. By considering this nonlinearity, one could expect that the optimized laser pulse would be as short as possible. However, it appears that by phase shaping the laser pulse, intermediate resonances can be exploited such that the transfer of population into the doubly ionized continuum is optimized. The results suggest that the adaptive laser pulse optimization has found an excitation scheme in which the highly nonlinear (12-photon) absorption process needed for double ionization is divided into several resonantly enhanced parts. In this way, a maximum nonlinearity of only fourth-order remains, which is consistent with findings from energy-variance experiments. The details of this excitation scheme are not known accurately at present, but it is evident that double-excited states are involved as intermediate resonances.

The optimizations of calcium single and double ionization constitute examples in which it should be possible to solve the so-called problem of inversion. We presented first (qualitative) steps toward explaining the observed behavior. Although the details are still somewhat speculative, it is clear from the optimized pulse shapes that bandwidth-limited laser pulses are not optimal for either single or double ionization, and that nontrivial excitation pathways can be exploited by the shaped pulses. Detailed quantum-mechanical calculations are necessary to obtain a quantitative understanding of all relevant mechanisms.

VI. Many-Parameter Quantum Control in the Liquid Phase

A. INTRODUCTION

It is evident from the previous experimental examples that adaptive feedback pulse shaping holds a great deal of promise as a general methodology for controlling the photochemistry of molecules. This is particularly true in the case of complex systems and solution-phase systems, where knowledge of the excited-state Hamiltonian is difficult to obtain. From this, it is exciting to speculate that this technique might someday find its way into the toolbox of synthetic chemists as a means of controlling the production of reactive intermediates or the formation of specific products. It is clear at the outset that if adaptive pulse shaping can have promise as a general synthetic method, it must be shown to be viable in the medium where most preparative chemistry is done. This medium is of course the solution phase.

In addition to synthetic photochemical motivations, physical motivations present themselves as both challenges and goals for achieving control in the solution phase. The primary motivation addresses the question of feasibility. In the experiments discussed previously on Na_2 in the gas phase, control of photochemical branching ratios using sequences of short laser pulses relied on the existence and persistence of vibrational coherence. It was then shown that optimal control of photochemical branching ratios for the organometallic species $CpFe(CO)_2Cl$ in the gas phase required a substantially more complex electric field. Presumably, this is in part because the time-dependent distribution of energy that takes place within the larger molecule is itself a complex process. We should expect that introduction of a solvent environment strongly increases the complexity of the time-dependent distribution of energy in comparable systems as well as providing an efficient mechanism for loss of vibrational coherence. For coherent control to be viable in the solution phase, these difficulties must be overcome. It should be stated that adaptive pulse shaping, because of the implementation of feedback as a learning mechanism and the use of large parameter spaces, is currently the only feasible way of approaching this difficult physical problem. What one stands to gain if control in solution is viable is information about what types of excited-state dynamics are amenable to control in the presence of solvent/solute interactions. It is also enticing to speculate that electric fields that are found to be optimal for particular excited-state dynamics will include information about the time-dependent response of the solvent throughout the evolution of the molecule.

Experimental challenges are not limited to the issues of feasibility just discussed. In contrast to experiments in the gas phase, propagation effects change the shape of the laser pulse as it travels through the solvent. This effect can be accounted for by the learning algorithm, which again searches for optimal results under given experimental conditions. However, the question remains which feedback signal has to be used for adaptive quantum control in the liquid phase. In contrast to the gas-phase experiments discussed previously, mass spectrometry is of course not possible and alternative feedback signals have to be found. The choice of feedback signal depends on the individual problem, since different techniques probe different aspects of the induced photoprocesses. Potential feedback implementations include the detection of emission [as used by Bardeen *et al.* (1997) on a laser dye], Raman-type techniques [as used by Weinacht *et al.* (1999b) on molecular liquids], differential transmission in a pump-probe setup [as used by Kunde *et al.* (2000) on semiconductors], or four-wave mixing [as used by Hornung *et al.* (2000b) on K_2 in a heat pipe].

B. EXPERIMENT

Our starting place is a technique that requires only a single laser beam (the shaped pulse) and relies on a photophysical observable for feedback. We have chosen to consider a molecular chromophore [Ru(dpb)$_3$](PF$_6$)$_2$ (where dpb = 4,4'-diphenyl-2,2'-bipyridine) dissolved in methanol which emits light following formation and evolution of excited-state population (Juris *et al.*, 1988; Damrauer *et al.*, 1997a). Because the emission occurs from a lower-energy excited state accessed via nonradiative relaxation, it is sensitive only to the excited-state population and is not coupled directly to the excitation. It can therefore provide information about the excited-state yield and can be used as feedback for controlling aspects of the excitation process (Brixner *et al.*, 2001b).

The chromophore used is shown in Fig. 26a along with important photophysical properties measured in methanol (Fig. 26b). It is a member of the ruthenium(II) polypyridyl family, which has received widespread attention for reasons which

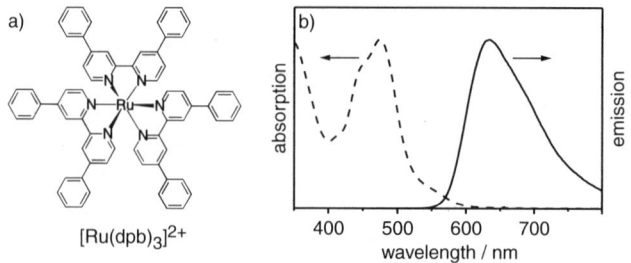

FIG. 26. Structure of [Ru(dpb)₃]²⁺ (a) and linear absorption/emission spectra (b).

include the use of these complexes as photoinduced electron-transfer sensitizers (Juris *et al.*, 1988; Ferraudi, 1988) and in applications leading to solar energy conversion (O'Regan and Gratzel, 1991). Like other members of its class, [Ru(dpb)₃]²⁺ has a strong visible metal-to-ligand charge transfer (MLCT) absorption band (Juris *et al.*, 1988). When this band is excited, formal oxidation of the metal center takes place (II → III) concomitant with reduction of a single ligand species. Following excitation of this molecule in typical solution-phase conditions, emission is observed on a microseconds time scale in a broad spectral region red-shifted from the MLCT excitation (Damrauer *et al.*, 1997a). A simplified picture of the excited-state evolution is shown in Fig. 27a. Here, absorption within the MLCT band results in Franck–Condon excitation of the singlet surface. The molecule then undergoes rapid intersystem crossing (~200 fs)

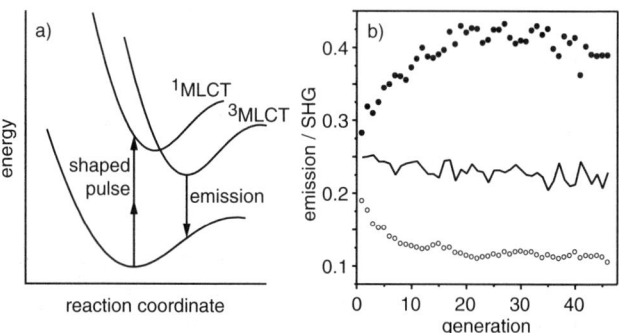

FIG. 27. Optimization of [Ru(dpb)₃]²⁺ emission in a methanol solution. Two-photon excitation of the ¹MLCT state with shaped pulses is followed by rapid intersystem crossing to the ³MLCT state and spontaneous emission (a). The emission/SHG ratio is plotted as a function of generation number (b) for the case of maximization (solid circles), minimization (open circles), and unmodulated laser pulses (solid line).

and vibrational relaxation to the lowest-energy triplet state (Damrauer and McCusker, 1999; Damrauer et al., 1997b). Because of weak coupling to the singlet ground-state surface, the ^{3}MLCT is long-lived and emits light with relatively large radiative quantum yield (Damrauer et al., 1997a). Because 800-nm light is used in our pulse shaper, excitation of this species requires at least two-photon absorption. Emission from the ^{3}MLCT can be collected at wavelengths disparate from the excitation. In this way, the relative emission intensity provides information about the molecular excitation process to be used in the feedback step. Specific details about our experimental setup are described in Section II.B.

There are a number of examples in the literature—mostly from the groups of Wilson (Bardeen et al., 1997, 1998a) and Shank (Bardeen et al., 1995, 1998b; Cerullo et al., 1996)—where phase-shaped pulses have been used to influence excited-state population of complex molecules in solution. In these experiments, photophysical emission is also used as the observable sensitive to excited-state population. In this context, Bardeen et al. (1997) have also demonstrated automated feedback control using the fluorescent dye IR125 and a five-parameter search space that includes pulse energy, central wavelength, spectral width, linear chirp, and quadratic chirp. In that experiment, the shaped pulse is able to excite the chromophore under control with a one-photon absorption. The influence of the pulse shape comes into play because of the possibility of a second photon interaction that can stimulate emission from the evolving wavepacket. The results then show that for intense laser fields, the genetic algorithm finds laser pulses with strong positive linear chirp. This result is consistent with previous one-parameter control experiments (Bardeen et al., 1998a,b; Cerullo et al., 1996).

In the experiment we have pursued, at least two photons are required for excitation of the chromophore. The use of a 128-pixel pulse shaper then provides access to extremely complex time-dependent electric fields. In the control experiment of Bardeen et al. (1997), only two parameters influencing the time-dependent electric field (linear and quadratic chirp) were varied and searched.

C. RESULTS AND DISCUSSION

Our initial experiments focused on two optimization goals, maximization of molecular emission and maximization of SHG. We know that in the optimization of SHG, the end result is a transform-limited pulse. The emission yield during the SHG optimization (not shown here) increases and reaches its highest level when the shortest possible laser pulse has been found. When emission itself is maximized, a very similar evolution curve is observed. This suggests that multiphoton excitation is the dominant nonlinear process explored using this optimization goal, and the adaptive learning algorithm finds solutions which primarily satisfy this excitation mechanism.

With these optimization goals, we cannot reasonably distinguish qualities of pulse shapes which discriminate between the physical processes of MLCT excitation of a molecule in solution and SHG of the laser beam in a nonlinear crystal. Because of this, we have explored two additional optimization goals: maximization and minimization of the ratio emission/SHG. The idea behind this choice of optimization goals is to cancel the dominating effect of the nonlinear excitation mechanism because of its presence in both the numerator and denominator of the ratio. If the mechanisms leading to the emission or SHG are identical, then optimization of their ratio should be impossible. Optimization curves for these experiments are shown in Fig. 27b. What is seen is very clear evolution of these ratios from the initial generation of random-phased individuals to maximization as well as minimization conditions. This shows that the learning algorithm is able to find laser fields which distinguish between the two excitation processes involved. Also included within the figure is the value of the ratio throughout the optimization procedure which results from an unmodulated (short and intense) laser pulse. It is seen that for both optimization goals, intense laser pulses are not satisfactory for achieving an optimal condition.

A useful way of looking at the full data set is presented in Fig. 28. Here we plot the values of the two observables considered, emission and SHG, for all individuals in all generations. The space defined by these axes is obviously very relevant for the ratio being maximized and minimized. Within this space, any ratio explored by the algorithm can be described by a line of a particular slope. The

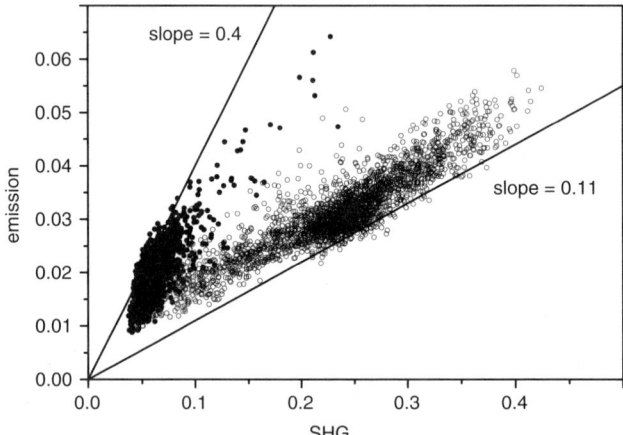

FIG. 28. Correlation diagram of emission versus SHG efficiency. The values of the two observables are plotted for all individual laser pulse shapes during maximization (solid circles) and minimization (open circles) of emission/SHG. The two straight lines correspond to the maximum and minimum emission/SHG ratios achieved in the end of the optimization.

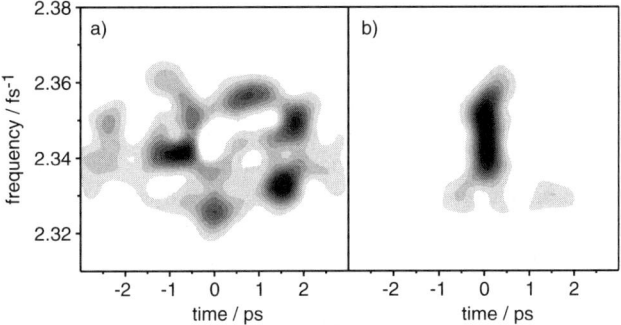

FIG. 29. Husimi transforms of the optimal laser pulse shapes leading to emission/SHG maximization (a) and minimization (b). The electric fields are calculated by Fourier-transforming the measured laser spectrum and the spectral phases applied by the pulse shaper.

two solid lines shown in Fig. 28 have slopes of 0.4 and 0.11, corresponding to the values of the ratio achieved in the maximization and minimization procedures (vide supra). It is seen nicely from the two distributions of data points that the learning algorithm is evaluating pulse shapes which explore distinct regions of this space. This observation supports the idea that the mechanisms involved in the two excitation processes can indeed be separated and selectively explored.

To visualize the pulse shapes found by the learning algorithm for the particular optimization goals, we plot the electric fields in time and frequency space using the Husimi representation (Husimi, 1940; Lalović et al., 1992). These are shown for the best individuals of the last generation of the emission/SHG maximization (Fig. 29a) and minimization (Fig. 29b) optimizations. What is observed are complex electric fields—most notable in the maximization field—as well as marked differences in the fields suitable for their particular goals. At this time, an interpretation of why these field shapes favor their respective optimization goals is difficult due to the electronic and physical complexity of the solution phase charge-transfer system itself. Nonetheless, certain brief comments can be made. For example, in the maximization experiment it is intriguing that we observe a discretization of pulse energy into smaller pulselike features in both time and frequency space. It seems reasonable that these pulselike features in time are necessary for achieving two-photon excitation of the molecule. We are also exploring the possibility that the discretization in frequency space indicates a sensitivity of the pulse to vibrational structure of the charge-transfer molecule or to vibrational dynamics occurring concomitant with the electronic excitation. In summary, with the large parameter space of 128 spectral phase values, electric fields can be found which exploit the differences in the physical mechanisms leading to charge-transfer excitation versus SHG.

VII. Summary and Outlook

In this contribution we have reviewed a number of experiments from our laboratories aiming at femtosecond quantum control. We started our discussion of experimental examples with one-parameter open-loop control schemes. There, manipulation of a single control variable (pulse energy, pulse duration, pump-probe delay, phase-sensitive excitation, or linear chirp) led to variations in the response of the sample molecules Na_2, Na_3, and $Fe(CO)_5$. The outcome of these photoreactions was determined by time-of-flight (TOF) mass spectrometry or by photoelectron kinetic energy analysis. In all examples, it was possible to control branching ratios of different photoreactive channels. These one-parameter experiments can therefore serve as a toolkit to study basic quantum control mechanisms.

In general, however, optimal control of complex systems requires more flexible laser fields. For this purpose we built a femtosecond laser pulse shaper and combined it with an evolutionary algorithm to find optimum pulse shapes according to a given target. As an initial example we have shown that it is possible to generate bandwidth-limited oscillator and amplified laser pulses in an automated fashion without requiring accurate pulse phase characterization. This is done by maximizing second-harmonic generation (SHG). Specific (non-transform-limited) pulse shapes can be produced as well in an optimal way if adequate and more complete pulse characterization methods provide the feedback for the learning loop.

Tailored femtosecond laser pulses were then used for many-parameter quantum control of gas-phase photoreactions. Different photofragmentation and ionization channels in $Fe(CO)_5$ and $CpFe(CO)_2Cl$ were investigated. The adaptive scheme "learned" pulse shapes, which led to a maximization or a minimization of different branching ratios as well as to the control of relative versus absolute product yields. No information about the system Hamiltonian was used in this procedure. In the case of $Fe(CO)_5$, it was even possible to gain insight into the dissociation time by analyzing the optimal laser field. In the case of $CpFe(CO)_2Cl$, more complex laser pulses were found by the optimization algorithm, and the problem complexity was analyzed by reducing the search parameter space as well as by performing correlation experiments. According to those findings, the $CpFe(CO)_2Cl$ optimization results cannot be explained by a simple "intensity" variation of the applied laser fields. This complexity analysis led us to the consideration of an example where the "problem of inversion" could be accessed more directly. For that purpose, we have optimized the single and double ionization in atomic calcium. Analysis of the resulting laser fields gave insight into the relevant photophysical processes. Surprisingly, the maximization of single or double ionization leads to a specific phase shape rather than to a bandwidth-limited laser pulse. It appears that dynamical stark shifting of levels and doubly excited states as intermediate resonances are exploited for optimum population transfer.

Finally, we employed the adaptive pulse shaping technique for many-parameter quantum control in the liquid phase. As a first implementation, the photophysical process of efficient metal-to-ligand charge transfer (MLCT) in [Ru(dpb)$_3$](PF$_6$)$_2$ was studied. Feedback-optimized control of the ratio emission/SHG is sensitive to the electronic structure of the charge-transfer chromophore in solution. This is because the two-photon intensity dependencies of molecular excitation and second-harmonic generation in the numerator and in the denominator of the emission/SHG ratio cancel each other. Successful optimizations were achieved with no prior knowledge of the vibronic surfaces involved in the photophysics. The implications of these results are not yet fully known; however, maximization and minimization of the emission/SHG ratio can be viewed as successful demonstrations of optimizing a competition between two photophysical mechanisms, one of which is a solution-phase process.

These results are promising for the next generation of control experiments in the solution phase, where the competition to be controlled will involve different excited-state dynamical processes in the same molecule, leading to different photochemical outcomes. Also promising is the possibility of using adaptive femtosecond pulse shaping to find complex pulse shapes which preferentially excite one type of molecule among several noninteracting species in a single solution. This achievement could be useful in a number of applications. For example, in two-photon microscopy, specifically shaped laser pulses could be used to selectively excite different fluorophores in addition to precompensating for pulse propagation effects. Furthermore, excitation of selected molecules could be helpful for purposes of chemical analysis and synthesis.

We conclude with some more general ideas about the future directions for quantum control. In writing this review, we have tried to outline our efforts in the 1990s, and we hope that from this work and the cited literature it is clear that rapid progress has been made in this field. With adaptive femtosecond pulse shaping we now have the techniques available to guide the temporal evolution of complex quantum systems. Future developments are likely to take advantage of this situation in many more and increasingly challenging chemical scenarios. One possibility, which is currently receiving considerable theoretical attention (Shapiro and Brumer, 1991; Fujimura *et al.*, 1999; Shapiro *et al.*, 2000), concerns the use of light to influence the chirality of molecular samples. It would be very useful if general methods could be developed either to separate racemic mixtures or to generate enantiomerically pure species. Progress in this area is likely to benefit from efforts made to achieve control in the solution phase. Another direction of interest which has not been extensively explored is the control of chemical reactions on surfaces. Specifically, we point to heterogeneous catalysis where product yields are influenced by both the electronic properties of the reactants and the catalytic surfaces (Bonn *et al.*, 1999; Petek *et al.* 2000). It should be possible to use shaped light fields to alter the dynamical behavior of the system in the vicinity of the catalytic sites and thereby to exert control over the ensuing chemistry. A final direction concerns the important scientific goal of using adaptive femtosecond pulse shaping techniques not only to control complex quantum systems but also to learn from the conditions leading to control. We are optimistic that it will soon be possible to analyze relevant electric field shapes to obtain information about the temporal evolution of the system leading to product states specified by the experimenter. Experimental and theoretical efforts (Rabitz and Zhu, 2000) are actively underway.

VIII. Acknowledgments

Pursuing quantum control as a major research topic in our laboratory would not have been possible without the dedicated efforts of many co-workers. We especially thank T. Baumert for the ideas, continuous motivation, and guidance he provided during this research. Throughout this chapter we have described experiments with essential contributions made by the following workers: A. Assion, L. Bañares, M. Bergt, T. Frohnmeyer, M. Grosser, J. Helbing, M. Hofmann, B. Kiefer, A. Oehrlein, V. Seyfried, M. Strehle, R. Thalweiser, U. Weichmann, V. Weiss, and E. Wiedenmann. We have also greatly benefitted from collaborations with the groups of V. Engel and P. Lambropoulos. Valuable scientific interactions in addition to financial support have come from the "Sonderforschungsbereich 276 (Universität Freiburg): Korrelierte Dynamik hochangeregter atomarer und molekularer Systeme," the "Sonderforschungsbereich 347 (Universität Würzburg): Selektive Reaktionen Metall-aktivierter Moleküle," the "Schwerpunktprogramm der Deutschen Forschungsgemeinschaft: Femtosekunden-Spektroskopie elementarer Anregungen in Atomen, Molekülen und Clustern," the "European Coherent Control Network (COCOMO): HPRN-CT-1999-00129" and the "Fonds der Chemischen Industrie." N.D. thanks the Alexander von Humboldt-Stiftung for generous support of a postdoctoral position.

IX. References

Ambartsumian, R. V., and Letokhov, V. S. (1977). *In* "Chemical and Biochemical Applications of Lasers" (C. B. Moore, Ed.), Vol. 3, Chapter 2. Academic Press, New York.

Assion, A., Baumert, T., Seyfried, V., Weiss, V., Wiedenmann, E., and Gerber, G. (1996a). Femtosecond spectroscopy of the (2) $^1\Sigma_u^+$ double minimum state of Na_2: Time domain and frequency spectroscopy. *Z. Phys. D* **36**, 265–271.

Assion, A., Baumert, T., Helbing, J., Seyfried, V., and Gerber, G. (1996b). Coherent control by a single phase shaped femtosecond laser pulse. *Chem. Phys. Lett.* **259**, 488–494.

Assion, A., Baumert, T., Seyfried, V., Weichmann, U., and Gerber, G. (1998a). Photodissociation of Na_2^+ in intense femtosecond laser fields. *In* "Ultrafast Phenomena XI" (T. Elsaesser, J. G. Fujimoto, D. A. Wiersma, and W. Zinth, Eds.), Springer Series in Chemical Physics, Vol. 63, pp. 453–455. Springer, Berlin.

Assion, A., Baumert, T., Bergt, M., Brixner, T., Kiefer, B., Seyfried, V., Strehle, M., and Gerber, G. (1998b). Automated coherent control of chemical reactions and pulse compression by an evolutionary algorithm with feedback. In "Ultrafast Phenomena XI" (T. Elsaesser, J. G. Fujimoto, D. A. Wiersma, and W. Zinth, Eds.), Springer Series in Chemical Physics, Vol. 63, pp. 471–473. Springer, Berlin.

Assion, A., Baumert, T., Bergt, M., Brixner, T., Kiefer, B., Seyfried, V., Strehle, M., and Gerber, G. (1998c). Control of chemical reactions by feedback-optimized phase-shaped femtosecond laser pulses. *Science* **282**, 919–922.

Assion, A., Baumert, T., Weichmann, U., and Gerber, G. (2001). Photofragmentation of Na_2^+ in intense femtosecond laser fields: From photodissociation on light-induced potentials to field ionization. *Phys. Rev. Lett.* **86**, 5695–5698.

Balling, P., Maas, D. J., and Noordam, L. D. (1994). Interference in climbing a quantum ladder system with frequency-chirped laser pulses. *Phys. Rev. A* **50**, 4276–4285.

Bañares, L., Baumert, T., Bergt, M., Kiefer, B., and Gerber, G. (1997). Femtosecond photodissociation dynamics of $Fe(CO)_5$ in the gas phase. *Chem. Phys. Lett.* **267**, 141–148.

Bañares, L., Baumert, T., Bergt, M., Kiefer, B., and Gerber, G. (1998). The ultrafast photodissociation of $Fe(CO)_5$ in the gas phase. *J. Chem. Phys.* **108**, 5799–5811.

Bardeen, C. J., Wang, Q., and Shank, C. V. (1995). Selective excitation of vibrational wave packet motion using chirped pulses. *Phys. Rev. Lett.* **75**, 3410–3413.

Bardeen, C. J., Yakovlev, V. V., Wilson, K. R., Carpenter, S. D., Weber, P. M., and Warren, W. S. (1997). Feedback quantum control of molecular electronic population transfer. *Chem. Phys. Lett.* **280**, 151–158.

Bardeen, C. J., Yakovlev, V. V., Squier, J. A., and Wilson, K. R. (1998a). Quantum control of population transfer in green fluorescent protein by using chirped femtosecond pulses. *J. Am. Chem. Soc.* **120**, 13023–13027.

Bardeen, C. J., Wang, Q., and Shank, C. V. (1998b). Femtosecond chirped pulse excitation of vibrational wave packets in LD690 and bacteriorhodopsin. *J. Phys. Chem. A.* **102**, 2759–2766.

Bartels, R., Backus, S., Zeek, E., Misoguti, L., Vdovin, G., Christov, I. P., Murnane, M. M., and Kapteyn, H. C. (2000). Shaped-pulse optimization of coherent emission of high-harmonic soft X-rays. *Nature* **406**, 164–166.

Baumert, T., and Gerber, G. (1994). Fundamental interaction of molecules (Na_2, Na_3) with intense femtosecond laser pulses. *Isr. J. Chem.* **34**, 103–114.

Baumert, T., and Gerber, G. (1995). Femtosecond spectroscopy of molecules and clusters. In "Advances in Atomic, Molecular, and Optical Physics" (B. Bederson and H. Walther, Eds.), Advances in Atomic, Molecular, and Optical Physics Series, Vol. 35, pp. 163–208. Academic Press, London.

Baumert, T., Grosser, M., Thalweiser, R., and Gerber, G. (1991). Femtosecond time-resolved molecular multiphoton ionization: The Na_2 system. *Phys. Rev. Lett.* **67**, 3753–3756.

Baumert, T., Engel, V., Meier, C., and Gerber, G. (1992). High laser field effects in multiphoton ionization of Na_2: Experiment and quantum calculations. *Chem. Phys. Lett.* **200**, 488–494.

Baumert, T., Thalweiser, R., and Gerber, G. (1993). Femtosecond two-photon ionization spectroscopy of the B state of Na_3 clusters. *Chem. Phys. Lett.* **209**, 29–34.

Baumert, T., Helbing, J., and Gerber, G. (1997a). Coherent control with femtosecond laser pulses. In "Chemical Reactions and Their Control on the Femtosecond Time Scale. XXth Solvay Conference on Chemistry" (P. Gaspard and I. Burghardt, Eds.), Advances in Chemical Physics, Vol. 101, pp. 47–82. Wiley, New York.

Baumert, T., Brixner, T., Seyfried, V., Strehle, M., and Gerber, G. (1997b). Femtosecond pulse shaping by an evolutionary algorithm with feedback. *Appl. Phys. B* **65**, 779–782.

Bergmann, K., Theuer, H., and Shore, B. W. (1998). Coherent population transfer among quantum states of atoms and molecules. *Rev. Mod. Phys.* **70**, 1003–1025.

Bergt, M., Brixner, T., Kiefer, B., Strehle, M., and Gerber, G. (1999). Controlling the femtochemistry of $Fe(CO)_5$. *J. Phys. Chem. A* **103**, 10381–10387.

Bergt, M., Brixner, T., Kiefer, B., Strehle, M., and Gerber, G. (2000). Control of quantum dynamics by adaptive femtosecond pulse shaping. *In* "Ultrafast Phenomena XII" (T. Elsaesser, S. Mukamel, M. M. Murnane, and N. F. Scherer, Eds.), Springer Series in Chemical Physics, Vol. 66, pp. 19–23. Springer, Berlin.

Bjorkholm, J. E., and Liao, P. F. (1974). Resonant enhancement of two-photon absorption in sodium vapor. *Phys. Rev. Lett.* **33**, 128–131.

Blanchet, V., Bouchene, M. A., Cabrol, O., and Girard, B. (1995). One-color coherent control in Cs_2. Observation of 2.7 fs beats in the ionization signal. *Chem. Phys. Lett.* **233**, 491–499.

Bloembergen, N., and Yablonovitch, E. (1978). Infrared-laser-induced unimolecular reactions. *Phys. Today* **31** (May), 23–30.

Bloembergen, N., and Zewail, A. H. (1984). Energy redistribution in isolated molecules and the question of mode-selective laser chemistry revisited. *J. Phys. Chem.* **88**, 5459–5465.

Bonn, M., Funk, S., Hess, Ch., Denzler, D. N., Stampfl, C., Scheffler, M., Wolf, M., and Ertl, G. (1999). Phonon- versus electron-mediated desorption and oxidation of CO on Ru(0001). *Science* **285**, 1042–1045.

Brixner, T., and Gerber, G. (2001). Femtosecond polarization pulse shaping. *Opt. Lett.* **26**, 557–559.

Brixner, T., Strehle, M., and Gerber, G. (1999). Feedback-controlled optimization of amplified femtosecond laser pulses. *Appl. Phys. B* **68**, 281–284.

Brixner, T., Oehrlein, A., Strehle, M., and Gerber, G. (2000). Feedback-controlled femtosecond pulse shaping. *Appl. Phys. B* **70**(Suppl.), S119–S124.

Brixner, T., Kiefer, B., and Gerber, G. (2001a). Problem complexity in femtosecond quantum control. *Chem. Phys.* **267**, 241–246.

Brixner, T., Damrauer, N. H., Kiefer, B., and Gerber, G. (2001b). In preparation.

Broers, B., van Linden van den Heuvell, H. B., and Noordam, L. D. (1992). Efficient population transfer in a three-level ladder system by frequency-swept ultrashort laser pulses. *Phys. Rev. Lett.* **69**, 2062–2065.

Brumer, P., and Shapiro, M. (1986). Control of unimolecular reactions using coherent light. *Chem. Phys. Lett.* **126**, 541–564.

Cerullo, G., Bardeen, C. J., Wang, Q., and Shank, C. V. (1996). High-power femtosecond chirped pulse excitation of molecules in solution. *Chem. Phys. Lett.* **262**, 362–368.

Chen, C., Yin, Y.-Y., and Elliott, D. S. (1990). Interference between optical transitions. *Phys. Rev. Lett.* **64**, 507–510.

Christian, J. F., Broers, B., Hoogenraad, J. H., van der Zande, W. J., and Noordam, L. D. (1993). Rubidium electronic wavepackets probed by a phase-sensitive pump-probe technique. *Opt. Commun.* **103**, 79–84.

Damrauer, N. H., Boussie, T. R., Devenney, M., and McCusker, J. K. (1997a). Effects of intraligand electron delocalization, steric tuning, and excited-state vibronic coupling on the photophysics of aryl-substituted bipyridyl complexes of Ru(II). *J. Am. Chem. Soc.* **119**, 8253–8268.

Damrauer, N. H., Cerullo, G., Yeh, A., Boussie, T. R., Shank, C. V., and McCusker, J. K. (1997b). Femtosecond dynamics of excited-state evolution in $[Ru(bpy)_3]^{2+}$. *Science* **275**, 54–57.

Damrauer, N. H., and McCusker, J. K. (1999). Ultrafast dynamics in the metal-to-ligand charge transfer excited-state evolution of $[Ru(4,4'-diphenyl-2,2'-bipyridine)_3]^{2+}$. *J. Phys. Chem. A* **38**, 8440–8446.

Demiralp, M. and Rabitz, H. (1993). Optimally controlled quantum molecular dynamics: A perturbation formulation and the existence of multiple solutions. *Phys. Rev. A* **47**, 809–816.

Efimov, A., Moores, M. D., Beach, N. M., Krause, J. L., and Reitze, D. H. (1998). Adaptive control of pulse phase in a chirped-pulse amplifier. *Opt. Lett.* **23**, 1915–1917.

Efimov, A., Moores, M. D., Mei, B., Krause, J. L., Siders, C. W., and Reitze, D. H. (2000). Minimization of dispersion in an ultrafast chirped pulse amplifier using adaptive learning. *Appl. Phys. B* **23**(Suppl.), S133–S141.

Elsaesser, T., and Kaiser, W. (1991). Vibrational and vibronic relaxation of large polyatomic molecules in liquids. *Annu. Rev. Phys. Chem.* **42**, 83–107.
Engel, V., and Metiu, H. (1994). Two-photon wave-packet interferometry. *J. Chem. Phys.* **100**, 5448–5458.
Ferraudi, G. J. (1988). "Elements of Inorganic Photochemistry." Wiley, New York.
Fittinghoff, D. N., Bowie, J. L., Sweetser, J. N., Jennings, R. T., Krumbügel, M. A., DeLong, K. W., Trebino, R., and Walmsley, I. A. (1996). Measurement of the intensity and phase of ultraweak, ultrashort laser pulses. *Opt. Lett.* **21**, 884–886.
Frohnmeyer, T., Hofmann, M., Strehle, M., and Baumert, T. (1999). Mapping molecular dynamics (Na_2) in intense laser fields: another dimension to femtochemistry. *Chem. Phys. Lett.* **312**, 447–454.
Fujimura, Y., González, L., Hoki, K., Manz, J., and Ohtsuki, Y. (1999). Selective preparation of enantiomers by laser pulses: Quantum model simulation for H_2POSH. *Chem. Phys. Lett.* **306**, 1–8.
Gaubatz, U., Rudecki, P., Becker, M., Schiemann, S., Külz, M., and Bergmann, K. (1988). Population switching between vibrational levels in molecular beams. *Chem. Phys. Lett.* **149**, 463–468.
Gaubatz, U., Rudecki, P., Schiemann, S., and Bergmann, K. (1990). Population transfer between molecular vibrational levels by stimulated Raman scattering with partially overlapping laser fields. A new concept and experimental results. *J. Chem. Phys.* **92**, 5363–5376.
Geremia, J. M., Zhu, W., and Rabitz, H. (2000). Incorporating physical implementation concerns into closed loop quantum control experiments. *J. Chem. Phys.* **113**, 10841–10848.
Goldberg, D. E. (1993). "Genetic Algorithms in Search, Optimization, and Machine Learning." Addison-Wesley, Reading.
Gordon, R. J., and Rice, S. A. (1997). Active control of the dynamics of atoms and molecules. *Annu. Rev. Phys. Chem.* **48**, 601–641.
Gruebele, M., and Bigwood, R. (1998). Molecular vibrational energy flow: beyond the Golden Rule. *Int. Rev. Phys. Chem.* **17**, 91–145.
Hartke, B., Kosloff, R., and Ruhmann, S. (1989). Large amplitude ground state vibrational coherence induced by impulsive absorption in CsI—a computer simulation. *Chem. Phys. Lett.* **158**, 238–244.
Herek, J. L., Materny, A., and Zewail, A. H. (1994). Femtosecond control of an elementary unimolecular reaction from the transition-state region. *Chem. Phys. Lett.* **228**, 15–25.
Heritage, J. P., Weiner, A. M., and Thurston, R. N. (1985). Picosecond pulse shaping by spectral phase and amplitude manipulation. *Opt. Lett.* **10**, 609–611.
Hornung, T., Meier, R., Zeidler, D., Kompa, K.-L., Proch, D., and Motzkus, M. (2000a). Optimal control of one- and two-photon transitions with shaped femtosecond pulses and feedback. *Appl. Phys. B* **71**, 277–284.
Hornung, T., Meier, R., and Motzkus, M. (2000b). Optimal control of molecular states in a learning loop with a parameterization in frequency and time domain. *Chem. Phys. Lett.* **326**, 445–453.
Huang, G. M., Tarn, T. J., and Clark, J. W. (1983). On the controllability of quantum-mechanical systems. *J. Math. Phys.* **24**, 2608–2618.
Husimi, K. (1940). Some formal properties of the density matrix. *Proc. Phys. Math. Soc. Jpn.* **22**, 264–314.
Jortner, J., Levine, R. D., and Rice, S. A., Eds. (1981). "Photoselective Chemistry," Advances in Chemical Physics, Vol. 47. Wiley, New York.
Judson, R. S., and Rabitz, H. (1992). Teaching lasers to control molecules. *Phys. Rev. Lett.* **68**, 1500–1503.
Juris, A., Baragelleti, S., Campagna, S., Balzani, V., Belser, P., and von Zelewski, A. (1988). Ru(II) polypyridine complexes: Photophysics, photochemistry, electrochemistry, and chemiluminescence. *Coord. Chem. Rev.* **84**, 85–277.
Kan, R. O. (1966). "Organic Photochemistry." McGraw-Hill, New York.

Kohler, B., Yakovlev, V. V., Che, J., Krause, J. L., Messina, M., Wilson, K. R., Schwentner, N., Whitnell, R. M., and Yan, Y. (1995). Quantum control of wave packet evolution with tailored femtosecond pulses. *Phys. Rev. Lett.* **74,** 3360–3363.
Kosloff, R., Rice, S. A., Gaspard, P., Tersigni, S., and Tannor, D. J. (1989). Wavepacket dancing: Achieving chemical selectivity by shaping light-pulses. *Chem. Phys.* **139,** 201–220.
Kunde, J., Baumann, B., Arlt, S., Morier-Genoud, F., Siegner, U., and Keller, U. (2000). Adaptive feedback control of ultrafast semiconductor nonlinearities. *Appl. Phys. Lett.* **77,** 924–926.
Lalović, D., Davidović, D. M., and Bijedić, N. (1992). Quantum mechanics in terms of non-negative smoothed Wigner functions. *Phys. Rev. A* **46,** 1206–1212.
Lepetit, L., Chériaux, G., and Joffre, M. (1995). Linear techniques of phase measurement by femtosecond spectral interferometry for applications in spectroscopy. *J. Opt. Soc. Am. B* **12,** 2467–2474.
Letokhov, V. S. (1977). Photophysics and photochemistry. *Phys. Today* **30** (May), 23–32.
Liao, P. F., and Bjorkholm, J. E. (1975). Direct observation of atomic energy level shifts in two-photon absorption. *Phys. Rev. Lett.* **34,** 1–4.
Meshulach, D., and Silberberg, Y. (1998). Coherent quantum control of two-photon transitions by a femtosecond laser pulse. *Nature* **396,** 239–242.
Meshulach, D., and Silberberg, Y. (1999). Coherent quantum control of multiphoton transitions by shaped ultrashort optical pulses. *Phys. Rev. A* **60,** 1287–1292.
Meshulach, D., Yelin, D., and Silberberg, Y. (1998). Adaptive real-time femtosecond pulse shaping. *J. Opt. Soc. Am. B* **15,** 1615–1619.
Nesbitt, D. J., and Field, R. W. (1996). Vibrational energy flow in highly excited molecules: Role of intramolecular vibrational redistribution. *J. Phys. Chem.* **100,** 12735–12756.
O'Regan, B., and Gratzel, M. (1991). A low-cost, high-efficiency solar cell based on dye-sensitized colloidal TiO_2 films. *Nature* **353,** 737–740.
Park, S. M., Lu, S.-P., and Gordon, R. J. (1991). Coherent laser control of the resonance-enhanced multiphoton ionization of HCl. *J. Chem. Phys.* **94,** 8622–8624.
Paye, J. (1992). The chronocyclic representation of ultrashort light pulses. *IEEE J. Quantum Electron.* **28,** 2262–2273.
Peirce, A., Dahleh, M., and Rabitz, H. (1988). Optimal control of quantum-mechanical systems: Existence, numerical approximation, and applications. *Phys. Rev. A* **37,** 4950–4964.
Petek, H, Nagano, H., Weida, M. J., and Ogawa, S. (2000). Quantum control of nuclear motion at a metal surface. *J. Chem. Phys. A* **104,** 10234–10239.
Potter, E. D., Herek, J. L., Pedersen, S., Liu, Q., and Zewail, A. H. (1992). Femtosecond laser control of a chemical reaction. *Nature* **355,** 66–68.
Rabitz, H., and Zhu, W. (2000). Optimal control of molecular motion: Design, implementation, and inversion. *Acc. Chem. Res.* **33,** 572–578.
Rabitz, H., de Vivie-Riedle, R., Motzkus, M., and Kompa, K. (2000). Chemistry—Whither the future of controlling quantum phenomena? *Science* **288,** 824–828.
Ramakrishna, V., Salapaka, M. V., Dahleh, M., Rabitz, H., and Peirce, A. P. (1995). Controllability of molecular systems. *Phys. Rev. A* **51,** 960–966.
Rice, S. A., and Zhao, M. (2000). "Optical Control of Molecular Dynamics." Wiley, New York.
Scherer, N. F., Carlson, R. J., Matro, A., Du, M., Ruggiero, A. J., Romero-Rochin, V., Cina, J. A., Fleming, G. R., and Rice, S. A. (1991). Fluorescence-detected wave packet interferometry: Time resolved molecular spectroscopy with sequences of femtosecond phase-locked pulses. *J. Chem. Phys.* **95,** 1487–1511.
Schwefel, H.-P. (1995). "Evolution and Optimum Seeking." Wiley, New York.
Shapiro, M., and Brumer, P. (1991). Controlled photon induced symmetry breaking: Chiral molecular products from achiral precursors. *J. Chem. Phys.* **95,** 8658–8661.

Shapiro, M., Hepburn, J. W., and Brumer, P. (1988). Simplified laser control of unimolecular reactions: Simultaneous (ω_1, ω_3) excitation. *Chem. Phys. Lett.* **149**, 451–454.
Shapiro, M., Frishman, E., and Brumer, P. (2000). Coherently controlled asymmetric synthesis with achiral light. *Phys. Rev. Lett.* **84**, 1669–1672.
Shi, S., Woody, A., and Rabitz, H. (1988). Optimal control of selective vibrational excitation in harmonic linear chain molecules. *J. Chem. Phys.* **88**, 6870–6883.
Strehle, M., and Gerber, G. (2001). In preparation.
Tannor, D. J., and Bartana, A. (1999). On the interplay of control fields and spontaneous emission in laser cooling. *J. Phys. Chem. A* **103**, 10359–10363.
Tannor, D. J., and Rice, S. A. (1985). Control of selectivity of chemical reaction via control of wavepacket evolution. *J. Chem. Phys.* **83**, 5013–5018.
Tannor, D. J., Kosloff, R., and Rice, S. A. (1986). Coherent pulse sequence induced control of selectivity of reactions: Exact quantum mechanical calculations. *J. Chem. Phys.* **85**, 5805–5820.
Tannor, D. J., Kosloff, R., and Bartana, A. (1999). Laser cooling of internal degrees of freedom of molecules by dynamically trapped states. *Faraday Disc.* **113**, 365–383.
Trebino, R., DeLong, K. W., Fittinghoff, D. N., Sweetser, J. N., Krumbügel, M. A., Richman, B. A., and Kane, D. J. (1997). Measuring ultrashort laser pulses in the time-frequency domain using frequency-resolved optical gating. *Rev. Sci. Instrum.* **68**, 3277–3295.
Warren, W. S., Rabitz, H., and Dahleh, M. (1993). Coherent control of quantum dynamics: The dream is alive. *Science* **259**, 1581–1589.
Weber, Th., Weckenbrock, M., Staudte, A., Spielberger, L., Jagutzki, O., Mergel, V., Afaneh, F., Urbasch, G., Vollmer, M., Giessen, H., and Dörner, R. (2000). Recoil-ion momentum distributions for single and double ionization of helium in strong laser fields. *Phys. Rev. Lett.* **84**, 443–446.
Wefers, M. M., and Nelson, K. A. (1995a). Generation of high-fidelity programmable ultrafast optical waveforms. *Opt. Lett.* **20**, 1047–1049.
Wefers, M. M., and Nelson, K. A. (1995b). Analysis of programmable ultrashort waveform generation using liquid-crystal spatial light modulators. *J. Opt. Soc. Am. B* **12**, 1343–1362.
Weinacht, T. C., Ahn, J., and Bucksbaum, P. H. (1999a). Controlling the shape of a quantum wavefunction. *Nature* **397**, 233–235.
Weinacht, T. C., White, J., and Bucksbaum, P. H. (1999b). Toward strong field mode-selective chemistry. *J. Phys. Chem. A* **103**, 10166–10168.
Weiner, A. M. (2000). Femtosecond pulse shaping using spatial light modulators. *Rev. Sci. Instrum.* **71**, 1929–1960.
Weiner, A. M., Leaird, D. E., Patel, J. S., and Wullert II, J. R. (1990). Programmable femtosecond pulse shaping by use of a multielement liquid-crystal phase modulator. *Opt. Lett.* **15**, 326–328.
Weiner, A. M., Leaird, D. E., Patel, J. S., and Wullert II, J. R. (1992). Programmable shaping of femtosecond optical pulses by use of 128-element liquid crystal phase modulator. *IEEE J. Quantum Electron.* **28**, 908–920.
Wong, V., and Gruebele, M. (1999). How does vibrational energy flow fill the molecular state space? *J. Phys. Chem. A* **103**, 10083–10092.
Yelin, D., Meshulach, D., and Silberberg, Y. (1997). Adaptive femtosecond pulse compression. *Opt. Lett.* **22**, 1793–1795.
Zare, R. N. (1998). Laser control of chemical reactions. *Science* **279**, 1875–1879.
Zeek, E., Maginnis, K., Backus, S., Russek, U., Murnane, M. M., Mourou, G., Kapteyn, H. C., and Vdovin, G. (1999). Pulse compression by use of deformable mirrors. *Opt. Lett.* **24**, 493–495.
Zeek, E., Bartels, R., Murnane, M. M., Kapteyn, H. C., Backus, S., and Vdovin, G. (2000). Adaptive pulse compression for transform-limited 15-fs high-energy pulse generation. *Opt. Lett.* **25**, 587–589.

Zeidler, D., Hornung, T., Proch, D., and Motzkus, M. (2000). Adaptive compression of tunable pulses from a non-collinear-type OPA to below 16 fs by feedback-controlled pulse shaping. *Appl. Phys. B* **70**(Suppl.), S125–S131.

Zewail, A. H. (1980). Laser selective chemistry—Is it possible? *Phys. Today* **33** (November), 27–33.

Zhao, M., and Rice, S. A. (1991). Comment concerning the optimal control of transformations in an unbounded quantum system. *J. Chem. Phys.* **95**, 2465–2472.

Zhu, L. C., Kleiman, V., Li, X. N., Lu, S. P., Trentelman, K., and Gordon, R. J. (1992). Coherent laser control of the product distribution obtained in the photoexcitation of HI. *Science* **270**, 77–80.

COHERENT MANIPULATION OF ATOMS AND MOLECULES BY SEQUENTIAL LASER PULSES

N. V. VITANOV
Helsinki Institute of Physics, PL 9, 00014 University of Helsinki, Finland

M. FLEISCHHAUER
Fachbereich Physik, Universität Kaiserslautern, 67653 Kaiserslautern, Germany

B. W. SHORE
Lawrence Livermore National Laboratory, Livermore, California 94550[1]

K. BERGMANN
Fachbereich Physik, Universität Kaiserslautern, 67653 Kaiserslautern, Germany

I. Introduction	57
A. Early Days of Laser State Selection	59
B. Incoherent Population Transfer Schemes	62
1. Incoherent Excitation of Two-State Systems	62
2. Optical Pumping	63
3. Stimulated Emission Pumping	64
C. Coherent Population Transfer	64
1. Resonant Coherent Excitation: Rabi Oscillations	65
2. Three-State Systems	66
II. Principles of Coherent Excitation	67
A. Coherent Excitation	67
B. Partial Coherence: The Density Matrix	68
C. Two States	69
1. The Rotating Wave Approximation (RWA)	69
2. Rabi Cycling	70
3. Probability Loss	71
4. Adiabatic States and Adiabatic Following	72
5. Rapid Adiabatic Passage	73
6. Estimating Transition Probabilities	75

N. V. Vitanov is also affiliated with the Department of Physics, Sofia University, James Boucher 5 blvd., 1126 Sofia, Bulgaria, and Institute of Solid State Physics, Bulgarian Academy of Sciences, 1784 Sofia, Bulgaria.
[1]Retired.

D. Three States 76
   1. The Three-State RWA Hamiltonian 76
   2. Pulse Sequences 77
   3. Adiabatic Elimination 78
   4. Autler-Townes Splitting 79
III. Three-State STIRAP: Theory 82
  A. Basic Properties of STIRAP 82
    1. Basic Equations and Definitions 82
    2. Adiabatic States 83
    3. The STIRAP Mechanism 85
    4. Five-Stage Description of STIRAP 86
    5. Intermediate-State Population: Importance of the Null Eigenvalue .. 88
    6. Intuitive versus Counterintuitive Pulse Sequences 89
    7. Adiabatic Condition: Local and Global Criteria 90
    8. Nonadiabatic Transitions 92
  B. Sensitivity of STIRAP to Interaction Parameters 93
    1. Sensitivity to Delay 93
    2. Sensitivity to Rabi Frequency and Pulse Width 94
    3. Sensitivity to Single-Photon Detuning 94
    4. Sensitivity to Two-Photon Detuning 95
    5. Sensitivity to Losses from the Intermediate State 97
    6. Sensitivity to Beam Geometry 100
    7. Multiple Intermediate States 101
    8. STIRAP Beyond the RWA 102
IV. Three-State STIRAP: Experiments 103
  A. Experimental Demonstrations with CW Lasers 103
    1. Sodium Dimers 103
    2. Metastable Neon Atoms 104
  B. Experimental Demonstrations with Pulsed Lasers 107
    1. General Considerations for Pulsed Lasers 107
    2. Nitrous Oxide Molecules 108
    3. Sulfur Dioxide Molecules 109
    4. STIRAP in a Ladder System: Rubidium Atoms 111
  C. STIRAP with Degenerate or Nearly Degenerate States 111
V. STIRAP-Like Population Transfer in Multistate Chains 117
  A. Resonantly Driven Chains with Odd Number of States 118
  B. Resonantly Driven Chains with Even Number of States 120
  C. The Off-Resonance Case 121
  D. Optimization of Multistate STIRAP: Dressed-State Picture 124
VI. Adiabatic Momentum Transfer 126
  A. Coherent Matter-Wave Manipulation 127
    1. Atomic Mirrors 127
    2. Atomic Beam Splitters 130
    3. Atomic Interferometers 131
  B. Coherent Manipulation of Laser-Cooled and Trapped Atoms 133
  C. Measurement of Weak Magnetic Fields with Larmor Velocity Filter ... 134
VII. Branched-Chain Excitation 135
  A. Branched Linkage Patterns 135
  B. The Tripod Linkage 136
    1. Concept 136

2. The Tripod Linkage........................... 136
3. Adiabatic Evolution.......................... 137
C. Experimental Demonstration..................... 141
VIII. Population Transfer via a Continuum of Intermediate States........ 143
A. Laser-Induced Continuum Structure............... 145
1. LICS Equations........................... 145
2. Coherent and Incoherent Channels............. 146
3. Fano Profile.............................. 146
4. Population Trapping: Dark and Bright States..... 149
B. Population Transfer via a Continuum............. 150
1. Population Transfer in the Ideal Case........... 150
2. Satisfying the Trapping Condition.............. 151
3. Suppressing Incoherent Ionization.............. 152
C. Tripod Coupling via a Continuum................ 155
IX. Extensions and Applications of STIRAP............... 156
A. Control of Chemical Reactions.................. 156
B. Hyper-Raman STIRAP (STIHRAP)............... 156
C. Stark-Chirped Rapid Adiabatic Passage........... 157
1. Theory.................................. 158
2. Experimental Demonstration.................. 160
D. Adiabatic Passage by Light-Induced Potentials (APLIP)......... 161
E. Photoassociative STIRAP as a Source for Cold Molecules........ 161
X. Propagation Phenomena............................ 162
A. Electromagnetically Induced Transparency (EIT)...... 163
1. Complete Decoupling of Light and Matter......... 163
2. Elimination of Absorption; Slow Light............ 165
B. Adiabatons................................... 167
C. Matched Pulses............................... 168
D. Coherence Transfer between Matter and Light...... 170
XI. Applications of STIRAP in Quantum Optics and Quantum Information... 173
A. Single-Atom Cavity Quantum Electrodynamics....... 173
B. Quantum Logic Gates Based on Cavity QED......... 175
C. Quantum Networking........................... 177
D. Many-Atom Systems........................... 177
XII. Summary and Outlook............................. 179
XIII. Acknowledgments............................... 180
XIV. References..................................... 180

I. Introduction

Many branches of contemporary physics require atoms or molecules prepared in specified quantum states—not only for traditional studies of state-to-state collision dynamics, isotope separation, or laser-controlled chemical reactions, but also in more recently developing research areas of atom optics and quantum information. Of greatest interest is the fraction of all atoms or molecules in a specific state, a time-varying probability here termed the *population P(t)*. Schemes for transferring population selectively (i.e., to a single predetermined quantum state), such as excitation with frequency-swept pulses and stimulated Raman adiabatic passage (STIRAP), have opened new opportunities for coherent control of atomic and molecular processes. With the growing interest in quantum information, there is also concern with creating and controlling specified coherent superpositions of quantum states. These more general properties of an ensemble of atoms or molecules are embodied in the time-varying state vector $\Psi(t)$.

This chapter describes the basic principles underlying a variety of techniques that can be used to control state vectors and, in particular, to transfer population, selectively, between quantum states of atoms or molecules. We also describe experimental demonstrations of the various principles. (Aspects of these population transfer schemes have been reviewed by Vitanov *et al.*, 2001.) All the methods share a common reliance on adiabatic time evolution, induced by a sequence of delayed, but partially overlapping, laser pulses. They begin with an ensemble of atoms or molecules in which the population is in a specified discrete quantum state. Then the sequence of laser pulses forces the population into a desired target state. Only highly monochromatized light can provide the selectivity needed to isolate a single final state—broadband light or charged particle pulses cannot so discriminate. The control of phase imposes further constraints; it requires coherent radiation, available only from a laser.

One goal of the theory of coherent excitation is to predict, for a given set of radiation pulses, the probability that atoms will undergo a transition between the initial state and the desired target state (the population *transfer efficiency*). More generally, theory can predict the changes of a state vector $\Psi(t)$ produced by specified radiation. Alternatively, theory can provide a prescription for pulses that will produce a desired population transfer or state vector change.

We begin our discussion, in Section I.A, with a brief summary of the historical background for the subsequent discussions of adiabatic transfer schemes. Although the excitation techniques described in this chapter require coherent radiation, incoherent light, such as that from filtered atomic vapor lamps or from broadband lasers with poor coherence properties, also has very useful applications for selective excitation, some of which are described briefly in Section I.B. Coherent excitation differs qualitatively from incoherent excitation. To emphasize this difference, Section I.C contrasts some simple examples.

Starting with Section II, we develop the general mathematical principles needed to describe coherent excitation and adiabatic time evolution of quantum systems. In Section III we apply this to the basic STIRAP process, wherein adiabatic evolution produces complete population transfer in a three-state Raman system. Section IV discusses various experimental demonstrations of the STIRAP technique. Sections V–VIII describe various theoretical and experimental extensions of the original three-state STIRAP. Although our primary concern is with the effect of prescribed fields on atoms, the atomic excitation creates localized polarization which alters the radiation as it propagates; Section X discusses some of the effects to be found by treating the field and the atoms together. Section XI

COHERENT MANIPULATION OF ATOMS AND MOLECULES 59

discusses some applications of the STIRAP principles in the rapidly growing area of quantum information and in the general area of quantum optics. Section XII offers a summary and comments on possible future work.

A. EARLY DAYS OF LASER STATE SELECTION

The use of lasers to address individual states in atoms or molecules dates back more than 30 years [for reviews see Bergmann (1988), Rubahn and Bergmann (1990)]. The early work, taking advantage of the small bandwidth and high spectral power density of laser radiation, employed lasers to populate individual rotational-vibrational levels in an electronically excited state of molecules, as preparation for collision studies (Kurzel and Steinfeld 1970; Bergmann and Demtröder, 1971) or for spectroscopic analysis (Demtröder *et al.*, 1969). Later, individual thermally populated states in the electronic ground state were labeled through population depletion by optical pumping (Bergmann *et al.*, 1978; Gottwald *et al.*, 1986). That work paved the way for detailed studies in crossed molecular beams involving molecules in preselected rotational (Hefter *et al.*, 1981) or individual vibrational states (Ziegler *et al.*, 1988) colliding with atoms (Gottwald *et al.*, 1987) or electrons (Ziegler *et al.*, 1987). By 1986 laser state selection by population depletion had even been developed sufficiently to allow collision studies of molecules in individual magnetic sublevels (Mattheus *et al.*, 1986; Hefter *et al.*, 1986) with high resolution of the scattering angle.

State selection by population depletion through optical pumping is limited to thermally populated levels. Access to higher lying vibrationally excited levels in the electronic ground state was gained by the Franck–Condon pumping method (Rubahn and Bergmann, 1990) whereby excitation, from thermally populated levels (j'', v''), into a suitably chosen rovibrational level (j', v') in the electronically excited state, followed by spontaneous decay back to the electronic ground state, establishes a distribution $f_{v'}(v'')$ of population over vibrationally excited levels v''. Within the limits given by the optical transition rates, the distribution $f_{v'}(v'')$ can be controlled by a suitable choice of the level v'.

In the early 1980s, high-power pulsed lasers became more readily available and a variety of other schemes for laser state selection, in particular for the population of vibrationally excited levels in the electronic ground state, such as overtone pumping (OTP) (Crim, 1984) or off-resonance stimulated Raman scattering (ORSRS) (Orr *et al.*, 1984; Meier *et al.*, 1986), were developed. In OTP, levels $v'' > 1$ are directly excited from $v'' = 0$ in a single photon transition. High laser power is needed because the transition probability decreases rapidly with $\Delta v''$. Only a small fraction of the molecules are typically excited to high-lying levels. In ORSRS the frequency difference between a strong (possibly fixed frequency) laser and a tunable laser matches the transition frequency between rotational levels in the vibrational states $v'' = 0$ and $v'' = 1$. A substantial fraction (<50%) of the molecules in a given state (j'', $v'' = 0$) can be excited to $v'' = 1$.

Another two-step process, *stimulated emission pumping* (SEP) (Kittrell et al., 1981), has proven to be a flexible and very successful method for population transfer. In SEP a suitable level in an electronic state is excited. Rather than allowing spontaneous emission to distribute the population over many vibrational levels, another laser, the frequency of which is tuned to resonance with the desired target state, forces as much as 50% of the electronically exited molecules into that state. However, about 50% of the population will remain in the excited electronic state and will subsequently be distributed by spontaneous emission over other vibrational levels v''.

In the context of this early work the coherence properties of the radiation were not essential. It was therefore natural, in seeking to improve the flexibility of state selection, to look for schemes which also exploit coherence properties of lasers. Of particular interest were methods that would efficiently and selectively populate high-lying vibrational levels of molecules, which otherwise were not accessible for detailed collision studies.

The attempts to *selectively* populate high-lying levels in the electronic ground state led to the development of a Raman laser that utilized a molecular beam as a gain medium (Jones et al., 1983; Hefter et al., 1985). In that work, a ring cavity was built around a molecular beam whose axis coincided with the waist of the cavity (Fig. 1). The directional flow carried molecules into and out of the active region of

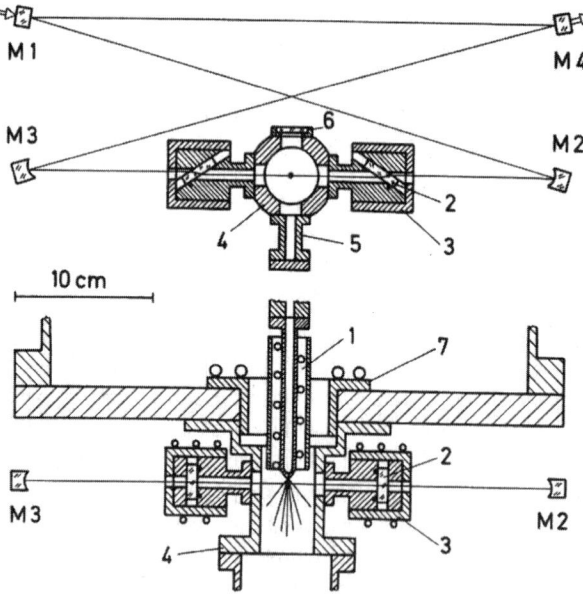

FIG. 1. Raman laser with molecular beam and ring cavity. Top and side view. (From P. L. Jones, U. Gaubatz, U. Hefter, B. Wellegehausen, and K. Bergmann. An optically pumped sodium-dimer supersonic beam laser. *Appl. Phys. Lett.* 1983;42:222–224.)

the laser cavity. A pump laser, propagating along the cavity axis, coupled thermally populated levels (j'', $v'' = 0$) to near resonance with levels v' in an electronically excited state. When implemented with a beam of sodium dimers, the rotational degree of freedom was cooled to a temperature of the order of $T_{rot} = 20$ K by supersonic expansion. Therefore the density of molecules in levels (j'', $v'' = 0$) with $j'' < 15$ was sufficiently high to provide a substantial gain. Thus, driven by the pump laser, Stokes radiation was generated in the cavity. Intracavity filters restricted laser oscillation to a single vibrational band $v' \to v''$, thereby providing the desired selectivity. Indeed, it was found (Becker *et al.,* 1987) that up to 70% of the molecules in (j'', $v'' = 0$) could be transferred to a specific level (j'', $v'' = 5$).

Further analysis of the scheme revealed, however, that the coincidence of the axis of the pump beam and the generated Stokes beam limited the achievable transfer efficiency for the following reason (see Fig. 2). The transit time of the molecules across the waist of the cavity was about an order of magnitude longer than the radiative lifetime in the electronically excited state. Near the axis of the cavity, the Stokes radiation (established during the prior buildup of laser oscillation) was sufficiently strong to compete successfully with spontaneous emission and thus to force the molecules into the desired state v'' by a Raman process. However, detrimental processes are unavoidable as the molecules enter or leave the cavity. When entering the cavity, the molecules are exposed to pump radiation while the Stokes radiation is still weak. Therefore, the latter cannot yet compete with the spontaneous emission and a fraction of the molecules is lost by optical pumping to other vibrational levels. Detuning the pump laser from resonance with the level v' would not cure the problem: although detuning would reduce the loss rate, the gain would be reduced as well. As the molecules, placed into a level $v'' \gg 1$ near the center of the pump and Stokes beams, leave the cavity, they are still exposed to the Stokes radiation. This, again, leads to some loss of molecules due to optical pumping.

Obviously, the detrimental losses due to spontaneous emission could be reduced when the axis of the Stokes laser beam, provided by an external source rather than generated in the cavity, would be shifted upstream of the axis of the pump laser,

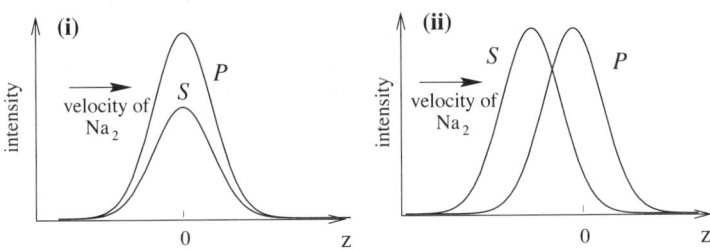

FIG. 2. (i) Profile of pump and Stokes laser intensities. The z axis is parallel to the axis of the molecular beam; $z = 0$ corresponds to the intersection with the cavity waist. (ii) Same as (i) but with Stokes field shifted somewhat upstream.

while retaining some overlap of the beams. Although that reasoning aimed at the reduction of the losses due to incoherent processes, it led quickly to the development of the stimulated Raman adiabatic passage process (STIRAP). Encouraging results were obtained from numerical studies, including incoherent processes, and were supported by first experimental evidence (Gaubatz et al., 1988), before the basic theory and experimental technique was fully developed (Gaubatz et al., 1990).

It is important to note that already in 1984, Oreg et al. had published theoretical work that included the essential elements of the consequences of sequential pulse interaction, namely, 100% transfer efficiency when the Stokes radiation interacts first. However, the practical relevance of the result of that work remained unrecognized until the phenomenon was independently rediscovered and the method developed in Kaiserslautern.

B. INCOHERENT POPULATION TRANSFER SCHEMES

1. Incoherent Excitation of Two-State Systems

Until the advent of laser light sources, theoretical descriptions of radiative excitation followed the lead of Einstein, who treated two-state atoms within a radiation-filled cavity (Einstein, 1917). With this approach one postulates a set of equations for the rate of change in atomic populations exposed to beams of light, the *radiative rate equations*,

$$\frac{d}{dt} P_n(t) = \sum_m R_{n,m}(t) P_m(t), \tag{1}$$

linking the various changes in probabilities $P_n(t)$ by a matrix of rate coefficients $R_{n,m}(t)$ in which excitation and stimulated emission rates for two-state atoms are proportional to the instantaneous radiation intensity $I(t)$ (power per unit area) and to the Einstein–Milne B coefficient, and in which spontaneous emission from the excited state (level 2) takes place at a rate A (cf. Shore, 1990, sect. 2.2). When all the two-level atoms are initially (at time $t \to -\infty$) unexcited (in state 1), when the radiation is sufficiently intense (the *saturated* regime in which spontaneous emission has a negligible effect, $BI \gg A$), and when the two levels have the same degeneracy, then the excited-state population at time t is

$$P_2(t) = \tfrac{1}{2} \left[1 - e^{-BF(t)} \right], \tag{2}$$

where $F(t) = \int_{-\infty}^{t} I(t')\, dt'$ is the pulsed radiation *fluence* (energy per unit area) up to time t. As this expression shows, an increase of pulse fluence always increases the excited-state population, which approaches monotonically the saturation value of 50%. This is the best population-transfer efficiency achievable with incoherent light. Once the radiation ceases, the atoms will spontaneously emit radiation and return (with exponential decay at rate A) to lower-lying levels. Eventually no excitation will remain; any excitation must be maintained by radiation.

2. Optical Pumping

The presence of spontaneous emission hinders direct excitation by providing an ever-present deexcitation rate. However, when one deals with more than two levels, spontaneous emission offers a mechanism for creating complete population transfer. The procedure can be understood from Fig. 3a: a radiation beam resonantly couples the initially populated state ψ_1 to the excited state ψ_2, from which spontaneous emission occurs. This uncontrolled radiative decay not only returns population to state ψ_1, it also takes some population into a third state ψ_3 that is unaffected by the radiation beam. (Either its energy lies far from resonance with the initial state or it is prevented from interacting by selection rules based on polarization of the light.) Each time an atom in state ψ_1 absorbs a photon, there is a chance that the resulting decay of excited state ψ_2 will carry population into state ψ_3. Once population is in state ψ_3 it is immune to further action by the radiation. If there are no other decay options than those to ψ_1 and ψ_3, the resulting population transfer, *optical pumping,* will eventually place all population into state ψ_3.

The simplicity of optical pumping has led to widespread use as a means of preparing atoms or molecules in a well-defined ground or metastable state—it requires only a single light source, which need not be a laser. Its main limitation is the lack of selectivity: the spontaneous emission step will generally place population in a mixture of final states—all the states into which the pump-excited state ψ_2 can decay. The distribution of final populations is determined by the relative decay rates that link each final state with the excited state. For vibrational transitions in molecules these rates are proportional to Franck–Condon factors. These rarely exceed a few percent, and so the selectivity is correspondingly low. Furthermore,

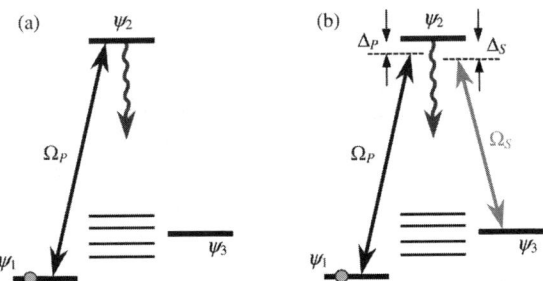

FIG. 3. (a) Linkage diagram for optical pumping. A pump field (not necessarily a laser) excites state ψ_2, which spontaneously decays either back to state ψ_1, or to state ψ_3, or possibly to some other states. (b) Linkage diagram for stimulated emission pumping. A pump field populates state ψ_2, and a subsequent Stokes (or dump) field populates state ψ_3.

optical pumping has limited use in optically thick vapors: subsequent absorption of the spontaneous-emission photons will reexcite the atoms, via the transition ψ_3 to ψ_2. Finally, the buildup of population in the target state takes place gradually, at a rate fixed by the A coefficient. It is therefore not possible to use optical pumping to move population more rapidly than this.

3. Stimulated Emission Pumping

Although optical pumping uses only a single light source, the overall population transfer involves two photons, as a Raman process: a *pump* photon from the imposed light source followed by a spontaneously emitted *Stokes* photon. It is natural to consider a *stimulated* Raman process in which externally supplied fields drive both transitions. A strong stimulated emission field will induce a transition to a selected final state, rather than to the mixture that would occur with spontaneous emission.

In one form of this two-photon process of population transfer, *stimulated emission pumping* (SEP) (Dai and Field, 1995), a pump field first places population from the initial state ψ_1 into the excited state ψ_2. Some time later a Stokes (or dump) field transfers population into the desired final state ψ_3 (hence the names *pump and dump*). Figure 3b depicts the relevant Λ-type linkage.

Because of its simplicity, the SEP technique has enjoyed widespread application in collision dynamics and spectroscopy (Hamilton *et al.*, 1986; Dai and Field, 1995). Its main limitation is the low efficiency: typically the transfer efficiency does not exceeds 10%, but this is quite adequate for many spectroscopic studies.

The reason for low efficiency is readily understood. If the intensity of the pump laser is sufficiently strong to saturate the $\psi_1 \leftrightarrow \psi_2$ transition, then, as suggested by the solution to the two-level rate equations, at most 50% of the population will be transferred from state ψ_1 to state ψ_2. If the Stokes laser is also sufficiently strong to saturate its transition, then half of the population in state ψ_2 will be subsequently transferred to the target state ψ_3. Therefore at most one-quarter of the population can be transferred to the target state. Half of the population remains in the initial state, and the remaining quarter is distributed according to the branching of spontaneous emission from state ψ_2.

SEP efficiency can be improved slightly if the pump and Stokes pulses are applied simultaneously, rather than successively. If they are sufficiently strong to saturate the transitions, thereby equalizing the populations, then one-third of the population can be transferred to the target state ψ_3.

C. COHERENT POPULATION TRANSFER

The response of a quantum system to coherent (laser) radiation differs significantly from its response to light from a lamp, even a very monochromatic lamp. Whereas the sudden application of *incoherent* radiation to an atom or molecule typically results in a monotonic approach to some equilibrium excitation, the sudden application of steady *coherent* radiation typically produces oscillating populations. These differences are clearly seen in the behavior of two-state systems, as noted below. They are equally evident in multilevel systems.

1. Resonant Coherent Excitation: Rabi Oscillations

The excited population $P_2(t)$ of a two-state system exposed to steady coherent radiation does not follow any monotonically increasing pattern, such as Eq. (2). Instead it oscillates sinusoidally (*Rabi oscillations*). When the radiation is resonant (the carrier frequency equal to the Bohr frequency), the result is

$$P_2(t) = \tfrac{1}{2}[1 - \cos(\Omega t)]. \tag{3}$$

The frequency of population oscillation, the *Rabi frequency* Ω, will be associated below with the strength of the atom–radiation interaction. As will be shown, it is proportional to the square root of the laser intensity.

When the radiation intensity varies, then so does the Rabi frequency $\Omega(t)$, and the cosine argument Ωt is replaced by the so-called *pulse area* $A(t)$, or time integral of the Rabi frequency,

$$\Omega t \rightarrow \int_{-\infty}^{t} \Omega(t')\,dt' \equiv A(t). \tag{4}$$

Unlike the monotonic approach to a steady saturation value observed for incoherent excitation, here the excited-state population oscillates between 0 and 1. At times when the pulse area $A(t)$ is an odd multiple of π (odd-π pulses) the population resides entirely in the excited state: the population is completely *inverted*. (Recall that with incoherent excitation, no more than half the population can be excited.) For pulse areas equal to even multiples of π (even-π pulses), the system returns to the initial state.

Although it is useful to consider single stationary atoms, in practice one must deal with an ensemble of atoms or molecules, often with a distribution of velocities. Atoms that move in the direction of a traveling wave experience a Doppler-shifted laser field, so their excitation is not exactly resonant. As noted in Eq. (18), their population oscillations are more rapid and have smaller peak values than do those of resonant atoms. Atoms that move across a laser beam will experience a pulse area that is dependent on the duration of their transit time across the beam, and hence on their velocity. These velocity-dependent interactions, and the presence of fluctuations in the laser intensity, require an averaging over excitation probabilities. The result is an effective excitation probability that has less pronounced oscillations; in extreme cases the averaging can bring the excitation probability to 0.5, the same as with incoherent excitation.

2. Three-State Systems

The oscillatory Rabi cycling characteristic of two-state systems can also be found in multistate systems. One example is a coherently driven three-state system subjected to the same pulse sequence as in SEP: first the pump pulse, followed after its completion (without overlap) by the Stokes pulse. Then the excitation can still be considered as a two-step process, but the probabilities for each step are different from those in SEP. In the case of exact single-photon resonances the transition probabilities P_{12} from state ψ_1 to state ψ_2 and P_{23} from state ψ_2 to state ψ_3 are

$$P_{12} = \tfrac{1}{2}(1 - \cos A_p), \qquad P_{23} = \tfrac{1}{2}(1 - \cos A_s). \tag{5}$$

where A_p and A_s are the pump and Stokes pulse areas [as in Eq. (3)]. If the system is initially in state ψ_1, then the population of state ψ_3 after the excitation is the product of the two probabilities,

$$P_{13} = \tfrac{1}{4}(1 - \cos A_p)(1 - \cos A_s). \tag{6}$$

Hence, when the pulse areas are both equal to odd multiples of π, there occurs complete population transfer from state ψ_1 to state ψ_3. However, the transfer efficiency depends strongly on the pulse areas, and it can even vanish (when A_p or A_s is an even multiple of π).

When the pump and Stokes pulses share a common time dependence, the population changes can no longer be separated into two consecutive independent two-state transitions. Nevertheless, an exact analytic expression can still be derived. If the system is initially in state ψ_1 and the two lasers are each resonantly tuned, then the population of state ψ_3 at the end of the excitation pulse is

$$P_{13} = \frac{A_p A_s}{A}\left(1 - \cos \tfrac{1}{2} A\right), \tag{7}$$

where $A = \sqrt{A_p^2 + A_s^2}$. Here again, the transfer efficiency depends on the pulse areas: complete population transfer from state ψ_1 to state ψ_3 occurs whenever $A = 2(2k + 1)\pi$ ($k = 0, 1, 2, \ldots$) and $A_p = A_s$, while complete population return to the initial state ψ_1 takes place when $A = 4k\pi$.

Just as with two-state Rabi oscillations, intensity fluctuations or a distribution of velocities will tend to average out the population oscillations and to lower the transfer efficiency. Moreover, because the population passes through the intermediate state ψ_2 during the transfer process, inevitable spontaneous emission will lead to population losses unless the excitation time is much shorter than the lifetime of ψ_2.

Rabi cycling is but one of the ways in which coherent laser pulses can induce population changes. Another class of change, the central theme of this chapter, *adiabatic evolution,* can occur when the Hamiltonian changes sufficiently slowly.

COHERENT MANIPULATION OF ATOMS AND MOLECULES   67

Section II explains the basic principles of adiabatic population transfer in terms of energy-level crossings, and Section III describes adiabatic population transfer by sequential laser pulses.

II. Principles of Coherent Excitation

A. COHERENT EXCITATION

The quantum mechanical description of the internal excitation of an atom (or molecule) is embodied in a time-dependent state vector, $\Psi(t)$, and in a fixed set of basis states $\psi_1, \psi_2, \ldots$, associated with likely energy states of the atom in the absence of any radiation pulses. These basis states serve as a fixed coordinate system (in a Hilbert space) wherein one can express the state vector as a time-varying superposition, which we take to be

$$\Psi(t) = \sum_{n=1}^{N} C_n(t) \exp[-i\zeta_n(t)]\hat{\psi}_n \equiv \sum_{n=1}^{N} C_n(t)\psi_n(t), \qquad (8)$$

where $\hat{\psi}_n$ is the static bare physical state. Hereafter, we write ψ_n for $\psi_n(t)$, the bare state with RWA phase factor $\exp[-i\zeta_n(t)]$. The real-valued phase $\zeta_n(t)$ is specified a priori, usually for mathematical convenience (as will be noted below), and the complex-valued function of time $C_n(t)$ is a *probability amplitude,* whose absolute square is the probability (or population) $P_n(t)$ that the atom will be found in state ψ_n at time t:

$$P_n(t) = |C_n(t)|^2 = |\langle\psi_n|\Psi(t)\rangle|^2. \qquad (9)$$

As the latter part of the equation indicates, the probability amplitude can be regarded (apart from a phase) as the projection of the state vector onto one of the coordinate axes of the Hilbert space.

Changes in the state vector are governed by the time-dependent Schrödinger equation,

$$\hbar \frac{\partial}{\partial t} \Psi(t) = -i\mathcal{H}(t)\Psi(t). \qquad (10)$$

For all the excitation processes discussed here (those produced by laser pulses whose frequencies lie in the optical or infrared region of the spectrum), the Hamiltonian operator $\mathcal{H}(t)$ varies with time as a result of the interaction energy $\mathbf{d} \cdot \mathbf{E}(t)$ of atomic dipole transition moment operator $\mathbf{d}$ with the electric field vector evaluated at the center of mass of the atom, $\mathbf{E}(t)$. In the simplest examples the electric field has the form of a periodic variation at a carrier frequency (the optical frequency ω) and a more slowly varying envelope, $\mathcal{E}(t)$:

$$\mathbf{E}(t) = \mathbf{e}\mathcal{E}(t)\cos(\omega t + \phi)$$
$$= \tfrac{1}{2}\mathbf{e}\mathcal{E}(t)[\exp(i\omega t + i\phi) + \exp(-i\omega t - i\phi)]. \tag{11}$$

Here $\mathbf{e}$ is a unit vector defining the direction of the electric field, i.e., the polarization direction. The connection between the electric field envelope and the intensity of the radiation $I(t)$ is

$$\mathcal{E}(t)[\text{V/cm}] = 27.4682\sqrt{I(t)[\text{W/cm}^2]}.$$

When the state-vector expansion of Eq. (8) is used with the time-dependent Schrödinger equation (10), there results a coupled set of ordinary differential equations. These may be written in vector form as

$$\hbar\frac{d}{dt}\mathbf{C}(t) = -i\mathsf{H}(t)\mathbf{C}(t), \tag{12}$$

where $\mathbf{C}(t) = [C_1(t), C_2(t), \ldots, C_N(t)]^T$ is an N-component column vector of (complex-valued) probability amplitudes. The elements of the $N \times N$ Hamiltonian matrix $\mathsf{H}(t)$ depend on how the phases are chosen in Eq. (8), as will be noted subsequently.

B. PARTIAL COHERENCE: THE DENSITY MATRIX

We shall be concerned in this chapter only with excitation by purely coherent radiation. Obviously, this is an idealization that cannot hold under all circumstances. In practice, a variety of uncontrollable stochastic events, ranging from spontaneous emission to atomic collisions and laser fluctuations, all cause irreversible changes. To treat such situations one must formulate quantum mechanics in terms of a *density matrix* $\rho(t)$ rather than a state vector $\Psi(t)$ (cf. Shore, 1990, chap. 6). In brief, probabilities are obtained as diagonal elements of this matrix,

$$P_n(t) = \rho_{n,n}(t) = \langle\psi_n|\rho(t)|\psi_n\rangle, \tag{13}$$

and the off-diagonal elements represent coherences. Instead of the Schrödinger equation, one deals with the equation of motion,

$$\hbar\frac{d}{dt}\rho(t) = -i\mathsf{H}(t)\rho(t) + i\rho(t)\mathsf{H}(t) - \mathcal{R}(t)\rho(t). \tag{14}$$

The Hamiltonian matrix H(t) is identical to what occurs with the Schrödinger equation; the operator $\mathcal{R}(t)$ incorporates the various stochastic processes which are not present with purely coherent excitation.

The density matrix can also be used to treat purely coherent excitation of an ensemble in which the initial state is not expressible as a single state vector, as in the example

$$\rho(0) = |\psi_1\rangle \cos\theta \langle\psi_1| + |\psi_2\rangle \sin\theta \langle\psi_2|. \tag{15}$$

In this case there is no stochastic operator, $\mathcal{R}(t) = 0$, and the evolution is equivalent to that predicted by the Schrödinger equation.

For simplicity we will not consider the density matrix, although many of the papers treating STIRAP do discuss it, and realistic modelings often require using it.

C. TWO STATES

For a simple two-state atom $\mathbf{C}(t) = [C_1(t), C_2(t)]^T$ is a two-component column vector whose elements are the probability amplitudes of the two states ψ_1 and ψ_2. The precise appearance of the matrix elements of the Hamiltonian H(t) depends on the choice of phases $\zeta_n(t)$. Were we to set the phases to zero (the Schrödinger picture), then the diagonal elements of H(t) would be the energies of the two states, E_1 and E_2, and the off-diagonal elements would contain the laser–atom dipole interaction energy, including the carrier-frequency oscillations from $\cos(\omega t + \phi)$.

1. The Rotating Wave Approximation (RWA)

It proves more convenient to incorporate the state energies and the carrier frequency ω as phases (the rotating-wave picture): $\zeta_1(t) = E_1 t$ and $\zeta_2(t) = \zeta_1(t) + \omega t$. By so doing we can make the first of the diagonal elements of the Hamiltonian matrix zero; the second diagonal element, the *detuning* Δ, is the difference between the Bohr frequency and the carrier frequency:

$$\hbar\Delta = E_2 - E_1 - \hbar\omega.$$

As presented here, the detuning is constant in time. More generally, the laser pulses induce time-varying Stark shifts of the energy levels (see Section II.D.3), and the detuning then becomes explicitly time-dependent.

The off-diagonal matrix element acquires a factor $\exp(-i\omega t)$ which cancels one part of the rapid carrier variation of the electric field—one of the two exponentials shown in Eq. (11). The remaining exponential is then $\exp(+i2\omega t)$. For most near-resonant excitation situations this factor varies much more rapidly than any changes of the probability amplitudes, and it is then permissible to neglect this term (thereby making the *rotating wave approximation,* RWA). When this is done, and the detuning is allowed to vary in time, the Hamiltonian matrix reads (Shore, 1990)

$$\mathsf{H}(t) = \hbar \begin{bmatrix} 0 & \frac{1}{2}\Omega(t) \\ \frac{1}{2}\Omega(t) & \Delta(t) \end{bmatrix}. \tag{16}$$

The off-diagonal element $\Omega(t)$, the *Rabi frequency*, parameterizes the strength of the atom–laser interaction; it is proportional to the component of the atomic transition dipole moment in the direction of the electric-field vector, d_{12}, and to the laser electric-field amplitude $\mathcal{E}(t)$; i.e.,

$$\hbar\Omega(t) = d_{12}\mathcal{E}(t). \tag{17}$$

As long as the phase ϕ of the electric field (11) remains constant, it is always possible to choose the expansion phases $\zeta_n(t)$ so that the Rabi frequency is real-valued and positive (Shore, 1990). The diagonal elements of $\mathsf{H}(t)$ are the two RWA energies: the zero element is the energy of state ψ_1 lifted (dressed) by the photon energy $\hbar\omega$ and used as the reference energy level, while the energy $\hbar\Delta$ is the frequency offset (detuning) of state ψ_2.

For the RWA to be valid it is necessary that both the frequencies $\Omega(t)$ and $\Delta(t)$ be much smaller than the carrier frequency ω, and that the pulse duration be many optical cycles. For a discussion of deviations from the RWA, see Section III.B.8.

We have presented the RWA as a combination of choosing phases and neglecting high-frequency oscillations. A more physical picture is possible (Series, 1978). The presence of the phases in the expansion (8) is equivalent to the introduction of a time-varying coordinate system in the two-dimensional Hilbert space spanned by the states ψ_1 and ψ_2. Indeed, this is a rotation at a steady rate ω, the carrier frequency of the laser. In this rotating coordinate system one term of the interaction energy is slowly varying (it varies only with the varying intensity envelope), while the second term oscillates at twice the carrier frequency. The neglect of this "counterrotating" term leads to the RWA.

2. Rabi Cycling

The simple periodic solution to the resonant RWA equations reveals the connection of the frequency of population oscillations and the strength of the interaction to be the Rabi frequency. For numerical estimates of this frequency the following formula is useful,

$$|\Omega[\text{rad/ns}]| = 0.22068 \left(\frac{d_{12}}{ea_0}\right) \sqrt{I[\text{W/cm}^2]},$$

where e is the electron charge and a_0 is the Bohr radius, so that ea_0 is the atomic unit of dipole moment.

An analytic solution for two-state excitation exists not only for resonant excitation but for detuned excitation, when the radiation remains steady. The excitation probability is then

$$P_e(t) = \frac{1}{2}\left(\frac{\Omega}{\tilde{\Omega}}\right)^2 [1 - \cos(\tilde{\Omega}t)], \tag{18}$$

where $\tilde{\Omega} = \sqrt{\Omega^2 + \Delta^2}$ is the *nonresonant flopping frequency*. As can be seen, when detuning grows, the population oscillations become more rapid and less complete—population never resides entirely in the excited state.

The solution (18) predicts that, at the termination of a pulse of constant-intensity radiation, some population may remain in the excited state. When one treats pulses that turn on and off more gradually, this tends not to be the case. Instead, a detuned, slowly varying pulse will return the population to the initial state—a process often termed *coherent population return* (Vitanov, 1995; Vitanov and Knight, 1995; Kuhn et al., 1998).

3. Probability Loss

The model of a two-state system is an idealization that, however useful, is never completely correct. Real atoms have an infinite number of bound states lying below the ionization continuum and molecules have numerous vibrational and rotational levels below each dissociation limit. When the radiation is very close to resonance between the ground state and a single excited state, the remaining levels hold little probability; primarily their influence comes indirectly, through providing the polarizability that appears as Stark shifts of the two energy levels of interest.

However, one effect of additional energy levels can become important: the excited state can always undergo spontaneous emission (over a time interval comparable to the inverse of the spontaneous emission time). When this emission leads to levels other than the ground state, then population is lost from the two-state system. To account correctly for this loss, one should really include not only additional levels in the description of the system, but one should base the mathematics on a density matrix rather than a state vector (cf. Section II.B). One simple way of accounting for probability loss out of the system of interest, occurring from state ψ_2 at a rate Γ, is to take the state energy to be a complex number. The resulting detuning is

$$\hbar\Delta = E_2 - \hbar\omega - i\tfrac{1}{2}\Gamma.$$

Under the influence of such a non-Hermitian Hamiltonian, probability is not conserved. The probabilities $P_n(t)$ will in general approach zero at long times.

4. Adiabatic States and Adiabatic Following

Theoretical discussion of time-evolving quantum systems is greatly facilitated by introducing instantaneous eigenstates $\Phi_k(t)$ of the time-varying Hamiltonian matrix,

$$\mathsf{H}(t)\Phi_k(t) = \hbar\varepsilon_k(t)\Phi_k(t). \tag{19}$$

Because the Hamiltonian changes with time, both the eigenvalues $\hbar\varepsilon_k(t)$ and the eigenvectors, the *adiabatic states* $\Phi_k(t)$, will change with time. The adiabatic states are time-dependent superpositions of the unperturbed states ψ_1 and ψ_2 (known also as *diabatic states*) (Shore, 1990),

$$\Phi_+(t) = \psi_1 \sin \Theta(t) + \psi_2 \cos \Theta(t), \tag{20a}$$

$$\Phi_-(t) = \psi_1 \cos \Theta(t) - \psi_2 \sin \Theta(t), \tag{20b}$$

where the mixing angle $\Theta(t)$ is defined (modulo π) as

$$\Theta(t) = \frac{1}{2} \arctan\left[\frac{\Omega(t)}{\Delta(t)}\right].$$

The energies of the adiabatic states are the two eigenvalues of $\mathsf{H}(t)$,

$$\varepsilon_\pm(t) = \tfrac{1}{2}[\Delta(t) \pm \sqrt{\Delta^2(t) + \Omega^2(t)}]. \tag{21}$$

What we here term *adiabatic states* are known also as *dressed states*, implying that the field interaction has clothed the atom in photons; the original (physical) states ψ_n are termed *bare states*.

The adiabatic states can serve as a moving coordinate system in which to place the state vector $\Psi(t)$ as it changes under the influence of the coherent radiation pulse. Such coordinates are most useful when the elements of the Hamiltonian—the Rabi frequency and the detuning—change sufficiently slowly (i.e., adiabatically); then the state vector remains fixed in the adiabatic coordinates space. Mathematically, adiabatic evolution requires that the coupling between the adiabatic states be negligible compared to the difference between their eigenfrequencies (Shore, 1990; Messiah, 1962; Crisp, 1973), viz.,

$$|\langle \dot{\Phi}_+|\Phi_-\rangle| \ll |\varepsilon_+ - \varepsilon_-|, \tag{22}$$

where the dot denotes a time derivative (note that $\langle \dot{\Phi}_+|\Phi_-\rangle = -\langle \Phi_+|\dot{\Phi}_-\rangle$). Explicitly, the two-state *adiabatic condition* reads

$$\tfrac{1}{2}|\dot{\Omega}\Delta - \Omega\dot{\Delta}| \ll (\Omega^2 + \Delta^2)^{3/2}. \tag{23}$$

COHERENT MANIPULATION OF ATOMS AND MOLECULES

According to Eq. (23), adiabatic evolution requires a smooth pulse, long interaction time, and large Rabi frequency and/or large detuning.

When the adiabatic condition holds, there are no transitions between the adiabatic states and their populations are conserved. That is, the state vector $\Psi(t)$ remains fixed in the time-varying coordinate system of adiabatic states, as the latter move with respect to the fixed basis ψ_1 and ψ_2. In particular, if the state vector $\Psi(t)$ coincides with a single adiabatic state $\Phi(t)$ at some time t, then it will remain in that adiabatic state as long as the evolution is adiabatic: the state vector $\Psi(t)$ will *adiabatically follow* the state $\Phi(t)$. The relationship of the single adiabatic state $\Phi(t)$ [and of the state vector $\Psi(t)$] to the diabatic (bare) states will change if the mixing angle $\Theta(t)$ changes, and so adiabatic evolution can produce population transfer between those diabatic states.

Generally speaking, successful population transfer means that the state vector $\Psi(t)$ connects to the initial state ψ_1 at early times, and connects to a specified target state ψ_N at late times. Under appropriate conditions, this connection can be provided by a single adiabatic state. Such a state is termed an *adiabatic transfer* (AT) state. One refers to these asymptotic connections with initial and target states as *connectivity*.

5. Rapid Adiabatic Passage

When discussing adiabatic time evolution it is useful to plot the values of the adiabatic energies as a function of time, and to view carefully the time intervals during which the curves are close together—it is during these intervals that the adiabatic condition is most likely to be violated. There are two distinct types of adiabatic population changes, distinguished by the behavior of the diabatic energies 0 and $\hbar \Delta(t)$. The *no-crossing* case is depicted in Fig. 4 (top left frame) in the particular case of constant detuning; then the diabatic energy curves are parallel to each other during the interaction. In the absence of interaction the adiabatic energies coincide with the diabatic ones, but the (pulsed) interaction $\Omega(t)$ pushes them away from each other. As Eqs. (20) show, at early and late times each adiabatic state is identified with the same diabatic state: $\Phi_-(t \to \pm\infty) = \psi_1$, $\Phi_+(t \to \pm\infty) = \psi_2$, while at intermediate times it is a superposition of diabatic states. Consequently, starting from the ground state ψ_1 initially, the population makes a partial excursion into the excited state ψ_2 at intermediate times and eventually returns to ψ_1 in the end (bottom left frame). Hence in the no-crossing case adiabatic evolution leads to *complete population return*.

A rather different situation occurs when the detuning $\Delta(t)$ sweeps slowly from some very large negative value to some very large positive value (or vice versa), as shown in Fig. 4 (top right frame). That is, the Hamiltonian at the end of the pulse differs from the Hamiltonian at the beginning, because of the detuning change. Large in this context means much larger than the Rabi frequency $\Omega(t)$. The two

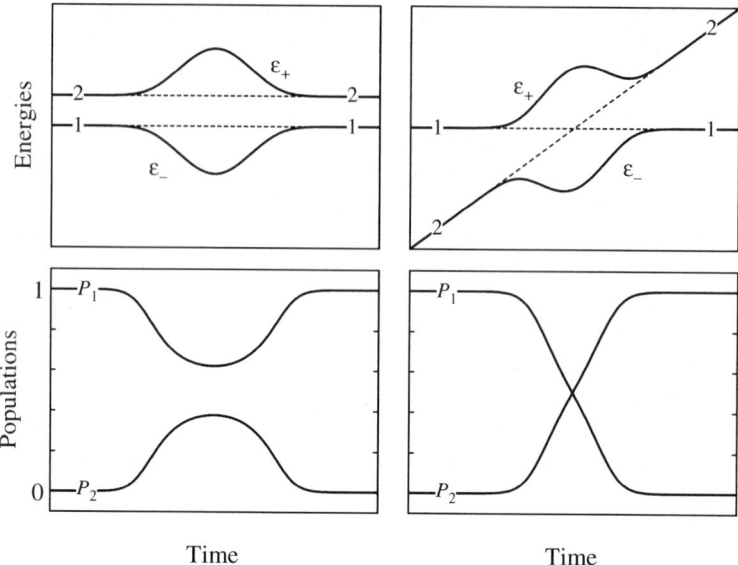

FIG. 4. Time evolution of the energies (upper frames) and the populations (lower frames) in a two-state system. In the upper plots, the dashed lines show the unperturbed (diabatic) energies, and the solid curves show the adiabatic energies. The left-hand frames are for the no-crossing case, while the right-hand frames are for the level-crossing case.

diabatic energies 0 and $\hbar\Delta(t)$ intersect at time t_0 when the detuning is zero. The adiabatic energies approach the diabatic energies when $\Delta(t)$ is large (at early and late times), but the presence of interaction prevents their intersection—the adiabatic energies have an *avoided crossing*. Indeed, as Eq. (21) shows, the eigenenergy separation $\hbar\varepsilon_+(t) - \hbar\varepsilon_+(t) = \hbar\sqrt{\Delta^2(t) + \Omega^2(t)}$ is equal to $\hbar\Omega(t_0)$ at the crossing. For constant $\Omega(t)$ this is the minimum value of the eigenvalue separation, while for pulse-shaped $\Omega(t)$ (as in Fig. 4), there are two minima near t_0. At very early and late times the ratio $\Delta(t)/\Omega(t) \stackrel{t\to\pm\infty}{\longrightarrow} \pm\infty$. Hence, during the excitation the mixing angle $\Theta(t)$ rotates clockwise from $\Theta(-\infty) = \pi/2$ to $\Theta(+\infty) = 0$ and the composition of the adiabatic states changes accordingly. Asymptotically, each adiabatic state becomes uniquely identified with a single unperturbed state,

$$\psi_1 \stackrel{-\infty \leftarrow t}{\longleftarrow} \Phi_+(t) \stackrel{t \to +\infty}{\longrightarrow} \psi_2, \tag{24a}$$

$$-\psi_2 \stackrel{-\infty \leftarrow t}{\longleftarrow} \Phi_-(t) \stackrel{t \to +\infty}{\longrightarrow} \psi_1. \tag{24b}$$

Consequently, starting from state ψ_1 initially, the system follows adiabatically the adiabatic state $\Phi_+(t)$ and eventually ends up in state ψ_2. The laser pulse, with detuning sweep, has produced *complete population transfer,* a process known as adiabatic passage or, because it must occur in a time shorter than the radiative lifetime of the excited state, as *rapid adiabatic passage.* We emphasize that adiabatic passage in a two-state system does not depend on the sign of the detuning slope: it takes place for both $\dot{\Delta}(t) > 0$ (as was assumed above) and $\dot{\Delta}(t) < 0$.

The adiabatic condition (23) expresses a constraint at each moment; it is a "local" condition. A "global" condition can be obtained from the expression $\langle \dot{\Phi}_+ | \Phi_- \rangle = \dot{\Theta}$. The average value of this angle, as it increases from 0 to $\pi/2$ is $\pi/4$. For resonant excitation the time-integrated difference of the two eigenvalues is just the time integral of the Rabi frequency, i.e., the pulse area. Hence one finds the commonly used rule of thumb that the pulse area must be very large,

$$A = \int_{-\infty}^{+\infty} dt\, \Omega(t) \gg 1.$$

Adiabatic passage offers significant advantages over Rabi cycling as a means of producing complete population transfer in an ensemble of atoms. Unlike Rabi cycling, adiabatic passage is robust against small-to-moderate variations in the laser intensity, detuning, and interaction time. Therefore it can produce uniform excitation for a broad range of Doppler shifts.

6. Estimating Transition Probabilities

A popular tool for estimating the transition probability between two crossing *diabatic* states is the Landau–Zener formula (Landau, 1932; Zener, 1932),

$$P = 1 - p, \qquad p = \exp\left[-\frac{\pi \Omega^2(t_0)}{2|\dot{\Delta}(t_0)|}\right], \qquad (25)$$

where $\dot{\Delta}(t_0)$ is the rate of change in the detuning evaluated at the crossing time t_0 and $\Omega(t_0)$ is the value of the Rabi frequency at t_0. This formula is exact only for a constant Rabi frequency and a linearly varying detuning over an infinite time interval, so it is only an approximation to any actual probability for adiabatic passage. Nevertheless, it correctly identifies the importance of the ratio of Ω^2 to $\dot{\Delta}$ as a measure of the likelihood of population transfer. We note here that the probability for (nonadiabatic) transition between the *adiabatic* states is $p = 1 - P$.

There exist other models of level-crossing excitation, more realistic because they allow pulsed interactions. The Allen–Eberly–Hioe model (Allen and Eberly, 1975; Hioe, 1984) assumes a hyperbolic-secant pulse and a hyperbolic-tangent chirp. The Demkov–Kunike model (Demkov and Kunike, 1969; Suominen and Garraway, 1992) adds a static detuning to this model.

D. THREE STATES

1. *The Three-State RWA Hamiltonian*

The usual situation with three discrete quantum states can be regarded as a linkage chain $\psi_1 \leftrightarrow \psi_2 \leftrightarrow \psi_3$ in which one of the states, the intermediate state ψ_2, is coupled by radiative interaction to two other states, which have no radiative coupling between them. (When the excitation is only by means of electric dipole radiation, the parity selection rule forbids the 3–1 linkage for free atoms or molecules.) Typically, each of the two linkages originates in a separate laser pulse, whose carrier frequency is close to the relevant Bohr frequency. The following variants of this chain occur, distinguished by the ordering of their unperturbed energies E_j (see Fig. 5):

- The states may form a *ladder*, in which successive energies lie higher than the predecessors. Population begins in state ψ_1. The ladder configuration occurs when one has interest in stepwise excitation toward ionization. It readily generalizes to multiple levels, each more highly excited.
- The linkage may form a *lambda*, in which the middle state, ψ_2, has unperturbed energy lying above either other state (the relative energy ranking of ψ_1 and ψ_3 is not significant). Again, population begins in state ψ_1. The lambda configuration exemplifies a Raman process.
- The linkage may form a *vee*, in which the middle state has lowest unperturbed energy, and initially holds the population. The vee configuration has interesting quantum-beat interference patterns between the two excitation branches leading to and from a single state.

In all of these cases the phases $\zeta_n(t)$ of the state vector expansion (8) can be chosen to place a zero as one of the diagonal elements of the Hamiltonian matrix, and to permit a generalized rotating wave approximation in which time variations

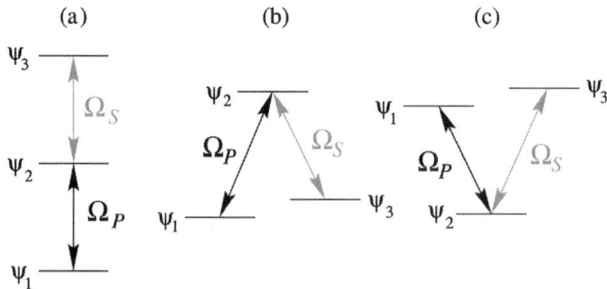

FIG. 5. Linkages for three-state coupling: (a) ladder configuration; (b) Λ configuration; (c) V configuration.

at the carrier frequencies have been eliminated. When the zero goes into the first element, the pattern is

$$\mathsf{H}(t) = \hbar \begin{bmatrix} 0 & \frac{1}{2}\Omega_1(t) & 0 \\ \frac{1}{2}\Omega_1(t) & \Delta_2 & \frac{1}{2}\Omega_2(t) \\ 0 & \frac{1}{2}\Omega_2(t) & \Delta_3 \end{bmatrix}. \quad (26)$$

The off-diagonal elements are the Rabi frequencies $\Omega_j(t)$, related to transition-dipole moments and electric-field envelopes by the relationships

$$\hbar\Omega_1(t) = d_{12} \cdot \mathcal{E}_1(t), \qquad \hbar\Omega_2(t) = d_{23} \cdot \mathcal{E}_2(t). \quad (27)$$

The expressions for the detunings $\Delta_j(t)$ depend on whether the Hamiltonian describes a ladder, a lambda, or a vee linkage. For the ladder arrangement one has

$$\hbar\Delta_2 = E_2 - E_1 - \hbar\omega_1, \quad (28a)$$
$$\hbar\Delta_3 = E_3 - E_1 - \hbar\omega_1 - \hbar\omega_2, \quad (28b)$$

while for the lambda the formulas read

$$\hbar\Delta_2 = E_2 - E_1 - \hbar\omega_1, \quad (29a)$$
$$\hbar\Delta_3 = E_3 - E_1 - \hbar\omega_1 + \hbar\omega_2. \quad (29b)$$

It is not difficult to obtain numerical solutions to the set of coupled ordinary differential equations, from which one can make plots of populations as a function of time or, for time fixed at the end of both pulses, as a function of other parameters. Some special limiting cases have properties that are of particular importance for population transfer.

2. Pulse Sequences

Much of the early theoretical work on three-state systems involved steady fields. Although population transfer can take place under such circumstances, all population changes are periodic, and so there is a regular return of population into the initial state, just as with the two-state atom. Of greater interest are various sequences of pulse pairs, one associated with each of the two transitions. These typically fall into two classes: when the first pulse introduces a coupling between the initial state and the next state of the chain, the sequence is often called "intuitive" ordering of pulses, because intuition suggests that one should start moving population by acting on a populated state. As will be noted in the sections on

STIRAP, a "counterintuitive" sequence, acting first on an unpopulated state, often is a better choice.

There are two possibilities for causing individual atoms (or molecules) to experience a sequence of pulses. The atoms may be relatively stationary, say in a vapor cell, and be exposed to pulsed lasers. Alternatively, the atoms may be moving together, say in an atomic beam, and pass across beams from steady lasers. What matters for excitation is the timing of the fields in the rest frame of each atom.

3. Adiabatic Elimination

An important special case of three-state excitation occurs when the detuning Δ from the intermediate state ψ_2 is very large, but the two-photon detuning δ between states ψ_1 and ψ_3 is small. This situation implies nearly resonant two-photon excitation without having a single-photon resonance. The RWA Hamiltonian of Eq. (26) provides the following equation for the amplitude in the intermediate state, ψ_2:

$$i\frac{d}{dt}C_2(t) = \tfrac{1}{2}\Omega_1 C_1(t) + \Delta_2 C_2 + \tfrac{1}{2}\Omega_2 C_3(t). \tag{30}$$

When Δ_2 is very large, rapid oscillations of amplitude $C_2(t)$ will occur, at this frequency. These variations are much more rapid than any changes of interest. One can average over many such oscillation periods (i.e., over a time interval $\Delta t \gg \Delta_2^{-1}$) to obtain more slowly varying amplitudes, $\bar{C}_n(t)$, for which the time variation of $\bar{C}_2(t)$ vanishes. By setting $(d/dt)\bar{C}_2 = 0$, we obtain from Eq. (30) that

$$\bar{C}_2(t) = \frac{-\Omega_1}{2\Delta_2}\bar{C}_1(t)\frac{-\Omega_2}{2\Delta_2}\bar{C}_3(t). \tag{31}$$

This approximation, known as *adiabatic elimination*, relates the intermediate-state amplitude $\bar{C}_2(t)$ to those of the other two states. Because Δ_2 is large (in order that adiabatic elimination be valid), the amplitude $\bar{C}_2(t)$ is small, and the population is confined to the two states ψ_1 and ψ_3. The two-state Hamiltonian that results from adiabatic elimination,

$$\mathsf{H}(t) = \hbar \begin{bmatrix} S_1(t) & \tfrac{1}{2}\bar{\Omega}(t) \\ \tfrac{1}{2}\bar{\Omega}(t) & \Delta_3(t) + S_3(t) \end{bmatrix}, \tag{32}$$

is similar to the simple two-state equation (16), but the interaction term—the Rabi frequency—now involves the product of two interactions,

$$\bar{\Omega}(t) = \frac{-\Omega_1(t)\Omega_2(t)}{2\Delta_2}, \tag{33}$$

and the diagonal elements have acquired dynamic Stark shifts,

$$S_1(t) = \frac{-|\Omega_1(t)|^2}{4\Delta_2}, \qquad S_3(t) = \frac{-|\Omega_2(t)|^2}{4\Delta_2}. \tag{34}$$

By incorporating $S_1(t)$ into the phase $\zeta_k(t)$, the Hamiltonian can be made to appear as in Eq. (16). The detuning then involves the difference between Stark-shifted Bohr frequencies and the carrier frequency,

$$\hbar\Delta(t) = E_3 + \hbar S_3(t) - E_1 - \hbar S_1(t) - \hbar\omega.$$

The preceding formulas give the essential properties of two-photon interactions. For more accurate results one should include a larger number of possible intermediate states (ideally an infinite number). In so doing one should avoid making the RWA at an early stage, and should treat the full Hamiltonian (cf. Shore, 1990, sect.14.8).

The two-photon Rabi frequency and the dynamic Stark shifts are both consequences of the presence of additional energy states that do not participate strongly in the excitation because they are far off resonance. The electric field interacts not only with any permanent dipole transition moments, as embodied in the operator **d**, but also with an induced dipole moment. The proportionality between this induced moment and the electric field is the complex-valued frequency-dependent *polarizability* tensor $\mathbf{Q}(\omega)$. The polarizability contributes to the Hamiltonian an interaction energy (Shore, 1990, sect. 14.9; Yatsenko *et al.*, 1998),

$$\mathcal{H}^{int}(t) = -\tfrac{1}{2}\mathbf{E}(t) \cdot \mathbf{Q}(\omega) \cdot \mathbf{E}(t). \tag{35}$$

The Hamiltonian matrix of such an interaction has both diagonal elements (acting as dynamic Stark shifts) and off-diagonal elements (acting as two-photon Rabi frequencies.)

4. Autler-Townes Splitting

Another important situation occurs when the coupling between two states (taken here to be Ω_2 between ψ_2 and ψ_3) is strong, whereas the coupling to the first state, Ω_1, is weak. For simplicity, let the fields be steady, not pulsed, and let the strong 2–3 transition be resonant with field of frequency ω_2 while the weak field has tunable frequency ω. It is natural to diagonalize the strong-coupling portion of the Hamiltonian, which is a two-level system. This leads to the dressed states Φ_- and Φ_+, whose composition is identical to that given above, and to the following expression for the state vector:

$$\Psi(t) = \psi_1 C_1(t)e^{-i\omega t} + \Phi_- B_-(t) + \Phi_+ B_+(t). \tag{36}$$

The RWA Hamiltonian takes the form

$$\mathsf{H}(t) = \hbar \begin{bmatrix} \Delta & \frac{1}{2}\Omega_- & \frac{1}{2}\Omega_+ \\ \frac{1}{2}\Omega_- & -\varepsilon & 0 \\ \frac{1}{2}\Omega_+ & 0 & +\varepsilon \end{bmatrix}, \tag{37}$$

where the dressed eigenvalues are $\pm\varepsilon = \pm\frac{1}{2}|\Omega_2|$, the detuning is $\Delta = (E_2 - E_1)/\hbar - \omega$, and the Rabi frequencies $\Omega_\pm$ are expressible in terms of the components of the dressed states. The situations of interest are when there occur two identical diagonal elements of this Hamiltonian, i.e., there is a degeneracy of diabatic energies; such a situation is one in which the excitation is dominated by Rabi oscillations between the degenerate states. There are two choices of probe frequency that make two diagonal element identical:

$$\hbar\omega = E_2 - E_1 \pm \hbar\varepsilon = E_2 - E_1 \pm \frac{1}{2}\hbar|\Omega_2|. \tag{38}$$

That is, there are two weak-field detunings for which there will be appreciable coupling between state ψ_1 and the strongly driven two-state system. In essence, the strong field has split the energy level accessed by the third state into two levels, separated by the *Autler–Townes* splitting $2\varepsilon = \hbar|\Omega_2|$ (Autler and Townes, 1955).

The Autler–Townes effect manifests itself in the absorption spectrum in the form of a splitting of the resonance. There is, however, another interesting feature when the lifetime of state ψ_3 is much longer than that of the excited state ψ_2. In this case one observes that the absorption exactly vanishes when the weak field is tuned midway between the two dressed energy levels (38). This is a result of an interference between the two absorption channels starting from the ground state ψ_1 via the two dressed states $\Phi_\pm$ with subsequent decay out of the system (Imamoglu et al., 1989; Lounis and Cohen-Tannoudji, 1992; Fleischhauer et al., 1992), as illustrated in Figs. 6 and 7. The interference of the decay channels becomes apparent if we include decay rates Γ_2 and Γ_3 in the above analysis. After the partial diagonalization we find

$$\mathsf{H}(t) = \frac{\hbar}{2} \begin{bmatrix} 2\Delta & \Omega_- & \Omega_+ \\ \Omega_- & -2\varepsilon - \frac{i}{2}(\Gamma_2 + \Gamma_3) & i(\Gamma_3 - \Gamma_2) \\ \Omega_+ & i(\Gamma_3 - \Gamma_2) & 2\varepsilon - \frac{i}{2}(\Gamma_2 + \Gamma_3) \end{bmatrix}. \tag{39}$$

COHERENT MANIPULATION OF ATOMS AND MOLECULES 81

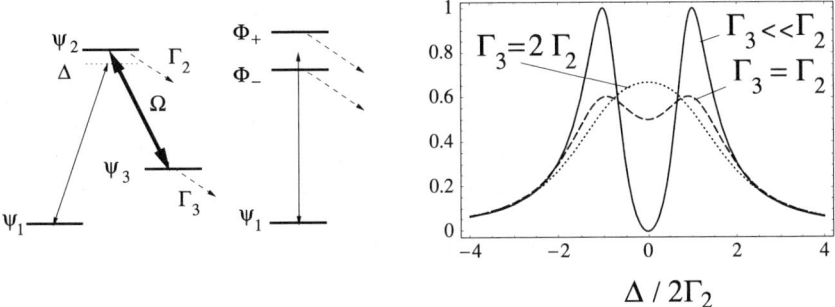

FIG. 6. *Left:* Diagonalization of strong, resonant coupling of levels ψ_2 and ψ_3 leads to symmetrically split Autler–Townes doublett $\Phi_\pm$. *Right:* Absorption of weak probe field as function of detuning Δ for $\Omega = 2\Gamma_2$. Destructive interference of absorption pathways for $\Gamma_3 \ll \Gamma_2$ leads to cancellation of absorption on resonance. Interference is absent for $\Gamma_3 = \Gamma_2$ and constructive for $\Gamma_3 > \Gamma_2$.

Three cases are of interest:

- If $\Gamma_3 \ll \Gamma_2$, i.e., if the decay out of state ψ_3 can be disregarded, the two absorption channels start and end in the same states and there is destructive interference. When $\Gamma_3 = 0$ there is complete cancellation of the absorption. This case, termed electromagnetically induced transparency (EIT) (Boller *et al.*, 1991), has a number of important implications (cf. Section XA).
- If $\Gamma_3 \gg \Gamma_2$, the interference is constructive and there is an enhanced absorption compared to a simple sum of two Lorentzian line profiles.
- Finally, if $\Gamma_3 \sim \Gamma_2$, the interference effect almost vanishes and only the splitting remains.

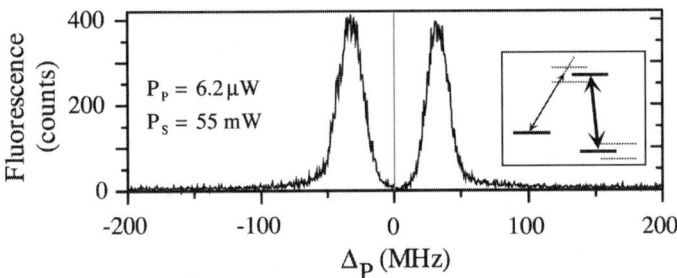

FIG. 7. Experimental demonstration of Autler–Townes splitting in neon. The Stokes (dressing) laser intensity is 55 mW and the probe laser intensity is 0.62 μW. From Bergmann *et al.* (1998).

III. Three-State STIRAP: Theory

A. BASIC PROPERTIES OF STIRAP

When the three-state linkage forms a lambda pattern (see Fig. 3b), one has a typical stimulated-Raman excitation scheme. The field acting on the initial state ψ_1 is termed the *pump* field, and the interaction leading to the final (target) state ψ_3 is termed the *Stokes*.

It is intuitively evident that population transfer between states ψ_1 and ψ_3 can take place, as in SEP, when the pump pulse precedes the Stokes pulse. Simultaneous and steady pulses can also produce complete population transfer, if the pulse areas are carefully chosen. What may not be obvious at first is that even more satisfactory population transfer can be produced if the Stokes pulse occurs first—the pulses then arrive in a "counterintuitive" ordering. This is the basis for a process now called stimulated Raman adiabatic passage (STIRAP) (Oreg et al., 1984; Kuklinski et al., 1989; Gaubatz et al., 1990; Bergmann and Shore, 1995; Shore, 1995; Bergmann et al., 1998, Vitanov et al., 2001).

The STIRAP technique uses the coherence of two pulsed laser fields to achieve a (nearly) complete population transfer from an initially populated state ψ_1 to a target state ψ_3 via an intermediate state ψ_2 (see Fig. 3b). If a two-photon resonance between ψ_1 and ψ_3 is maintained, if there is sufficient overlap of the two pulses, and if the pulses are sufficiently strong that the time evolution is adiabatic, then (almost) complete population transfer occurs between states ψ_1 and ψ_3. Furthermore, there is almost no population in the (usually decaying) intermediate state ψ_2 at any time. The following sections offer explanations for this remarkable result.

1. Basic Equations and Definitions

The mathematical description of STIRAP derives from the Schrödinger equation (12) with $\mathbf{C}(t)$ a three-component column vector. Initially the population resides entirely in state ψ_1, meaning $\mathbf{C}(-\infty) = [1, 0, 0]^T$. The objective is to transfer population into state ψ_3, meaning $\mathbf{C}(+\infty) = [0, 0, 1]^T$.

With the rotating wave approximation, the Hamiltonian $\mathsf{H}(t)$ for purely coherent excitation has the form

$$\mathsf{H}(t) = \hbar \begin{bmatrix} 0 & \frac{1}{2}\Omega_p(t) & 0 \\ \frac{1}{2}\Omega_p(t) & \Delta_p & \frac{1}{2}\Omega_s(t) \\ 0 & \frac{1}{2}\Omega_s(t) & \Delta_p - \Delta_s \end{bmatrix}. \quad (40)$$

Here $\Omega_p(t)$ and $\Omega_s(t)$ are the (real-valued) Rabi frequencies of the pump and Stokes pulses, respectively,

$$\hbar\Omega_p(t) = d_{12}\mathcal{E}_p(t), \qquad \hbar\Omega_s(t) = d_{23}\mathcal{E}_s(t).$$

The diagonal elements of this matrix involve the single-photon detunings of the pump and Stokes lasers from their respective transitions,

$$\hbar\Delta_p = E_2 - E_1 - \hbar\omega_p, \qquad \hbar\Delta_s = E_2 - E_3 - \hbar\omega_s. \tag{41}$$

An essential condition for STIRAP is that there be two-photon resonance between states ψ_1 and ψ_3, meaning $\Delta_p = \Delta_s \equiv \Delta$, or

$$\delta \equiv \Delta_p - \Delta_s = 0. \tag{42}$$

The single-photon detuning Δ has relatively little effect on STIRAP; Section III.B.3 will discuss its effect.

Although the definitions of Eq. (14) pertain to excitation in the absence of any incoherent processes, it is easy to include the possibility of loss from state ψ_2 at a rate Γ by making the replacement $E_2 \to E_2 - \frac{1}{2}i\Gamma$. In this way it is possible to model the effect of spontaneous emission out of state ψ_2 into states *other than* ψ_1 or ψ_2. As long as the excitation is adiabatic, the presence of a complex-valued detuning has no effect on the STIRAP process, because population never is found in state ψ_2. However, as the detuning and the loss rate increase, adiabaticity deteriorates, which eventually reduces the transfer efficiency; for further discussion see Sections III.B.3 and III.B.5.

In practice, a part of the spontaneous emission acts to repopulate states ψ_1 and ψ_3. Such effects cannot be treated within the Schrödinger equation; they require a density matrix treatment.

2. Adiabatic States

The population transfer mechanism in STIRAP is most easily understood in a Hilbert space whose coordinate basis vectors are instantaneous eigenstates of the time-varying Hamiltonian (i.e., a basis of adiabatic states). When the two-photon resonance condition (42) is fulfilled, the eigenvalues of $\mathbf{H}(t)$, which represent the energies of the adiabatic states, are $\hbar\varepsilon_-$, $\hbar\varepsilon_0$, and $\hbar\varepsilon_+$, where

$$\varepsilon_+(t) = \tfrac{1}{2}[\Delta + \sqrt{\Delta^2 + \Omega^2(t)}] = \tfrac{1}{2}\Omega(t)\cot\varphi(t), \tag{43a}$$

$$\varepsilon_0(t) = 0, \tag{43b}$$

$$\varepsilon_-(t) = \tfrac{1}{2}[\Delta - \sqrt{\Delta^2 + \Omega^2(t)}] = -\tfrac{1}{2}\Omega(t)\tan\varphi(t). \tag{43c}$$

The presence of a null eigenvalue follows from the choice of energy zero point [and the phases $\zeta_n(t)$], which here reckons all energies as excitation from the initial state ψ_1.

The corresponding eigenstates $\Phi_+(t)$, $\Phi_0(t)$, and $\Phi_-(t)$ of $\mathsf{H}(t)$ are connected to the bare (diabatic) states ψ_1, ψ_2, and ψ_3 by the relations

$$\Phi_+(t) = \psi_1 \sin\vartheta(t)\sin\varphi(t) + \psi_2 \cos\varphi(t) + \psi_3 \cos\vartheta(t)\sin\varphi(t), \quad (44\text{a})$$

$$\Phi_0(t) = \psi_1 \cos\vartheta(t) - \psi_3 \sin\vartheta(t), \quad (44\text{b})$$

$$\Phi_-(t) = \psi_1 \sin\vartheta(t)\cos\varphi(t) - \psi_2 \sin\varphi(t) + \psi_3 \cos\vartheta(t)\cos\varphi(t), \quad (44\text{c})$$

where the time-dependent mixing angles $\vartheta(t)$ and $\varphi(t)$ are defined as

$$\tan\vartheta(t) = \frac{\Omega_p(t)}{\Omega_s(t)}, \quad \tan 2\varphi(t) = \frac{\Omega(t)}{\Delta}, \quad (45)$$

with $\Omega(t)$ the root-mean-square (RMS) field

$$\Omega(t) = \sqrt{\Omega_p^2(t) + \Omega_s^2(t)}. \quad (46)$$

The adiabatic state $\Phi_0(t)$ associated with the null eigenvalue has particular importance: it has no component of the excited state ψ_2. The latter state can undergo spontaneous emission back to state ψ_1, state ψ_3 (in the Λ configuration), and in most cases it can also decay to other states. By avoiding the possibility of such loss, the state $\Phi_0(t)$ acts to trap population; it is known as a *trapped state* (Alzetta et al., 1976, 1979; Arimondo and Orriols, 1976; Gray et al., 1978; Arimondo, 1996).

We introduce the diabatic or adiabatic bases by writing one of the two expansions

$$\begin{aligned}\Psi(t) &= \sum_n \hat{\psi}_n C_n(t) \exp[-i\zeta_n(t)] \\ &= \sum_k \Phi_k(t) B_k(t).\end{aligned} \quad (47)$$

According to Eqs. (44), the probability amplitudes of the adiabatic states $\mathbf{B}(t) = [B_+(t), B_0(t), B_-(t)]^T$ are connected to the diabatic-state (or bare-state) amplitudes $\mathbf{C}(t)$ by the orthogonal transformation

$$\mathbf{C}(t) = \mathsf{R}(t)\mathbf{B}(t), \quad (48)$$

where the rotation matrix $\mathsf{R}(t)$ is given by

$$\mathsf{R}(t) = \begin{bmatrix} \sin\vartheta\sin\varphi & \cos\vartheta & \sin\vartheta\cos\varphi \\ \cos\varphi & 0 & -\sin\varphi \\ \cos\vartheta\sin\varphi & -\sin\vartheta & \cos\vartheta\cos\varphi \end{bmatrix}. \tag{49}$$

The Schrödinger equation in the adiabatic representation is obtained from Eqs. (12), (48), and (49) and is given by

$$\hbar\frac{d}{dt}\mathbf{B}(t) = -i\mathsf{H}_b(t)\mathbf{B}(t), \tag{50}$$

with the Hamiltonian given by $\mathsf{H}_b = \mathsf{R}^{-1}\mathsf{H}\mathsf{R} - i\hbar\mathsf{R}^{-1}\dot{\mathsf{R}}$, or explicitly,

$$\mathsf{H}_b = \hbar \begin{bmatrix} \tfrac{1}{2}\Omega\cot\varphi & i\dot\vartheta\sin\varphi & i\dot\varphi \\ -i\dot\vartheta\sin\varphi & 0 & -i\dot\vartheta\cos\varphi \\ -i\dot\varphi & i\dot\vartheta\cos\varphi & -\tfrac{1}{2}\Omega\tan\varphi \end{bmatrix}, \tag{51}$$

where an overdot means a time derivative.

3. The STIRAP Mechanism

STIRAP is based on tying the state vector $\Psi(t)$ to the zero-eigenvalue adiabatic state $\Phi_0(t)$, which is a coherent superposition of the initial state ψ_1 and the final state ψ_3 only. For the counterintuitive pulse ordering the relations $\Omega_p(t)/\Omega_s(t) \xrightarrow{t\to-\infty} 0$ and $\Omega_p(t)/\Omega_s(t) \xrightarrow{t\to+\infty} \infty$ apply; hence, as time progresses from $-\infty$ to $+\infty$, the mixing angle $\vartheta(t)$ rises from 0 to $\pi/2$. Consequently, the adiabatic state $\Phi_0(t)$ evolves from the bare state ψ_1 initially to a superposition of states ψ_1 and ψ_3 at intermediate times and finally to the target state ψ_3 at the end of the interaction; thus, state $\Phi_0(t)$ links adiabatically the initial state ψ_1 to the target state ψ_3. Since the Hamiltonian is explicitly time dependent, the derivative terms in Eq. (51) (the nonadiabatic couplings) are nonzero and, consequently, diabatic transitions between the adiabatic states will occur. The goal is to reduce the diabatic transition rates to negligibly small values, i.e., to ensure adiabatic evolution. Then the system can be forced to stay in the trapped state at all times, and a complete population transfer from ψ_1 to ψ_3 will be achieved, as shown in Fig. 8. Moreover, because the intermediate state ψ_2 does not participate in the trapped state Φ_0, it does not participate in the population transfer either and remains unpopulated throughout the interaction. Hence, as long as the excitation is adiabatic, its properties, such as radiative decay, do not influence STIRAP. From another viewpoint, when the Stokes pulse is stronger (in the beginning) the population is predominantly in state ψ_1, and when the pump pulse is stronger (in the end) the population is

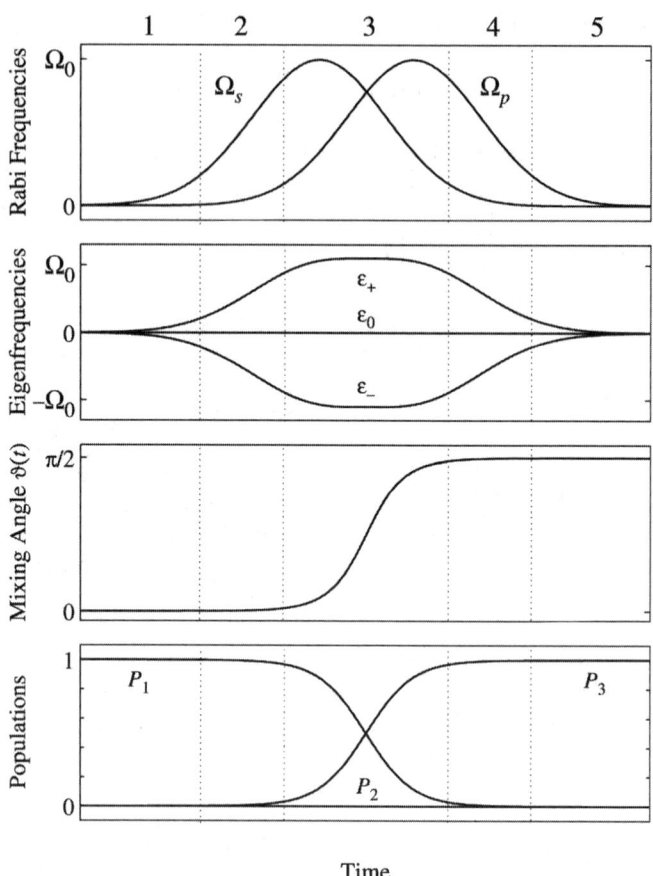

FIG. 8. Time dependencies of the pump and Stokes Rabi frequencies, the eigenfrequencies, the mixing angle and the populations in three-state STIRAP. Dotted lines separate five phases of STIRAP; see Section III.A.4.

predomonantly in state ψ_3; thus the intermediate state is always weakly coupled to the more populated state, which provides another explanation why this state is bypassed by STIRAP during the transfer. A vector picture of the STIRAP process is shown in Fig. 9.

4. Five-Stage Description of STIRAP

The STIRAP process can be viewed as comprising five stages, each defined by the relative strengths of the two fields (Fig. 8). For each stage, coherence is essential.

COHERENT MANIPULATION OF ATOMS AND MOLECULES

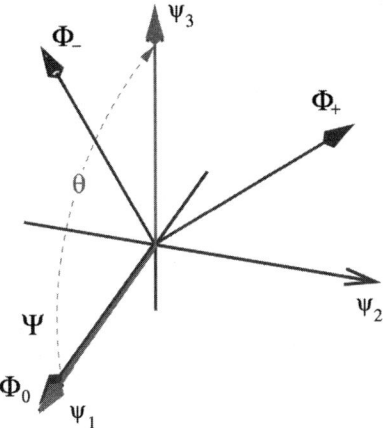

FIG. 9. Vector picture of STIRAP. ψ_1, ψ_2, and ψ_3 are bare atomic eigenstates or diabatic states. Φ_0, Φ_+, and Φ_- are the adiabatic dressed states. The trapped state Φ_0 is rotated from ψ_1 to ψ_3. Under adiabatic conditions, the state vector Ψ follows the evolution of Φ_0.

- *Phase 1:* Only the Stokes pulse is present; its intensity increases steadily. The pump laser does not act yet, i.e., the population in state ψ_1 is not perturbed yet, the mixing angle remains zero, and the state vector remains parallel to state ψ_1. The Stokes pulse prepares for the lossless transfer process in the sense that it provides the Autler–Townes splitting of levels, needed in phase 2. This phase is therefore the *Stokes-induced Autler–Townes phase*. Its purpose is to line up the state vector Ψ with the state Φ_0, i.e., $|\langle\Psi|\Phi_0\rangle| = 1$.
- *Phase 2:* The Stokes pulse has nearly reached the maximum intensity, but the pump pulse is still weak. The state vector deviates only by a very small angle from state ψ_1. One might ask: why is the pump laser radiation not absorbed? Here the Autler–Townes effect leads to a cancellation of the transition rate from the ground state to the two Autler–Townes states. This is the same mechanism that leads to EIT. The effect of the Stokes-induced Autler–Townes splitting is very obvious here. This phase is therefore the *Stokes-induced EIT phase*.
- *Phase 3:* The Stokes pulse decreases and the pump pulse increases. Now the essential part of the population dynamics starts. The pump pulse couples state ψ_1 strongly to the other levels. The mixing angle increases and the state vector departs from the state ψ_1 direction toward state $-\psi_3$, while remaining in the $\psi_1\psi_3$ plane and leaving state ψ_2 unpopulated. This phase is the *adiabatic passage (AP) phase*, with Stokes and pump acting on equal footing, because it is the ratio of these two Rabi frequencies which determines the dynamics.

- *Phase 4:* The population is now almost completely deposited into state ψ_3, but the Stokes pulse is still not zero. Why is there no loss due to optical pumping out of state ψ_3 by the Stokes field? The answer is that the pump field protects the population in state ψ_3 by inducing Autler–Townes splitting and interference (coupling of states ψ_1 and ψ_2). This is the *pump-induced EIT phase*.
- *Phase 5:* The Stokes pulse intensity is zero and the pump-induced Autler–Townes splitting must be reduced to zero. This phase is the *pump-induced Autler–Townes phase*.

We emphasize that the phenomena of Autler–Townes splitting, EIT, and adiabatic passage all depend on the coherence of the radiation. Phase fluctuations would cause the state vector to "jiggle arround," thereby causing strong nonadiabatic coupling.

5. Intermediate-State Population: Importance of the Null Eigenvalue

Once the conditions for STIRAP are fulfilled—two-photon resonance between states ψ_1 and ψ_3, counterintuitive pulse ordering, and adiabatic evolution—a complete population transfer from ψ_1 to ψ_3 is guaranteed. Moreover, because the adiabatic transfer (AT) state (44b) does not involve the intermediate state ψ_2, the latter remains unpopulated during the transfer: the AT state is a *dark state*. This means that its properties have little impact on the transfer efficiency. For example, this remarkable feature of STIRAP allows efficient population transfer on time scales exceeding the lifetime of the intermediate state, which usually can decay on the nanosecond scale.

For example, such a situation arises in the implementation of STIRAP with continuous lasers, when the atomic or molecular beam crosses two spatially displaced and partially overlapping continuous-wave (CW) laser beams at right angles (Gaubatz et al., 1990; Theuer and Bergmann, 1998). The time it takes for the atoms or molecules to cross the laser beams is often two orders of magnitude longer than the lifetime of the excited state. It would be impossible to achieve any population transfer by intermediate storage in the excited state, e.g., by stimulated emission pumping. STIRAP, however, produces a transfer efficiency of nearly 100%, because the upper state is never populated appreciably.

It is readily shown that only on two-photon resonance ($\delta = 0$) does there exist a trapped state, without a contribution from the intermediate state ψ_2. Indeed, it follows from the Hamiltonian (40) that the components of any of the eigenstates of H must obey the relation

$$\tfrac{1}{2}\Omega_s C_2 = (\varepsilon - \delta)C_3.$$

Because on two-photon resonance ($\delta = 0$) one of the eigenvalues is zero ($\varepsilon = 0$), the right hand side (RHS) of the above equation vanishes and the corresponding eigenstate—the trapped state—has no component from the intermediate state, $C_2 = 0$. Off two-photon resonance ($\delta \neq 0$), one can easily show that no eigenvalue is equal to δ ($\varepsilon \neq \delta$) and thus the RHS does not vanish; the implication is a nonzero component $C_2 \neq 0$ from the intermediate state.

6. Intuitive versus Counterintuitive Pulse Sequences

When both the pump and Stokes lasers are on resonance with their respective transitions, the two opposite pulse sequences lead to qualitatively different results. While, as explained above, the counterintuitive sequence induces complete population transfer to state ψ_3, the intuitive sequence produces generalized Rabi oscillations in the populations of all three states (He et al., 1990; Shore et al., 1992a,b; Vitanov and Stenholm, 1997b). This is readily seen by noting that for $\Delta = 0$, we have $\varphi \equiv \pi/4$. Because for the intuitive ordering $\vartheta(-\infty) = \pi/2$ and $\vartheta(+\infty) = 0$, we find that the adiabatic states behave as

$$\frac{1}{\sqrt{2}}(\psi_1 + \psi_2) \xleftarrow{-\infty} \Phi_+(t) \xrightarrow{+\infty} \frac{1}{\sqrt{2}}(\psi_2 + \psi_3), \quad (52a)$$

$$-\psi_3 \xleftarrow{-\infty} \Phi_0(t) \xrightarrow{+\infty} \psi_1, \quad (52b)$$

$$\frac{1}{\sqrt{2}}(\psi_1 - \psi_2) \xleftarrow{-\infty} \Phi_-(t) \xrightarrow{+\infty} \frac{1}{\sqrt{2}}(-\psi_2 + \psi_3). \quad (52c)$$

It follows from the above equations that initially both states $\Phi_+(t)$ and $\Phi_-(t)$ are populated. Because of the interference between the two different paths from state ψ_1 to state ψ_3, the final population of state ψ_3 will oscillate (Vitanov and Stenholm, 1997b),

$$P_1 = 0, \qquad P_2 = \sin^2 \tfrac{1}{2} A, \qquad P_3 = \cos^2 \tfrac{1}{2} A, \quad (53)$$

with $A = \int_{-\infty}^{+\infty} \Omega(t)\,dt$. Thus, only for certain values of the RMS pulse area (generalized π pulses) is it possible to obtain complete population transfer from state ψ_1 to state ψ_3 with a resonant intuitive pulse sequence. However, in such a setup, the intermediate state would receive appreciable transient population and considerable population losses would occur, unless the pulse durations are much shorter than the intermediate-state lifetime.

When the two lasers are tuned away from the respective single-photon resonances, while maintaining the two-photon resonance, adiabatic evolution produces complete population transfer from ψ_1 to ψ_3 for both pulse sequences (Shore et al., 1992b; Vitanov and Stenholm, 1997b). For the counterintuitive sequence (Stokes before pump), this occurs because, as emphasized above, the trapped state does not depend on the single-photon detuning Δ. For the intuitive sequence (pump before Stokes), the adiabatic transfer is carried out through the adiabatic state $\Phi_-(t)$. Indeed, for the intuitive ordering, the relations $\vartheta(-\infty) = \pi/2$, $\vartheta(+\infty) = 0$, and $\varphi(-\infty) = \varphi(+\infty) = 0$ apply; then the adiabatic state $\Phi_-(t)$ has the following asymptotic behavior:

$$\psi_1 \xleftarrow{-\infty} \Phi_-(t) \xrightarrow{+\infty} \psi_3. \tag{54}$$

Thus the adiabatic state $\Phi_-(t)$ provides adiabatic connection between states ψ_1 and ψ_3. However, unlike the counterintuitive ordering, here the intermediate state receives a significant transient population, $P_e = \sin^2 \varphi(t)$. Hence, if the lifetime of state ψ_2 is comparable or shorter than the excitation duration, then efficient population transfer can be achieved only with the counterintuitive pulse sequence.

The similarity of population transfer by the two pulse sequences in the off-resonant case ($\Delta \neq 0$) is easily explained when the single-photon detuning Δ is large ($|\Delta| \gg \Omega_p, \Omega_s$); then the intermediate state ψ_2 can be eliminated adiabatically (Shore et al., 1992b; Vitanov and Stenholm, 1997b). As shown in Section II.D.3, the resulting effective two-state model involves a coupling $\Omega_{\text{eff}} = -\Omega_p \Omega_s / 2\Delta$ and a detuning $\Delta_{\text{eff}} = (\Omega_p^2 - \Omega_s^2)/4\Delta$. Obviously, for sequential pulses the detuning $\Delta_{\text{eff}}(t)$ passes through resonance at the time t_0 when $\Omega_p(t_0) = \Omega_s(t_0)$; this level crossing leads, in the adiabatic limit, to complete population transfer for both pulse orderings, because the ordering reversal leads to the unimportant change of sign in Δ_{eff}, and does not affect Ω_{eff}.

Finally, another interesting feature of STIRAP is that the population P_1 of the initial state ψ_1 does not depend on the pulse order, as can be inferred from the symmetric dependence of P_1 on the pulse delay τ in Fig. 10. This symmetry can be deduced rigorously (Vitanov, 1999).

7. Adiabatic Condition: Local and Global Criteria

For adiabatic evolution, the coupling between each pair of adiabatic states should be negligible compared to the difference between the energies of these states. With respect to the trapped state $\Phi_0(t)$ the adiabatic condition reads (Messiah, 1962; Crisp, 1973; Shore, 1990)

$$|\langle \dot{\Phi}_0 | \Phi_\pm \rangle| \ll |\varepsilon_0 - \varepsilon_\pm|, \tag{55}$$

(note that $\langle \Phi_0 | \dot{\Phi}_\pm \rangle = -\langle \dot{\Phi}_0 | \Phi_\pm \rangle$) or, explicitly (Vitanov and Stenholm 1997a,b),

$$\left| \dot{\vartheta} \frac{\sin^2 \varphi}{\cos \varphi} \right| \ll \tfrac{1}{2}\Omega, \quad \left| \dot{\vartheta} \frac{\cos^2 \varphi}{\sin \varphi} \right| \ll \tfrac{1}{2}\Omega. \tag{56}$$

COHERENT MANIPULATION OF ATOMS AND MOLECULES   91

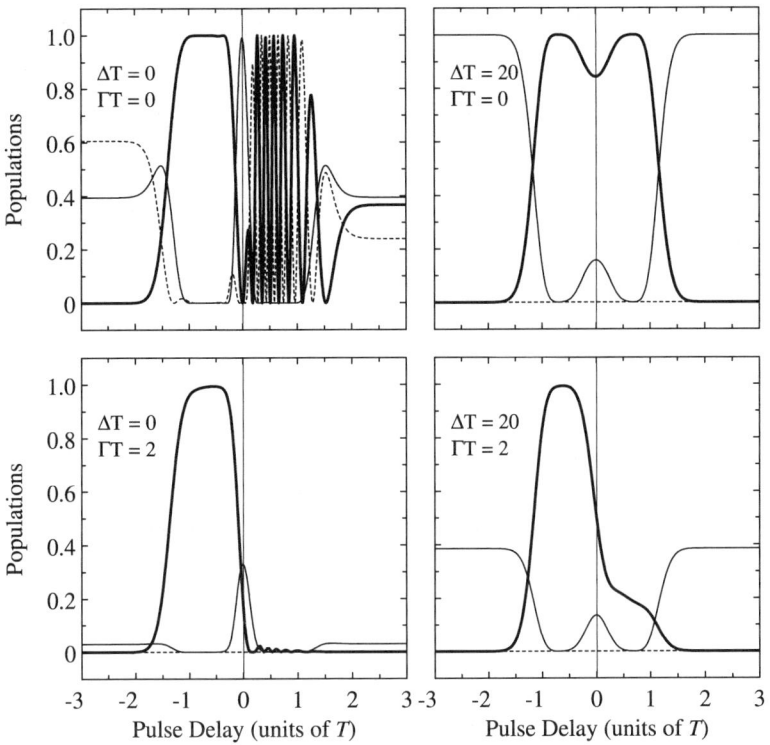

FIG. 10. Numerically calculated populations of the initial state (thin solid line), the intermediate state (dashed line), and the final state (thick solid line) plotted against the pulse delay for different single-photon detuning Δ and loss rate Γ, whose values are denoted in each frame. We have assumed Gaussian pulse shapes, $\Omega_p(t) = \Omega_0 \exp[-(t-\tau)^2/T^2]$, $\Omega_s(t) = \Omega_0 \exp[-(t+\tau)^2/T^2]$, with $\Omega_0 T = 40$.

On one-photon resonance ($\Delta = 0$), we have $\varphi = \pi/4$ and the adiabaticity condition simplifies (Gaubatz et al., 1990),

$$\Omega \gg |\dot{\vartheta}| \propto T^{-1},$$

where T is the pulse width. Assuming that the pump and Stokes pulses are sufficiently smooth and have the same peak Rabi frequency Ω_0, this condition can be written as

$$\Omega_0 T \gg 1. \quad (57)$$

On the left-hand side of this inequality we have the factor $\Omega_0 T$, which, up to an unimportant pulse-shape-dependent factor of the order of unity, is essentially the pulse area. Hence adiabaticity demands a large pulse area. In terms of incoherent excitation, the large pulse area means saturation of the transitions. In practical applications, the pulse area should exceed 10 to provide efficient population transfer, $\Omega_0 T > 10$.

The above adiabatic criteria (56) and (57) are directly applicable to STIRAP with continuous-wave lasers in the crossed-beam geometry (atomic beam crossing two spatially displaced CW laser beams), because CW lasers have almost perfect coherence properties. For pulsed lasers, the adiabatic conditions need to be modified. For perfectly coherent pulsed lasers, the adiabatic condition is essentially the same, but it is more conveniently written in the equivalent form

$$\Omega_0^2 T > \frac{100}{T}. \tag{58}$$

This condition imposes a lower limit on the pulse energy (which is proportional to $\Omega_0^2 T$) for a given pulse duration T; obviously, the needed laser energy grows rapidly when the pulse duration decreases.

Pulsed lasers, however, often suffer from phase fluctuations, i.e., the actual linewidth $\Delta\omega$ deviates from the transform limit $\omega_{TL} = 1/T$. A careful analysis of the effect of imperfect laser coherence on the STIRAP efficiency leads to the modified adiabaticity condition (Kuhn et al., 1992),

$$\Omega_0 T \gg \left[1 + \left(\frac{\Delta\omega}{\Delta\omega_{TL}}\right)^2\right]^{1/2}. \tag{59}$$

The adverse effect of imperfect laser coherence derives from the fact that both phase fluctuations and frequency chirp correspond to time-dependent changes in the laser frequencies. Unless these changes are correlated (e.g., if the pump and Stokes pulses are derived from the same laser), they will result in time-dependent detuning from two-photon resonance. The two-photon detuning induces nonadiabatic couplings between the dark state and the other adiabatic states that reduce the transfer efficiency. These population losses can be reduced by increasing the laser intensity, thereby suppressing nonadiabatic transitions. This problem is discussed in more detail in Section IV.B.

8. Nonadiabatic Transitions

An important subject of theoretical investigation is the behavior of the system away from the adiabatic limit. Of particular interest is here the question in what manner the adiabatic limit is approached.

For smooth, ramped pump, and Stokes pulses, diabatic corrections are exponentially small in the inverse adiabaticity parameter (Elk, 1995). This behavior is very similar to the classic result by Dykhne (1962) and Davis and Pechukas (1976) for two-level systems. For many STIRAP models this is not the case, however, and there is a power-law dependence (Laine and Stenholm, 1996; Vitanov and Stenholm, 1996; Drese and Holthaus, 1998).

The following section discusses various other factors that affect adiabaticity.

B. SENSITIVITY OF STIRAP TO INTERACTION PARAMETERS

1. Sensitivity to Delay

The following considerations affect the choice of an optimum delay between the Stokes and pump pulses.

- *Coincident pulses:* In this case the mixing angle ϑ is constant and the evolution is fully adiabatic because the nonadiabatic coupling vanishes ($\dot{\vartheta} = 0$). However, the initial projection of the state vector Ψ onto the trapped state $\Phi_0(t)$ is incomplete, i.e., the trapped state is not the only adiabatic state populated initially and the interference between different evolution paths from ψ_1 to ψ_3 lead to oscillations in the final population of the target state ψ_3 (Vitanov, 1998b), rather than to complete population transfer. Hence this regime presents the "good adiabaticity argument" but the "bad projection argument."

- *Small delay, very large overlap:* For delayed and counterintuitively ordered pulses, the initial projection of the state vector onto Φ_0 is unity. When the pulses are only slightly delayed, the mixing angle ϑ is nearly constant during most of the overlap and hence $\dot{\vartheta} \approx 0$ ("good adiabaticity argument"); if the pump and Stokes peak Rabi frequencies are equal, then $\vartheta \approx \pi/4$ during the overlap. However, ϑ rises too quickly from 0 to about $\pi/4$ during the short interval between the arrivals of the Stokes and pump pulses, and then again rises too quickly from $\pi/4$ to $\pi/2$ during the short interval between the disappearance of the two pulses. During these two time intervals, the nonadiabatic coupling (which is $\propto \dot{\vartheta}$) can reach significant values and can cause nonadiabatic transitions to the other adiabatic states Φ_+ and Φ_-. Thus when the system enters the pulse overlap region the state vector Ψ is not parallel to the trapped state Φ_0, i.e., we have "bad projection." The presence of two nonadiabatic zones eventually leads to interference and oscillations in the transfer efficiency, as for coincident pulses. Hence this regime presents the "good adiabaticity argument" during the pulse overlap, but the "bad projection argument" at the entry and the exit of the overlap region.

- *Large delay, very small overlap:* The initial projection of the state vector onto Φ_0 is unity. However, the mixing angle ϑ stays nearly constant for most of the excitation (≈ 0 early when only the Stokes pulse is present and $\approx \pi/2$ at late times when only the pump pulse is present) and rises from 0 to $\pi/2$ during only a very short period when the pulses overlap; during this period the nonadiabatic coupling ($\propto \dot{\vartheta}$) is very large and causes transitions to the other adiabatic states Φ_+ and Φ_-, resulting in loss of transfer efficiency. Hence this regime presents the "good projection argument" but the "bad adiabaticity argument" during the pulse overlap.

- *Optimum delay:* The optimal pulse delay, which must lead to maximal transfer efficiency, can be determined by maximizing adiabaticity. For maximal adiabaticity, the mixing angle $\vartheta(t)$ must change slowly and smoothly in time, so that the nonadiabatic coupling ($\propto \dot{\vartheta}$) remains small, without pronounced peaks. As follows from the above discussion, the optimal adiabaticity occurs for a certain range of moderate delays, when both the two-peak time dependence of $\dot{\vartheta}$ appearing for small delay and the sharp-single-peak time dependence appearing for large delay, are absent. The particular optimal value of τ depends on the pulse shapes. For Gaussian pulses, the optimum occurs when the delay is nearly equal to the pulse width, $\tau_{opt} \approx T$.

2. Sensitivity to Rabi Frequency and Pulse Width

It is best to have intensities adjusted such that when combined with the given dipole transition moments rates the two peak Rabi frequencies are about equal. If the maximum Rabi frequencies are very different, and the pulse widths are about the same, than the projection of the state vector onto the adiabatic transfer state is very good early (or late) but necessarily less good late (or early), i.e., "good projection" cannot be achieved both early *and* late, and consequently the transfer efficiency will be small. It is interesting to note that in this case the population which does not end up in state ψ_3 returns to state ψ_1, i.e., the transfer efficiency is not limited by spontaneous emission losses. The same reasoning leads to the conclusion that the pump and Stokes pulse widths should also be about equal.

It should be noted that with large pulse areas those prescriptions are not in conflict with the claim that the method is robust. For large pulse areas, small deviations from the optimum do not lead to significant drop in transfer efficiency. Using equal peak Rabi frequencies and equal pulse widths allows to reduce the necessary pulse areas and hence facilitates efficient population transfer.

3. Sensitivity to Single-Photon Detuning

Typical STIRAP experiments permit measurement of the population transfer probability for various choices of the two carrier frequencies. Because a plot of transfer probability versus frequency appears similar to a plot of emission or absorption intensity versus frequency (a *spectral line profile*), the full-width at half-maximum (FWHM) profile is termed the line width.

Two profiles are of interest. Variation of both the pump and Stokes frequencies, while maintaining the two-photon resonance condition, presents a *single-photon* profile, $P_3(\Delta)$. Alternatively, variation of either carrier frequency, while keeping the other fixed, will cause a change in the two-photon detuning, and will lead to a two-photon profile, $P_3(\delta)$.

As is evident from Eqs. (44), the single-photon detuning Δ does not affect the formation or the composition of the trapping state (as long as two-photon resonance is maintained), because the mixing angle ϑ does not depend on Δ. However, the other mixing angle φ depends on Δ and therefore the detuning affects the adiabatic conditions (56). Since $|\varphi| \leq \pi/4$, and hence $|\sin\varphi| \leq \cos\varphi$, the latter of conditions (56) is more stringent. Because φ is a decreasing function of Δ [cf. Eq. (45)], the LHS of this condition increases with Δ. This implies that STIRAP works best on single-photon resonance; when the single-photon detuning Δ increases, adiabaticity deteriorates and the transfer efficiency decreases.

We can easily derive the scaling properties of the FWHM $\Delta_{1/2}$ of the single-photon line profile $P_3(\Delta)$. For large Δ, when Δ begins to affect adiabaticity, we have $\varphi \approx \Omega/2\Delta$ and condition (56) becomes $\frac{1}{4}\Omega^2 \gg |\Delta\dot\vartheta|$. Because $\dot\vartheta$ depends only on the pulse delay τ and the pulse shapes [cf Eq. (45)], but not on Δ or on the peak Rabi frequency Ω_0, the width $\Delta_{1/2}$ must scale with Ω_0^2. Upon introducing a pulse-delay dependent proportionality factor $D(\tau)$, one may write (Vitanov and Stenholm, 1997a)

$$\Delta_{1/2} = D(\tau)\Omega_0^2. \tag{60}$$

Because the Rabi frequency is proportional to the electric field amplitude, the single-photon width $\Delta_{1/2}$ is proportional to the peak intensity.

4. Sensitivity to Two-Photon Detuning

It is also possible to derive a simple scaling relationship for the sensitivity of population transfer to the two-photon detuning $\delta = \Delta_p - \Delta_s$ (Danileiko *et al.*, 1994; Romanenko and Yatsenko, 1997; Fewell *et al.*, 1997; Vitanov, 2001). The detuning from two-photon resonance is much more crucial for STIRAP than the single-photon detuning, because the two-photon detuning prevents the exclusive population of the trapped state.

The sensitivity of the transfer efficiency to the two-photon detuning δ can be quantified by examining the behavior of the three eigenstates of the Hamiltonian (40), which are no longer given by Eqs. (44) (Danileiko *et al.*, 1994; Romanenko and Yatsenko, 1997; Fewell *et al.*, 1997). In particular, the trapped state $\Phi_0(t)$ is no longer an eigenstate of $H(t)$. For nonzero δ, each of the three eigenstates of the

Hamiltonian (40) connects to *the same* bare state at both $t = -\infty$ and $t = +\infty$, and hence there is no adiabatic transfer state providing an adiabatic connection from state ψ_1 to state ψ_3, as does the trapped state Φ_0 for $\delta = 0$. Thus adiabatic evolution leads to complete population return of the system to its initial state ψ_1, i.e., to zero transfer efficiency. The only mechanism by which some population transfer to state ψ_3 can occur is by nonadiabatic transitions between the adiabatic states. Such transitions can take place for small values of δ when there are narrow avoided crossings between the adiabatic eigenvalues. This is illustrated in Fig. 11. By using the Landau–Zener formula (Landau, 1932; Zener, 1932) to evaluate the nonadiabatic transitions at these avoided crossings, analytic expressions for the two-photon linewidth have been derived (Danileiko *et al.*, 1994; Romanenko and Yatsenko, 1997).

An alternative approach to estimating the two-photon line width makes use of the adiabatic condition (Vitanov, 2001). We assume for simplicity single-photon resonance, $\Delta = 0$. From the initial bare-state basis we make a transformation to the basis of the states (44), which are the eigenstates of $\mathsf{H}(t)$ for $\delta = 0$, rather than to the genuine $\delta \neq 0$ adiabatic basis. In the basis (44), the effect of nonzero two-photon detuning δ shows up in additional terms in the respective Hamiltonian proportional to δ (Vitanov 2001),

$$\mathsf{H}_b(\delta) = \mathsf{H}_b(\delta = 0)$$

$$+ \frac{1}{2}\hbar\delta \cos^2 \vartheta \begin{bmatrix} 1 & -\sqrt{2}\tan\vartheta & 1 \\ -\sqrt{2}\tan\vartheta & 2\tan^2\vartheta & -\sqrt{2}\tan\vartheta \\ 1 & -\sqrt{2}\tan\vartheta & 1 \end{bmatrix}, \quad (61)$$

where $\mathsf{H}_b(\delta = 0)$ is given by Eq. (51). If we now assume that the evolution is adiabatic for $\delta = 0$ (i.e., that we have unity transfer efficiency on two-photon resonance), we can neglect all nondiagonal terms in $\mathsf{H}_b(\delta = 0)$. We can also neglect

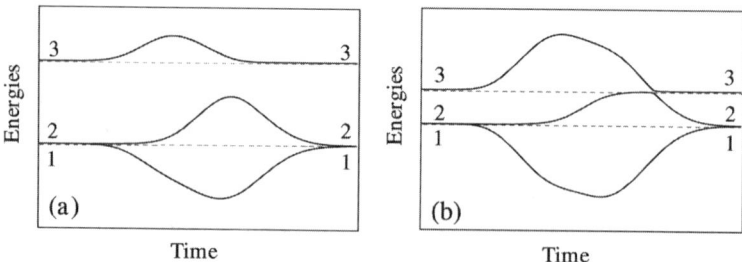

FIG. 11. Time evolution of the energies in a three-state system with two-photon detuning: (a) large two-photon detuning (no transfer); (b) narrow avoided crossings for small two-photon detuning (transfer through diabatic transitions).

all δ terms in the diagonal elements of $\mathbf{H}_b(\delta)$, unless the two-photon detuning is comparable with the eigenvalue separation (then the transfer efficiency would be virtually zero). Then the effect of the nonzero two-photon detuning δ shows up as additional nonadiabatic couplings (which do not vanish in the adiabatic limit) between the $\delta = 0$ adiabatic states (44). Considerable population transfer between states ψ_1 and ψ_3 can still be realized if the Λ system is forced to stay in the trapped state Φ_0. This requirement leads to the condition

$$\delta \sin \vartheta \cos \vartheta \ll \frac{1}{\sqrt{2}}\Omega.$$

Because, as we emphasized above, the mixing angle ϑ does not depend on the peak Rabi frequency Ω_0 but only on the ratio Ω_p/Ω_s, i.e., on the pulse delay τ and the pulse shapes, the width $\delta_{1/2}$ of the two-photon line profile $P_3(\delta)$ must scale with Ω_0. By introducing a pulse-delay dependent proportionality factor $d(\tau)$, one may write (Vitanov 2001)

$$\delta_{1/2} = d(\tau)\Omega_0. \qquad (62)$$

By contrast with the single-photon width $\Delta_{1/2}$, the two-photon width $\delta_{1/2}$ varies in proportion to the Rabi frequency, meaning the square root of the peak intensity.

In conclusion, STIRAP efficiency is much less sensitive to single-photon detuning (because the single-photon linewidth grows with the square of the pulse area, $\Delta_{1/2} \propto (\Omega_0 T)^2$) than to two-photon detuning (where the linewidth increases only linearly with the pulse area, $\delta_{1/2} \propto \Omega_0 T$). A numerical example is shown in Fig. 12 and experimental results in Fig. 13.

5. Sensitivity to Losses from the Intermediate State

As we emphasized above, in the adiabatic limit, no population resides in state ψ_2, and spontaneous emission from ψ_2 is not detrimental to successful population transfer. However, a strong decay from state ψ_2 may reduce adiabaticity and demands larger laser intensity for high transfer efficiency. The influence of spontaneous emission from the intermediate state ψ_2 both within the Λ system (back to states ψ_1 and ψ_3) (Band and Julienne, 1991a, 1992) and to other states (Glushko and Kryzhanovsky, 1992; Fleischhauer and Manka, 1996; Vitanov and Stenholm, 1997d) has been studied. Modeling spontaneous emission within the Λ system requires using the Liouville density-matrix equation and it is hard to derive analytic estimates, while population losses to states outside the Λ system allow analytic treatment (Fleischhauer and Manka, 1996; Vitanov and Stenholm, 1997d) using the Schrödinger equation.

The effect of irreversible losses is most conveniently estimated in the lossless ($\Gamma = 0$) adiabatic basis (51), where the effect of the loss rate shows as additional

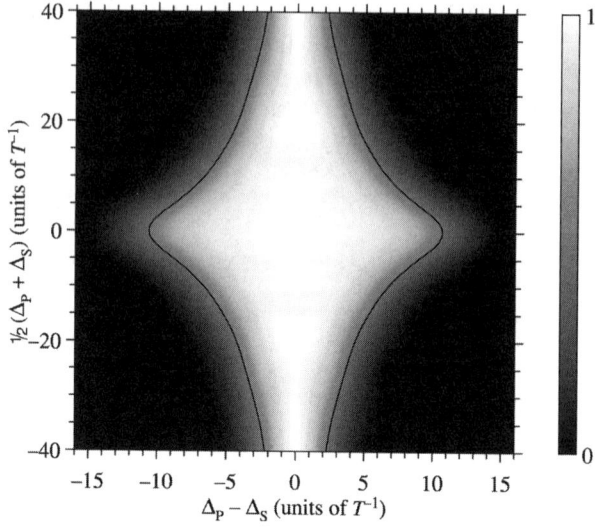

FIG. 12. Numerically calculated transfer efficiency in STIRAP plotted versus the sum and the difference of the pump and Stokes detunings (i.e., versus the single-photon and two-photon detunings) for Gaussian pulse shapes, $\Omega_p = \Omega_0 \exp[-(t-\tau)^2/T^2]$, $\Omega_s = \Omega_0 \exp[-(t+\tau)^2/T^2]$, with $\Omega_0 T = 20$, $\tau = 0.5T$. The curves show the $P_3 = 0.5$ value.

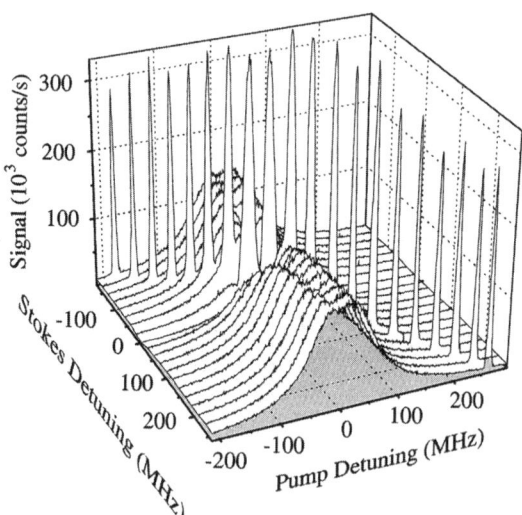

FIG. 13. Experimentally measured transfer efficiency of STIRAP plotted versus the pump and Stokes detunings. (From J. Martin, B. W. Shore, and K. Bergmann. Coherent population transfer in multilevel systems with magnetic sublevels. III. Algebraic analysis. *Phys. Rev. A* 1996;54:1556–1569.)

imaginary terms (Vitanov and Stenholm, 1997d) in the Hamiltonian

$$H_b(\Gamma) = H_b(\Gamma = 0) + \frac{1}{2}i\Gamma \begin{bmatrix} -\cos^2\varphi & 0 & \frac{1}{2}\sin 2\varphi \\ 0 & 0 & 0 \\ \frac{1}{2}\sin 2\varphi & 0 & -\sin^2\varphi \end{bmatrix}, \quad (63)$$

where $H_b(\Gamma = 0)$ is given by Eq. (51). Obviously, the losses affect much more strongly population transfer by the intuitive pulse ordering because then the decaying intermediate state may receive considerable transient population. Since then the population is transferred via the adiabatic state Φ_-, which decays with a rate $\frac{1}{2}\Gamma \sin^2\varphi$, the transfer efficiency decreases exponentially with Γ.

For the counterintuitive pulse ordering, the population losses occur by two mechanisms. The first mechanism, which dominates at small to medium decay rates, is by dissipation of population that visits state ψ_2 due to imperfect adiabaticity. In the adiabatic basis, imperfect adiabaticity leads to nonadiabatic transitions from the nondecaying dark state Φ_0 to the other, decaying adiabatic states Φ_+ and Φ_-, from which population losses occur. The second mechanism is quantum overdamping, which dominates for large Γ and leads to effective decoupling of the three-state system from the laser fields. These two mechanisms lead to different damping of the transfer efficiency with Γ: exponential at small Γ (but with a much smaller effective loss rate than for the intuitive pulse ordering) and polynomial at large Γ.

By using a similar approach as for the estimation of the single-photon line width $\Delta_{1/2}$ (Section III.B.3), one can show that the "linewidth" $\Gamma_{1/2}$, the loss rate value at which the transfer efficiency drops to $\frac{1}{2}$, is proportional to the squared pulse area (Vitanov and Stenholm, 1997d),

$$\Gamma_{1/2} \approx G\left(\frac{\tau}{T}\right)(\Omega_0 T)^2, \quad (64)$$

where $G(\tau/T)$ is a coefficient that depends on the pulse delay τ (and on the specific pulse shapes) but not on $\Omega_0 T$. In the case of Gaussians, an analytic approximation for the coefficient $G(\tau/T)$ reads (Vitanov and Stenholm, 1997d)

$$G(\tau/T) \approx \frac{3(\tau/T)\ln 2}{8(\tau/T)^2 + \pi/2} e^{-2(\tau/T)^2}. \quad (65)$$

Formula (65) suggests that $G(\tau/T)$ rises from zero at $\tau = 0$ to its maximum value of about 0.23 at $\tau \approx 0.302T$ and then decreases in a near-Gaussian fashion with τ. The decrease of $G(\tau/T)$ at small τ/T is because then the loss rates of states Φ_+ and Φ_- increase and larger pulse area is needed to suppress them. The decrease of $G(\tau/T)$ at large τ/T is due to the larger nonadiabatic coupling between the dark state Φ_0 and the other adiabatic states Φ_+ and Φ_-, which again requires larger pulse area to ensure sufficient adiabaticity.

6. Sensitivity to Beam Geometry

The optimum geometry of the laser beams (circular or cylindrical with different orientations of the ellipsoid) depends on the purpose of the excitation. For scattering experiments it is desirable to maximize the flux of atoms which crosses the laser beams in regions that allow adiabatic evolution. In other applications, one may want to manipulate a highly collimated beam, and the laser power may be at the limit of what is needed; in that case, one wants to increase the interaction time to improve the adiabatic evolution.

Given the nonspherical intensity distribution according to

$$\vec{E}(x,y,t) = \tfrac{1}{2}\vec{e}\mathcal{E}_0 e^{i\omega t} e^{-(x/w_x)^2 - (y/w_y)^2} + \text{c.c.}, \tag{66}$$

the intensity $I(x,y)$ and power P are given by $I(x,y) = \tfrac{1}{2}c\epsilon_0 \vec{E}^2(x,y)$ and $P = \int_{x,y} dx\, dy\, I(x,y) = \tfrac{\pi}{2} w_x w_y I_0$, respectively. From these one obtains the Rabi frequency,

$$\Omega = \frac{d}{\hbar}\sqrt{\frac{4}{c\pi\epsilon_0}}\sqrt{\frac{P}{w_x w_y}}. \tag{67}$$

The global adiabaticity criterion can be written in a form which contains the shape parameters of the laser beam,

$$\frac{4d}{\hbar v}\sqrt{\frac{1}{c\pi\epsilon_0}}\sqrt{\frac{w_x P}{w_y}} \gg 1, \tag{68}$$

where v is the atom velocity.

It is interesting to note that the adiabaticity criterion is independent of the laser beam diameter D for circular beam geometry! This is because the interaction time increases linearly with D, while the local (and maximum) Rabi frequency decreases inversely proportional to D. Thus, the dependence on D cancels.

As illustrated in Fig. 14, a laser beam focused cylindrically with the long axis perpendicular to the particle beam axis allows one to manipulate a relatively large flux, but the Rabi frequency increases less than the interaction time decreases (given the same power of the laser beam, of course). Thus the price for a high flux of manipulated (excited) atoms or molecules is a higher intensity.

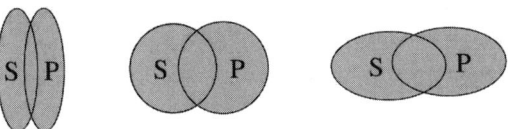

FIG. 14. Laser beam geometries for STIRAP.

A cylindrical focus with the long axis parallel to the particle beam axis reduces the flux of manipulated particles. However, since the interaction time increases more than the Rabi frequency decreases, the required intensity is smaller.

Given these considerations it is interesting to ask: how large can we allow the cylindrical focus to be? One obvious limitation is that the interaction time can not be increased indefinitely. Increasing the interaction time makes the transfer process more sensitive to the detrimental phase fluctuations during the adiabatic passage process. Thus the transit time through the overlap region should be small compared to the inverse of the laser linewidth.

7. Multiple Intermediate States

In the standard three-state STIRAP, the Raman linkage between the initial state ψ_1 and the final state ψ_3 takes place via a single intermediate state ψ_2. In real atoms, and particularly in molecules, it may happen that there are multiple intermediate states strongly coupled to ψ_1 and ψ_3 by the pump and Stokes fields, thus forming a parallel multi-Λ system. Such couplings may be present because, while very sensitive to the two-photon resonance, STIRAP is relatively insensitive to the single-photon detuning from the intermediate state. Coulston and Bergmann (1992) were the first to consider the effects of multiple intermediate states in the simplest case of $N = 2$ states and equal couplings $\Omega_p(t)$ to state ψ_1 and equal couplings $\Omega_s(t)$ to state ψ_3. Vitanov and Stenholm (1999) studied the general case of unequal couplings and unevenly distributed, N intermediate states. It has been concluded that the dark state (44b) remains a zero-eigenvalue eigenstate of the Hamiltonian only when, for each intermediate state ψ_k, the ratio $\Omega_p^{(k)}(t)/\Omega_s^{(k)}(t)$ between the couplings to the initial and target states is the same and does not depend on k. Then the multi-Λ system behaves very similarly to the single-Λ system in STIRAP, and complete population transfer with no transient population in any intermediate state can take place for adiabatic evolution. When this proportionality condition is not fulfilled, a dark state does not exist but a more general adiabatic transfer state, which links adiabatically the initial and target states, may exist under certain conditions on the single-photon detunings and the relative coupling strengths. This AT state, unlike the dark state, contains contributions from the intermediate states which therefore acquire transient populations during the transfer.

It has been shown (Vitanov and Stenholm, 1999) that when the pump and Stokes frequencies are scanned across a manifold of N intermediate states (while maintaining the two-photon resonance), the target-state population passes through N regions of high transfer efficiency (unity in the adiabatic limit) and $N-1$ regions of low efficiency (zero in the adiabatic limit). It is most appropriate to tune the pump and Stokes lasers either just below or just above all intermediate states because there the AT state always exists, the adiabatic regime is achieved more quickly, the transfer is more robust against laser fluctuations, and the transient intermediate-state populations, which are inversely proportional to squared single-photon detunings, can easily be suppressed.

8. STIRAP Beyond the RWA

Conventional STIRAP assumes applicability of the rotating-wave approximation. This approximation requires that the two Rabi frequencies and the single-photon detunings are much smaller than the Bohr transition frequencies. In most experiments, these conditions are well satisfied.

Several extensions of conventional STIRAP beyond the RWA have been explored. In the most extreme case, Guérin and Jauslin (1998) have examined, by using an adiabatic Floquet approach, the situation when the Rabi frequencies are comparable to the Bohr frequencies. Then the feasibility of adiabatic population transfer can be deduced by analyzing a plot of the eigenenergies of a Floquet Hamiltonian.

Yatsenko et al. (1998) have studied the case when the envelopes of the laser pulses are not smooth but are modulated periodically in time. In the adiabatic Floquet picture, a success or failure of adiabatic population transfer can be deduced, again, from a plot of the eigenenergies of a Floquet Hamiltonian. In such a plot, there occurs an infinite sequence of eigenenergy triplets. The distance between the triplets is proportional to the modulation frequency, while the splitting of the Floquet energies within each triplet is proportional to the peak Rabi frequency Ω_0. When the modulation frequency exceeds the peak Rabi frequency, high transfer efficiency can be achieved because the adjacent triplets are well separated and do not interfere with each other. When Ω_0 is comparable or larger than the modulation frequency, the success or failure of adiabatic population transfer is determined by an interplay between the splitting within each triplet and the separation of the triplets.

It may also occur that the two laser fields act on both the pump and Stokes transitions, e.g., when the two laser fields have the same polarization. If the two Bohr frequencies are sufficiently different, so that their difference is large compared to the Rabi frequencies of the pump and Stokes pulses ($|\omega_{12} - \omega_{32}| \gg \Omega_p, \Omega_s$), one can neglect the off-resonant channels, i.e., the action of the pump laser on the Stokes transition and the action of the Stokes laser on the pump transition. If the Rabi frequencies are comparable to or bigger than the difference of the Bohr frequencies (Ω_p, $\Omega_s \gtrsim |\omega_{12} - \omega_{32}|$), the off-resonant channels have to be accounted for (Unanyan et al., 2000c). Then efficient and robust population transfer can still be possible under certain conditions with the exchange of one or more (odd-number) photons between each laser field and the atom. With the multiple-photon scenario, however, the transient population of the intermediate level is no longer negligibly small, even when the evolution is adiabatic.

IV. Three-State STIRAP: Experiments

When continuous-wave lasers are used in combination with atomic or molecular beams, it is straightforward to expose the atoms or molecules to a delayed sequence of interactions by spatially displacing the axes of the laser beams (see Fig. 15). When pulsed lasers are used, the axes of the laser beams need to coincide but the pulses must be delayed in time.

A. Experimental Demonstrations with CW Lasers

1. Sodium Dimers

After preliminary, incomplete, results (Gaubatz et al., 1988), the first convincing experimental demonstration of STIRAP was achieved by Bergmann and co-workers in studies of Na$_2$ (Gaubatz et al., 1990). A beam of sodium molecules crossed two spatially displaced but partially overlapping CW laser beams. When

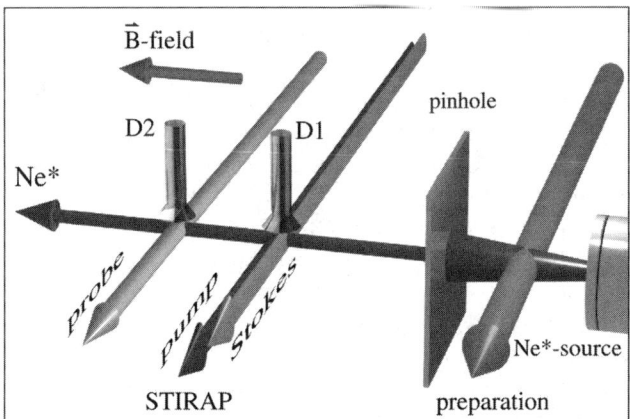

FIG. 15. Experimental setup for the Ne experiment. (From K. Bergmann, H. Theuer, and B. W. Shore. Coherent population transfer among quantum states of atoms and molecules. Rev. Mod. Phys. 1998;70:1003–1025.)

the molecules interacted first with the Stokes laser (counterintuitive ordering), complete population transfer was observed from the initial level ($v = 0$, $J = 5$) to the final level ($v = 5$, $J = 5$) of the molecules in their electronic ground state $X^1\Sigma_g^+$ via an intermediate level ($v = 7$, $J = 6$) of the excited electronic state $A^1\Sigma_u^-$. The time required for a molecule to traverse the two laser beams was about 200 ns. Although this interaction time was much longer than the excited-state lifetime ($\approx$15 ns), efficient population transfer was achieved because STIRAP does not populate the excited state appreciably. Because the interaction time was relatively long and because the sodium dimers have relatively strong transition moments, only moderate laser intensities were needed to induce large pulse areas. Typical intensities in the range of 100 W/cm^2 were sufficient to produce adiabatic passage. This radiation was provided by CW lasers having of the order of 100 mW power, with radiation mildly focused to a spot diameter of a few hundred micrometers into the molecular beam.

2. Metastable Neon Atoms

STIRAP has been studied in detail in metastable neon in a similar crossed-beam geometry (Rubahn et al., 1991; Martin et al., 1996; Lindinger et al., 1997; Theuer and Bergmann, 1998). In the experiment by Theuer and Bergmann (1998), the population was transferred from state $2p^53s\,^3P_0$ to state $2p^53s\,^3P_2$ via the intermediate state $2p^53p\,^3P_1$. In the experiment, a beam of Ne* atoms emerged from a discharge source (see Fig. 15). A preparation laser depleted the population of the $^3P_2$ metastable level by optical pumping. Excitation of the $^3P_2$ level to the $^3D_2$ level resulted in spontaneous emission to the short-lived levels of the $2p^53s$ configuration, followed by decay to the ground state. After passing through a collimating slit, the atoms crossed the STIRAP zone, composed of the Stokes and pump lasers, with their axes suitably displaced. In the STIRAP region, the population of the $^3P_0$ state was transferred to the other metastable level $^3P_2$. If the population transfer was incomplete, some transient population would reside in the intermediate level and VUV fluorescence (photon energy 16.7 eV) would be observed at the channeltron detector D1. Downstream from the STIRAP zone a probe laser excited atoms in the $^3P_2$ level and the subsequent VUV radiation was monitored by a channeltron detector D2, with almost no background signal. The pump and Stokes intensities used in the experiment, typically a few W/cm^2, were provided by CW radiation, focused into the atomic beam by cylindrical lenses.

Figure 16 displays typical evidence for STIRAP. The upper frame shows laser-induced fluorescence from the final state $^3P_2$ (monitored by detector D2) as a function of the pump laser frequency, while the Stokes laser frequency is held fixed slightly off-resonance. The signal shows a broad feature, widened by strong saturation to exceed the natural linewidth by an order of magnitude. At a specific pump detuning, the two-photon resonance condition is met and coherent population

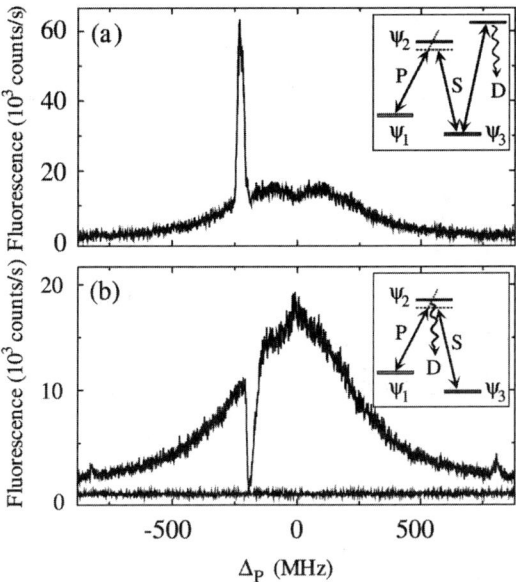

FIG. 16. Laser-induced fluorescence from the final state $^3P_2$ (upper frame) and fluorescence from the intermediate state $^3P_1$ in the Ne* experiment plotted against the pump laser frequency, while the Stokes laser frequency, tuned slightly off-resonance, remains unchanged. (From K. Bergmann, H. Theuer, and B. W. Shore. Coherent population transfer among quantum states of atoms and molecules. Rev. Mod. Phys. 1998;70:1003–1025.)

transfer occurs. Then very little (if any) transient population will reside in the intermediate state $^3P_1$ and the fluorescence from there (monitored by detector D1) will disappear, as shown in Fig. 16b. Thus, efficient population transfer is accompanied by a pronounced dark resonance. Indeed, although the typical time required for passage of atoms across the laser beams was more than 20 times longer than the radiative lifetime ($\approx$20 ns) of the intermediate state, no more than 0.5% of the population was detected in this state at the center of the dark resonance (i.e., at two-photon resonance).

Figure 17 displays another characteristic signature for STIRAP. This figure shows the transfer efficiency plotted versus the spatial displacement between the pump and Stokes laser beams, i.e., versus the pulse delay from the viewpoint of the Ne* atoms. Positive displacement (on the left-hand side) corresponds to counterintuitive pulse ordering (Stokes before pump) and negative displacement (on the right-hand side) to intuitive ordering (pump before Stokes). When the Stokes beam was shifted too far upstream (far left in the figure), it was excluded from the interaction because there was no overlap with the pump pulse; in this case about 25% of the population was optically pumped into the target state because the pump laser excited the atoms to the intermediate state from which they decayed radiatively.

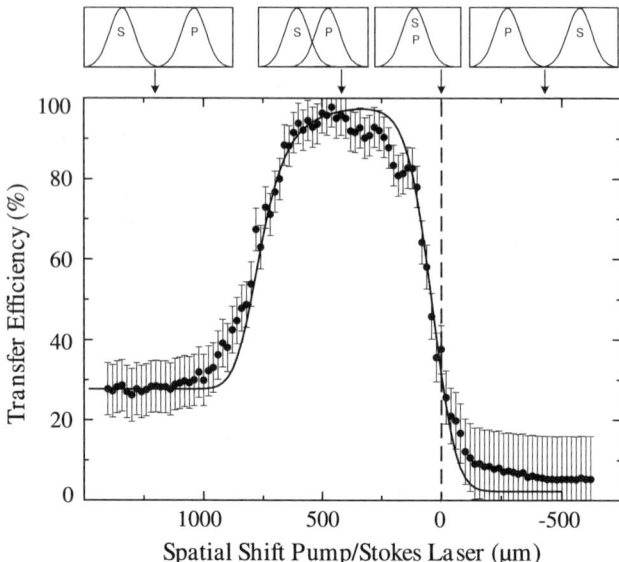

FIG. 17. Transfer efficiency versus displacement between pump and Stokes pulses in Ne* experiment. The broad plateau, showing nearly complete population transfer for counterintuitive pulse sequence, is a typical STIRAP signature, as contrasted with the low efficiency for intuitively ordered pulses. The dots are experimental data and the solid curve shows numeric simulation. (From K. Bergmann, H. Theuer, and B. W. Shore. Coherent population transfer among quantum states of atoms and molecules. Rev. Mod. Phys. 1998;70:1003–1025.)

As the Stokes beam was shifted toward the pump beam, while still preceding it, the transfer efficiency increased dramatically and reached almost unity. When the axes of the two lasers coincided, the transfer efficiency dropped to about 25%. Virtually no transfer was observed when the Stokes beam was moved farther downstream so that the Ne* atoms encountered the pump laser first (intuitive pulse ordering). In this configuration the atoms were transferred to the intermediate state by the pump laser and reached the final state by spontaneous emission from there; toward the end of the interaction they were exposed to the Stokes laser only and were thus lost from state ψ_3 by optical pumping and subsequent spontaneous emission of VUV radiation in a two-step radiative-decay cascade to the neon ground state. The broad plateau for counterintuitive pulse ordering is a characteristic feature of STIRAP and indicates the robustness of the population transfer.

In the previously discussed setups the propagation direction of pump and Stokes fields was chosen parallel. This is not necessary, however, as was demonstrated experimentally again using Ne* atoms (Theuer et al., 1999). Here circularly (σ) polarized Stokes radiation couples the $^3P_2 \leftrightarrow {}^3P_1$ transition and a linear (π) polarized pump laser the $^3P_1 \leftrightarrow {}^3P_0$ transition. As will be shown in Fig. 36, pump and Stokes are at 90° with respect to each other and both intersect the beam of metastable Ne* at right angles to minimize the effect of Doppler shifts. They are properly displaced in order to guarantee a counterintuitive coupling in the frame of the moving atoms. After careful preparation of the initial state by depopulation optical pumping, a two-photon resonance width of 6 MHz (FWHM) was measured, which is clearly below the single-photon value. This type of coupling using σ–π polarizations and orthogonal propagation directions is of particular importance for the realization of a variable coherent atomic beam splitter, which will be discussed in Section VI.

B. EXPERIMENTAL DEMONSTRATIONS WITH PULSED LASERS

1. General Considerations for Pulsed Lasers

A very interesting and important application of STIRAP is the selective excitation of high-lying vibrational levels in molecules. In most molecules the first electronically excited states have energies more than 4 eV above the ground state. Therefore a Raman-type linkage from the vibrational ground level to a high vibrational level requires ultraviolet lasers. Strong ultraviolet radiation is most readily provided by frequency-conversion techniques involving high-intensity pulsed lasers. Furthermore, since the molecular transition dipole moments are usually considerably smaller than for atoms, the adiabaticity condition is difficult to satisfy with CW lasers. Sufficiently strong light intensity, and hence large enough couplings, can be delivered only by pulsed lasers. Pulsed lasers, however, often have inferior coherence properties, e.g., they suffer from phase fluctuations and frequency chirping; both of these effects increase the pulse bandwidth.

As follows from the adiabatic condition (59), the required laser energy increases quadratically with the pulse bandwidth. For conventional nanosecond lasers that are not specially designed to yield nearly transform-limited pulses, the ratio $\Delta\omega/\Delta\omega_{TL}$ is typically bigger than 10. According to the above estimate, the intensity needed for STIRAP has to be increased by a factor of 100 with respect to transform-limited pulses. While in principle it is possible to obtain higher intensities by focusing the laser beam, this would be detrimental for most applications, where large volumes of the molecular jet have to be excited, e.g., in reactive scattering experiments. Therefore, it is very difficult to satisfy the adiabaticity criterion for laser pulses whose bandwidth clearly exceeds the transform-limited bandwidth.

In short-pulse implementations of STIRAP, one should also account for the rapidly increasing laser energy required to ensure adiabatic evolution, as prescribed by condition (58). Thus, if a pulse energy of 1 mJ were sufficient to ensure adiabatic evolution for a 1-ns laser pulse, then the same degree of adiabaticity would require energy of 1 J for a 1-ps laser pulse, even if the bandwidth of the latter is transform limited. Still, this does not mean that long-pulse or CW lasers offer the best possibilities to implement STIRAP. Indeed, the product $\Omega_0 T$, which is proportional to the square root of the pulse energy times the pulse duration T, is typically largest for tunable laser sources of intermediate pulse duration, i.e., nanosecond lasers. Conventional tunable picosecond and femtosecond lasers typically cannot compensate for the shorter pulse duration by an adequate increase in pulse energy. High peak intensity leads to detrimental multiphoton couplings and ionization. On the other hand, CW lasers may provide longer interaction times, but suffer for the weak intensities available.

2. Nitrous Oxide Molecules

STIRAP has been successfully demonstrated with nanosecond pulses in NO molecules (Schiemann et al., 1993; Kuhn et al., 1998), where highly efficient and selective population transfer has been achieved in the electronic ground state from the $X\,^2\Pi_{1/2}(v=0, J=\frac{1}{2})$ rovibrational state to the $X\,^2\Pi_{1/2}(v=6, J=\frac{1}{2})$ state via the intermediate state $A\,^2\Sigma(v=0, J=\frac{1}{2})$. The NO molecule provides an example of complications that may arise due to hyperfine structure. What seems to be a three-level system is actually a system of 18 sublevels (see Fig. 18). Because $^{14}N^{16}O$ has a nuclear spin of $I=1$, each of the three levels is split into a pair of sublevels with $F=\frac{1}{2}$ and $F=\frac{3}{2}$, which in turn possess magnetic sublevels. For linearly polarized light, and when the pump and Stokes polarizations are parallel, the 18-state system decomposes into two independent three-state systems (one with $F=\frac{3}{2}, m_F=\frac{3}{2}$ and another with $F=\frac{3}{2}, m_F=-\frac{3}{2}$) and two six-state systems for $m_F=\frac{1}{2}$ and $m_F=-\frac{1}{2}$. Because the hyperfine splittings of the initial and final levels (214 MHz) is large enough to be resolved experimentally, the complexity of the system can be further reduced. Thus, despite the complications, a nearly complete transfer has been achieved, as shown in Fig. 19.

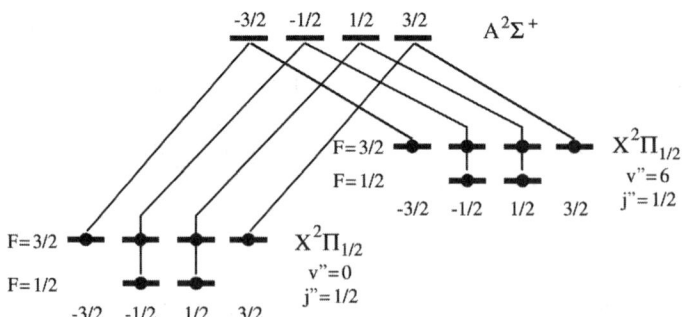

FIG. 18. Linkage scheme in the NO experiment. With hyperfine levels included, the seemingly three-level system becomes an 18-level system, because each of the three levels is split into two hyperfine levels with $F=\frac{1}{2}$ and $F=\frac{3}{2}$, which in turn have magnetic sublevels. The splitting of the excited hyperfine components of 15 MHz is too small to be resolved with pulsed lasers of a few nanoseconds duration and is not shown.

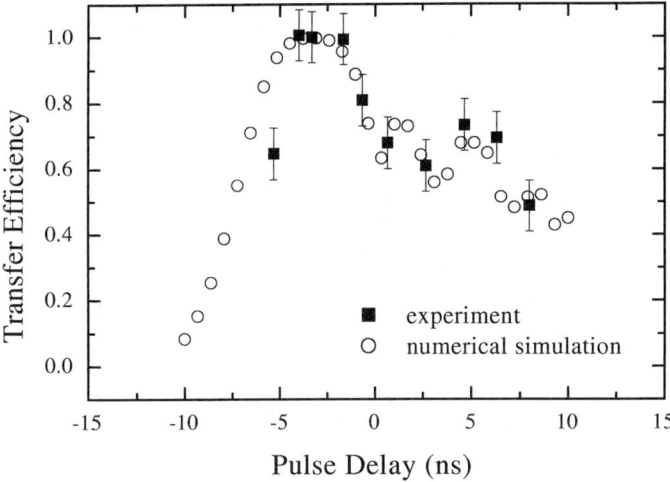

FIG. 19. Experimental demonstration of STIRAP in NO molecules versus the pulse delay. (From S. Schiemann, A. Kuhn, S. Steuerwald, and K. Bergmann. Efficient coherent population transfer in NO molecules using pulsed lasers. *Phys. Rev. Lett.* 1993;71:3637–3640.)

3. Sulfur Dioxide Molecules

The population transfer achieved in SO_2 molecules (Halfmann and Bergmann, 1996) is an example of STIRAP in a polyatomic molecule. The enormously increased density of levels, as compared to atoms or diatomic molecules, results in much smaller transition dipole moments. Nevertheless, efficient population transfer becomes possible with adequate laser power, when the level density in the final state is not too high. Figures 20 and 21 show examples of population transfer from the rotational state 3_{03} of the vibrational ground state (0,0,0) to the same rotational level of the (9,1,0) overtone in the electronic ground state X $^1A_1$ via the vibrational level (1,1,0) of the excited electronic state C^1B_2. The wavelengths were 227 nm for the pump and 300 nm for the Stokes lasers (Halfmann and Bergmann, 1996), with pulse durations of 2.7 ns for the pump and 3.1 ns for the Stokes pulse. Typical laser intensities were 10 MW/cm^2, yielding Rabi frequencies of about 10^{10} s^{-1}. The population in the target state was probed by laser-induced fluorescence. The lower trace in Fig. 20 displays a signal of the probe laser-induced fluorescence from the final state (magnified 10 times) in the case when only the pump pulse was present; then the final state was populated by spontaneous emission from the intermediate state. When the Stokes pulse was turned on and applied before the pump pulse (with an appropriate overlap between them) the final-state population increased by more than two orders of magnitude. When the delay between the pump and Stokes pulses was varied, the typical plateau region of complete population transfer for

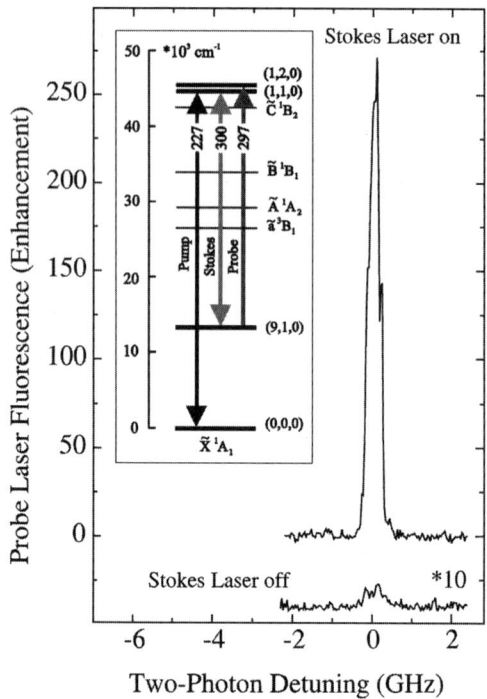

FIG. 20. Experimental demonstration of STIRAP in SO_2 molecules: fluorescence versus two-photon detuning. The inset shows the level linkage. (From T. Halfmann and K. Bergmann. Coherent population transfer and dark resonances in SO_2. *J. Chem. Phys.* 1996;104:7068–7072.)

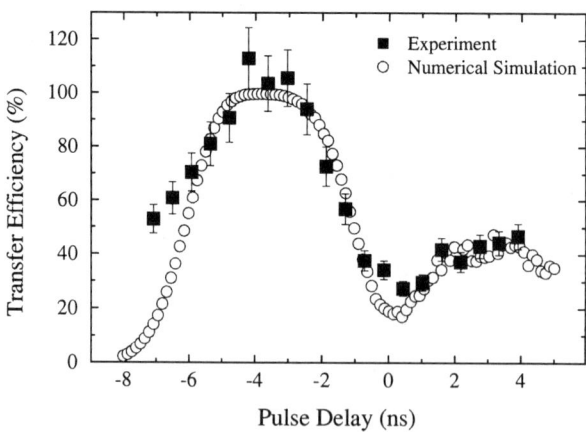

FIG. 21. Experimental demonstration of STIRAP in SO_2 molecules: efficiency versus pulse delay (right plot). (From T. Halfmann and K. Bergmann. Coherent population transfer and dark resonances in SO_2. *J. Chem. Phys.* 1996;104:7068–7072.)

COHERENT MANIPULATION OF ATOMS AND MOLECULES 111

negative pulse delay was observed, as shown in Fig. 21. For positive pulse delay (pump before Stokes), i.e., the case of SEP, a transfer efficiency of about 25% was observed, as one expects from rate equation calculations.

4. STIRAP in a Ladder System: Rubidium Atoms

STIRAP has been successfully implemented to produce samples of ultracold highly excited rubidium atoms by using the ladder transition $5S_{1/2} \to 5P_{3/2} \to 5D_{5/2}$ (Süptitz *et al.*, 1997). The Rb atoms in their ground state $5S_{1/2}$ were initially laser cooled to less than 500 μK and trapped in a magnetooptical trap with a diameter of 7 mm. Then the excitation was performed by illuminating the trapped sample with laser pulses from two injection-locked high-power (100-mW) diode lasers tuned near resonance with the $5S_{1/2}(F = 3) \to 5P_{3/2}(F' = 4)$ and $5P_{3/2}(F' = 4) \to 5D_{5/2}(F' = 5)$ transitions at 780 nm and 776 nm, respectively. The pulses were generated by acustooptic modulators and the delay was controlled electronically. The pulses had nearly Gaussian shapes and were linearly polarized in the same direction. The laser pulse durations of 33 ns (FWHM) were short compared to the lifetime of the target $5D_{5/2}$ state (241 ns), but longer than the lifetime of the intermediate state $5P_{3/2}$ (27 ns). Transfer efficiencies exceeding 90% have been achieved for counterintuitively ordered pulses (the $5P_{3/2} \to 5D_{5/2}$ transition driven first) with peak laser intensities about 10 W/cm^2 (corresponding to peak Rabi frequencies about 2×10^8 s^{-1}). There are no principal limitations to use STIRAP for efficient excitation of even higher-lying (Rydberg) states in this and other atoms.

C. STIRAP WITH DEGENERATE OR NEARLY DEGENERATE STATES

A problem that often arises when implementing STIRAP in real atoms and molecules is the existence of multiple intermediate and final states. These states may be present due to fine and/or hyperfine structure, Zeeman sublevels, or closely spaced rovibrational levels in polyatomic molecules. A multistate system has multiple eigenenergies, which may present a very complicated picture when plotted in time. For example, narrow avoided crossings between the eigenenergies may appear; if such avoided crossings involve the eigenstate that provides the adiabatic linkage for population transfer, then the adiabatic path will be blocked and STIRAP may fail.

A detailed numerical, analytical, and experimental investigation of this problem has been presented in a series of papers on STIRAP in metastable neon atoms (Shore *et al.*, 1995; Martin *et al.*, 1995, 1996). It has been concluded and demonstrated that the presence of closely spaced levels near the intermediate and final state may pose a problem and even be detrimental for STIRAP.

The level scheme in the neon experiment (Martin *et al.*, 1996), shown in Fig. 22, involves states with $J = 0$, 1, and 2, and thus there are $1 + 3 + 5 = 9$ magnetic

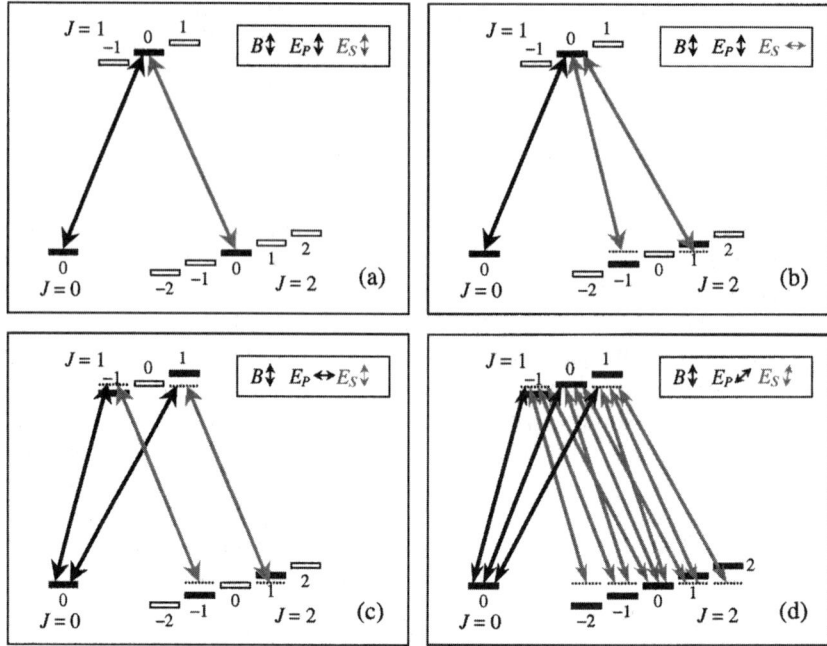

FIG. 22. Linkage patterns in Ne* for various choices of pump and Stokes polarizations with respect to the direction of the magnetic field B.

sublevels that may be coupled by the pump and Stokes lasers. A uniform magnetic field **B** can be used to remove the Zeeman degeneracy in the intermediate and final levels. On the other hand, the optical selection rules allow control of the number of levels participating in the process by an appropriate choice of the laser polarizations with respect to the direction of the uniform magnetic field, which sets the quantization axis. Four particular cases are shown in Fig. 22. When the pump and Stokes fields are linearly polarized along the direction of **B**, the selection rule $\Delta M = 0$ applies. Then only the three $M = 0$ sublevels (one in each of the initial, intermediate and final level) are coupled by the laser fields, as shown in Fig. 22a. When only the pump polarization is parallel to **B**, while the Stokes polarization is perpendicular to **B**, four states are coupled, as seen in Fig. 22b. By contrast, when the Stokes polarization is parallel to **B** and the pump polarization is perpendicular to **B**, five states are coupled, as illustrated in Fig. 22c. Finally, in the most general case when the pump and the Stokes are polarized in arbitrary directions, all nine magnetic sublevels are coupled by the laser fields, as seen in Fig. 22d.

Figure 23 shows the behavior of the final-state population for a set of incrementally increasing values of the magnetic field **B**. The pump laser polarization

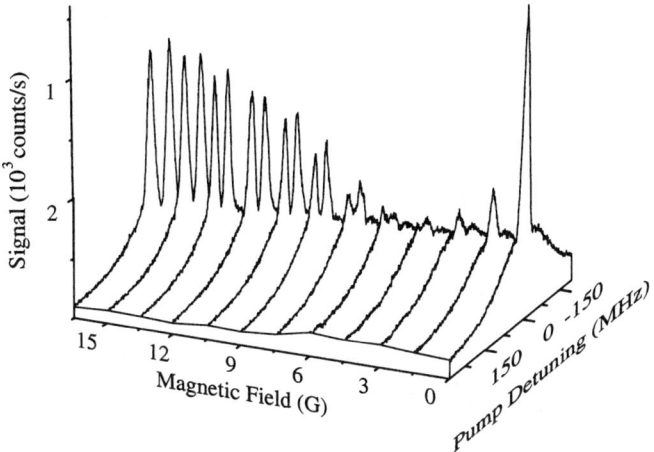

FIG. 23. Population transfer in Ne* versus magnetic field strength for the polarization choice in Fig. 22c. The Stokes laser frequency is held fixed on resonance, while the pump frequency is varied. (From J. Martin, B. W. Shore, and K. Bergmann. Coherent population transfer in multilevel systems with magnetic sublevels. III. Algebraic analysis. *Phys. Rev. A* 1996;54:1556–1569.)

is perpendicular to the magnetic field, while the Stokes polarization is parallel (Martin et al., 1996); the ensuing linkage pattern is that in Fig. 24c and involves five states. The Stokes frequency is tuned to resonance with the Bohr frequency of the degenerate $(B = 0)$ $^3P_2 \leftrightarrow {}^3P_1$ transition, while the pump laser frequency is scanned across the resonance. For small magnetic field, a single peak in the target-state population is observed near resonance ($\Delta_p = 0$), because the $M = +1$ and $M = -1$ sublevels are too close to be resolved. For large magnetic field, the Zeeman splitting increases and a symmetric two-peaked structure emerges, indicating populations of the $M = +1$ and $M = -1$ sublevels (depending in which sublevel is on two-photon resonance). A significant drop in the transfer efficiency is observed at intermediate magnetic field strengths. This drop was identified (Martin et al., 1996) as due to lack of adiabatic connectivity between the initial and final states, i.e., the adiabatic path between them was blocked because of coupling to neighboring adiabatic states. Figure 23 also demonstrates the possibility of orienting the atomic angular momentum simply by tuning a laser frequency to the respective two-photon resonance.

The success of STIRAP at zero and large magnetic field and its failure at some intermediate values can be understood by examining the evolution of the energies of the adiabatic states, shown in Fig. 24. With five coupled states we also have five dressed eigenstates and five adiabatic energies. For $\mathbf{B} = 0$, the pattern looks similar to the one for three states. Because the pump and Stokes laser frequencies are tuned to the respective one-photon resonances, the five eigenenergies are degenerate at

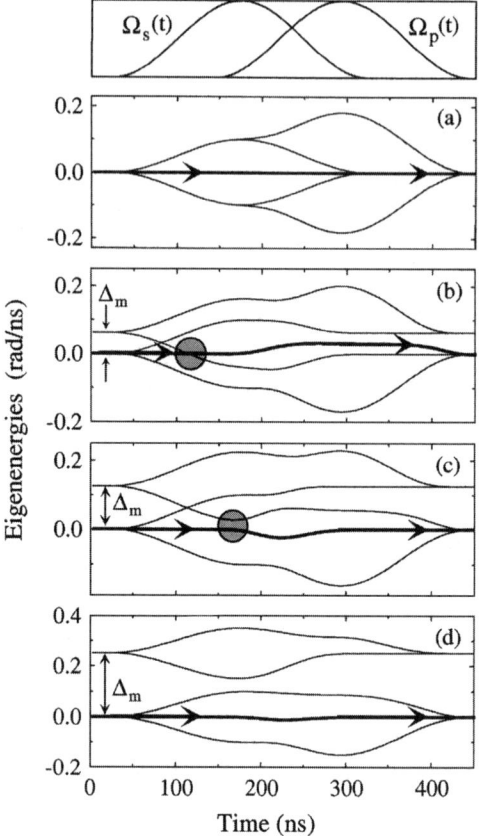

FIG. 24. Time evolution of the eigenenergies in Ne* for different magnetic fields. The circles mark avoided crossings of the eigenenergies, which block the adiabatic path (thick curves) and impede population transfer. (From J. Martin, B. W. Shore, and K. Bergmann. Coherent population transfer in multilevel systems with magnetic sublevels. III. Algebraic analysis. *Phys. Rev. A* 1996;54:1556–1569.)

early and late times, as seen in Fig. 24a. As in the case of three states, we have a zero adiabatic energy at all times, which is the adiabatic path linking state ψ_1 to state ψ_3. Thus STIRAP is possible, as verified in Fig. 23.

When the magnetic field is nonzero, the Zeemen splittings remove the M degeneracy. Two-photon resonance can be established between the initial state ψ_1 and only one of the final states. As a consequence, the eigenenergies are no longer degenerate at early and late times. Therefore, at early times we have a triple degeneracy, which is lifted as soon as the Stokes pulse arrives, and a double degeneracy of states, separated from the former ones by the detuning from the two-photon resonance, i.e., by the Zeeman splitting. The triplet is associated with the three bare states that are resonantly coupled by the laser field, while the doublet is associated with the other two bare states. These latter states also show an Autlor-Townes splitting, and, since at early times their energy is separated from that of the other three states, a crossing of the states occur, provided the Rabi frequency exceeds the Zeeman detuning.

One can show that a coupling between states, related to the crossing energy levels, is induced by the pump laser. When the Zeeman splitting is small, this crossing occurs while the pump laser is still weak and the system, while following the zero-energy path, passes (diabatically) through this crossing (Fig. 24b). At intermediate times, the pump laser induces interaction between the states and forces the energy of the transfer path to deviate from zero. At later times, the transfer path does not connect to the state with zero energy when the Stokes laser is turned off, as is needed for successful completion of the transfer. Therefore the adiabatic transfer path is blocked and coherent population transfer is not possible.

With increasing magnetic field strength, leading to larger Zeeman splitting, the curve crossing with the zero-energy path occurs later, when the pump Rabi frequency is already large (Fig. 24c). Then the avoided crossing is sufficiently broad and does not impede the population transfer. Although a transient deviation from zero energy is observed (meaning that some population will transiently reside in the intermediate states and may be lost by radiative decay), the transfer path connects to the zero-energy eigenvalue as the Stokes laser is turned off; population transfer is again possible. At even larger Zeeman splitting, the avoided crossing is barely noticeable and a (nearly) zero-energy transfer path is again established (Fig. 24d). Moreover, when the Zeeman splitting is larger than the Rabi

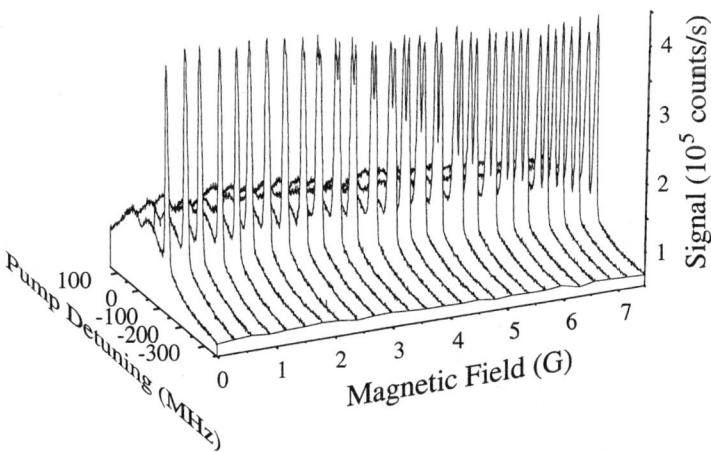

FIG. 25. Population transfer in Ne* versus magnetic field strength for the polarization choice in Fig. 22c. Unlike Fig. 23, here the Stokes laser frequency is held fixed off resonance ($\Delta_s = 200$ MHz), while the pump frequency is varied. (From J. Martin, B. W. Shore, and K. Bergmann. Coherent population transfer in multilevel systems with magnetic sublevels. III. Algebraic analysis. *Phys. Rev. A* 1996;54:1556–1569.)

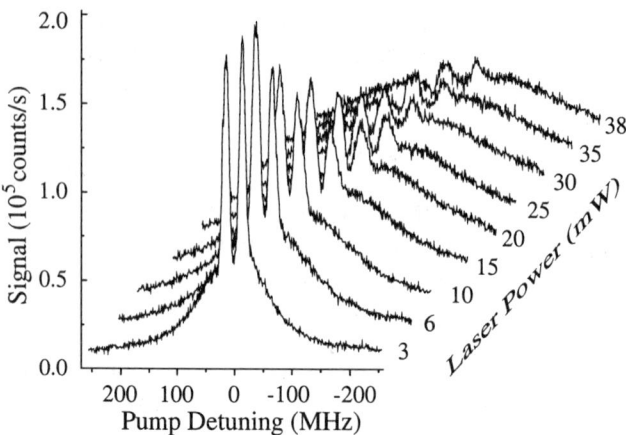

FIG. 26. Top: linkage patterns in Ne* for various choices of pump and Stokes polarizations with respect to the direction of the magnetic field B. Bottom: population transfer in Ne* versus magnetic field strength. (From J. Martin, B. W. Shore, and K. Bergmann. Coherent population transfer in multilevel systems with magnetic sublevels. III. Algebraic analysis. *Phys. Rev. A* 1996;54:1556–1569.)

frequencies, the relevant bare states can again be considered a (nearly) isolated three-state system.

Detuning of the Stokes laser from the one-photon resonance may eliminate the connectivity problem recognized in Fig. 24b. Figure 25 shows the fluorescence signal from the $^3P_2$ state for the same polarizations as in Fig. 23. Again, the figure shows traces for incrementally increasing magnetic field as the pump laser frequency is scanned across resonance. However, unlike Fig. 23, the Stokes laser frequency is held fixed off resonance by 200 Mhz. This detuning prevents the crossing at early times (Fig. 24b) from appearing and ensures high transfer efficiency for all magnetic field values.

Finally, Fig. 26 shows an example from Martin *et al.* (1996) in which population transfer is possible for small Rabi frequencies but fails at higher laser power. The reason is again the blocking of the adiabatic path and can be cured, again, by detuning the lasers off their single-photon resonances. Thus, while in three-state systems STIRAP gets more robust and efficient as the Rabi frequency increases, in multistate systems this is not garanteed in general.

In conclusion, coherent population transfer in multilevel systems depends on the availability of an adiabatic path that connects the initial and final states with only a small deviation from the zero eigenvalue at intermediate times when both laser fields are nonzero. Whether this is possible or not depends on the level structure and coupling scheme. In most cases, an analysis like the one shown in Fig. 24 is needed in order to understand the details of the transfer process and to make an appropriate choice of Rabi frequencies and detunings in order to control the crossing of dressed-state eigenvalues. In many cases, the connectivity problems can be avoided if the laser frequencies are tuned sufficiently far from any one-photon resonance with an intermediate state. Moreover, for efficient population transfer the two-photon linewidth should not exceed the intermediate level separation, if the laser frequencies are close to one-photon resonance. An important conclusion, predicted theoretically and confirmed experimentally, is that one-photon resonances should be avoided (while the two-photon resonance is still essential) when more than three states are involved. Although the laser–atom coupling is strongest for one-photon resonance, this resonance may lead to blocking of the adiabatic connection and at least to non-negligible transient population in the decaying intermediate states.

Finally, we point out that some other aspects of the influence of multiple nearly degenerate final states in STIRAP have been explored theoretically (Band and Magnes, 1994; Kobrak and Rice, 1998b).

V. STIRAP-Like Population Transfer in Multistate Chains

The simplest extensions of the three-state coherent excitation chain are those dealing with multistate chainwise excitation, i.e., linkages of the type

$$\psi_1 \leftrightarrow \psi_2 \leftrightarrow \psi_3 \leftrightarrow \ldots \leftrightarrow \psi_N$$

in which each state is connected to at most two other states (see Fig. 27). For steady-amplitude fields, analytic solutions to the Schrödinger equation are available in a variety of cases (cf. Shore, 1990, chap. 15). The ordering of the energy levels is not important: the linkage pattern may appear as a simple ladder, or as some bent chain such as has the appearance of the letter N for four states or, for five states, the letter M or W. Chainwise ladder excitation is of considerable interest particularly for producing dissociation or ionization, whereas bent chains have interesting applications described in this section and in Section VI.

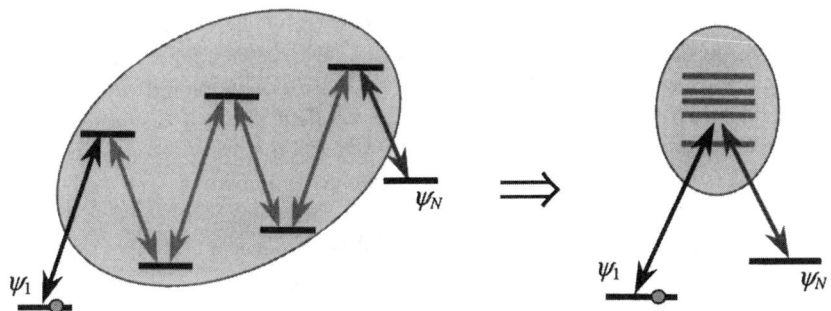

FIG. 27. Linkage pattern for chain-STIRAP (serial multi-Λ system) and equivalent parallel multi-Λ system, obtained by diagonalization of the subsystem comprising the intermediate states in the original chainwise system.

The extension of STIRAP to multistate chains has drawn considerable attention (Shore *et al.*, 1991; Marte *et al.*, 1991; Smith, 1992; Oreg *et al.*, 1992; Pillet *et al.*, 1993; Valentin *et al.*, 1994; Goldner *et al.*, 1994a,b; Malinovsky and Tannor, 1997; Theuer and Bergmann, 1998; Vitanov, 1998a, Vitanov *et al.*, 1998; Nakajima 1999). Chain-STIRAP has proved to be a viable technique of great potential, particularly for momentum transfer in atom optics (Section VI). In this section we will describe the basic features and problems of chain STIRAP and will provide a few illustrations.

The RWA Hamiltonian of a multistate chain is a tridiagonal matrix which has the diabatic energies (the detunings) on its diagonal, and the laser-induced couplings as off-diagonal elements,

$$\mathsf{H} = \frac{\hbar}{2} \begin{bmatrix} 0 & \Omega_{1,2} & 0 & \cdots & 0 & 0 \\ \Omega_{1,2} & 2\Delta_2 & \Omega_{2,3} & \cdots & 0 & 0 \\ 0 & \Omega_{2,3} & 2\Delta_3 & \cdots & 0 & 0 \\ \vdots & \vdots & \vdots & \ddots & \vdots & \vdots \\ 0 & 0 & 0 & \cdots & 2\Delta_{N-1} & \Omega_{N-1,N} \\ 0 & 0 & 0 & \cdots & \Omega_{N-1,N} & 0 \end{bmatrix}. \qquad (69)$$

The zeros in the first and last diagonal elements indicate that the initial state and the final state of the chain are on $(N-1)$-photon resonance, a condition that generalizes the two-photon resonance in three-state STIRAP, whereas the intermediate states may be off-resonance in general. It has been discovered that such multistate chains behave differently when they involve odd and even numbers of states.

A. RESONANTLY DRIVEN CHAINS WITH ODD NUMBER OF STATES

The key for STIRAP-like population transfer in multistate chainwise connected systems is the existence of a multilevel dark state, which generalizes the usual dark state (44b) of three-state Λ systems and links adiabatically the initial state ψ_1 to the final state ψ_N of the chain. Such a multilevel dark state exists only when the multistate chain comprises an odd number of states ($N = 2n + 1$). It also requires that all lasers are on resonance with the corresponding transitions or only the even states in the chain are detuned from resonance (Shore *et al.*, 1991; Marte *et al.*, 1991; Smith, 1992). Such a chain can be viewed as a sequence of serially connected Λ systems, each of which is on two-photon resonance.

The Hamiltonian describing such a multistate chain has a zero eigenvalue. The multilevel dark state $\Phi_0(t)$ is the corresponding zero-eigenvalue eigenstate and it is a time-dependent coherent superposition of the odd states in the chain

$\psi_1, \psi_3, \ldots, \psi_{2n+1}$. For example, the Hamiltonian of a five-state chain is given by

$$H = \frac{\hbar}{2}\begin{bmatrix} 0 & \Omega_{1,2} & 0 & 0 & 0 \\ \Omega_{1,2} & 2\Delta & \Omega_{2,3} & 0 & 0 \\ 0 & \Omega_{2,3} & 0 & \Omega_{3,4} & 0 \\ 0 & 0 & \Omega_{3,4} & 2\Delta & \Omega_{4,5} \\ 0 & 0 & 0 & \Omega_{4,5} & 0 \end{bmatrix}. \quad (70)$$

States ψ_1 and ψ_3 are on two-photon resonance, and so are states ψ_3 and ψ_5. The even states in the chain ψ_2 and ψ_4 may be off the respective single-photon resonances by the same detuning Δ. The multilevel dark state of this system reads (Morris and Shore, 1983; Hioe and Carroll, 1988; Shore et al., 1991; Marte et al., 1991; Smith, 1992; Milner and Prior, 1998)

$$\Phi_0(t) = \frac{1}{\mathcal{N}(t)}[\Omega_{2,3}(t)\Omega_{4,5}(t)\psi_1 - \Omega_{1,2}(t)\Omega_{4,5}(t)\psi_3 + \Omega_{1,2}(t)\Omega_{3,4}(t)\psi_5], \quad (71)$$

where $\mathcal{N}(t)$ is a normalization factor.

A particularly suitable system for multistate STIRAP is the chainwise transition formed by the magnetic sublevels of a degenerate two-level system with excited-level angular momentum $J_e = J_g$ or $J_g - 1$, driven by two sequential pulses with opposite circular polarizations. For example, if the system is prepared initially in the $M_g = -J_g$ ground-state sublevel (e.g., by optical pumping), then a STIRAP-like transfer to the $M_g = J_g$ sublevel can be achieved by applying a pulse of σ^- polarization (the Stokes) before a pulse of σ^+ polarization (the pump), i.e., in the counterintuitive ordering.

In the five-state example, the Rabi frequencies $\Omega_{1,2}(t)$ and $\Omega_{3,4}(t)$ will follow the time dependence $f_+(t)$ of the σ^+ pulse [and the difference in their peak values is determined by the respective Clebsch–Gordan coefficients (cf. Shore, 1990, chap. 19.1)], while the Rabi frequencies $\Omega_{2,3}(t)$ and $\Omega_{4,5}(t)$ will follow the time dependence $f_-(t)$ of the σ^- pulse. Then the dark state (71) takes the form

$$\Phi_0(t) = \frac{1}{\mathcal{N}(t)}[\Omega_{2,3}^0\Omega_{4,5}^0 f_-^2(t)\psi_1 - \Omega_{1,2}^0\Omega_{4,5}^0 f_-(t)f_+(t)\psi_3 + \Omega_{1,2}^0\Omega_{3,4}^0 f_+^2(t)\psi_5]. \quad (72)$$

Hence, if the σ^- pulse precedes the σ^+ pulse (counterintuitive pulse ordering), state $\Phi_0(t)$ is equal to ψ_1 ($M = -J_g$) at early times and to ψ_5 ($M = +J_g$) at late times, i.e., it provides an adiabatic connection between the initial state ψ_1 and the last state ψ_5 of the chain. As in STIRAP, the sublevels of the excited level (ψ_2 and ψ_4 in this example) remain unpopulated throughout the transfer if the interaction is adiabatic; however, the intermediate odd states in the chain

(ψ_3 here)—the sublevels of the ground state—do acquire some transient populations. In this particular system these transient intermediate-state populations do not pose a problem because these sublevels do not decay and there are no population losses. Indeed, multistate STIRAP in such chainwise systems has been demonstrated experimentally by several groups (Pillet et al., 1993; Valentin et al., 1994; Goldner et al., 1994a; Goldner et al., 1994b; Theuer and Bergmann 1998). Sections VI and XI discuss some applications of STIRAP in multistate chains.

B. RESONANTLY DRIVEN CHAINS WITH EVEN NUMBER OF STATES

The chains with an even number of states ($N = 2n$) behave very differently in the resonant case (when all intermediate-state detunings vanish, $\Delta_2 = \Delta_3 = \ldots = \Delta_{N-1} = 0$) compared with the chains with an odd number of states. For even-N systems we have det $\mathsf{H} = (-)^{N/2} \Omega_{1,2}^2 \Omega_{3,4}^2 \ldots \Omega_{N-1,N}^2 \neq 0$, which means that the Hamiltonian does not have a zero eigenvalue, in contrast to the case of odd N. More important, $\mathsf{H}(t)$ does not possess any adiabatic state that provides an adiabatic connection between the initial state ψ_1 and the final state ψ_N of the chain. Consequently, even when such a system is driven adiabatically by counterintuitively ordered resonant pulses, a STIRAP-like population transfer between the initial and final states of the chain cannot occur. Instead, the final-state population exhibits Rabi-like oscillations as the pulse intensities increase (Oreg et al., 1992; Vitanov, 1998a; Band and Julienne, 1991b). These oscillations occur because at early times the initial state is equal to a superposition of adiabatic states (rather than to a single adiabatic state as in STIRAP) and so is the final state at late times; hence interference between the different paths from the initial state to the final state takes place.

For example, in a resonantly driven four-state chain the four eigenvalues are $\pm \varepsilon_+$ and $\pm \varepsilon_-$, where

$$\varepsilon_{\pm} = \frac{1}{2}\sqrt{\frac{1}{2}\left(\Omega^2 \pm \sqrt{\Omega^4 - 4\Omega_p^2 \Omega_s^2}\right)} \tag{73}$$

and $\Omega^2 = \Omega_p^2 + \Omega_i^2 + \Omega_s^2$, where Ω_i is the coupling between the two intermediate states. The initial state ψ_1 cannot be identified with a single adiabatic state at $-\infty$, but is rather given by a superposition of the two adiabatic states corresponding to the smallest eigenenergies $-\varepsilon_-$ and ε_-. The same applies to the final state ψ_4 at $+\infty$:

$$\psi_1 = \tfrac{1}{\sqrt{2}}[\Phi_{-\varepsilon_-}(-\infty) + \Phi_{\varepsilon_-}(-\infty)], \tag{74a}$$

$$\psi_4 = \tfrac{1}{\sqrt{2}}[\Phi_{-\varepsilon_-}(+\infty) + \Phi_{\varepsilon_-}(+\infty)]. \tag{74b}$$

If the system starts its evolution in state ψ_1 initially, then the adiabatic solution for the final populations of the bare states is (Vitanov 1998a)

$$P_1(\infty) \approx \cos^2 \vartheta \cos^2 \Theta, \qquad (75a)$$

$$P_2(\infty) \approx 0, \qquad (75b)$$

$$P_3(\infty) \approx \sin^2 \vartheta \cos^2 \Theta, \qquad (75c)$$

$$P_4(\infty) \approx \sin^2 \Theta. \qquad (75d)$$

Here $\tan \vartheta = \lim_{t \to +\infty} [\Omega_p(t)/\Omega_i(t)]$ and $\Theta = \int_{-\infty}^{\infty} \varepsilon_-(t) dt$. Hence, as we approach the adiabatic limit, $P_4(t)$ oscillates rather than tend to unity. Note that $P_3(\infty) = 0$ if the pump pulse $\Omega_p(t)$ vanishes before $\Omega_i(t)$ [$\Omega_p(t)/\Omega_i(t) \stackrel{t \to +\infty}{\longrightarrow} 0$], while $P_1(\infty) = 0$ if the pump pulse $\Omega_p(t)$ vanishes after $\Omega_i(t)[\Omega_p(t)/\Omega_i(t) \stackrel{t \to +\infty}{\longrightarrow} +\infty]$.

We also point out that if the pulse $\Omega_i(t)$, coupling the intermediate states, is much stronger than the pump and Stokes pulses at all times, then $\vartheta \approx 0$. Hence, although not being of STIRAP-type, the excitation process in an on-resonance four-state system involving a strong intermediate pulse (Malinovsky and Tannor, 1997), is quite interesting by itself because it demonstrates how the population can flip between states ψ_1 and ψ_4, bypassing the intermediate states ψ_2 and ψ_3 despite the fact that the latter states are on resonance with the corresponding lasers.

C. THE OFF-RESONANCE CASE

When the intermediate states are off resonance while the initial state and the final state of the chain are still on $(N - 1)$-photon resonance, chains with odd and even number of states behave quite similarly. In both cases a STIRAP-like transfer is only possible if there exists an adiabatic path—adiabatic transfer state $\Phi_T(t)$—linking the initial and final states of the chain. This depends on the laser parameters, particularly on the intermediate detunings.

Let us consider the case when all intermediate detunings are nonzero, $\Delta_k \neq 0$ ($k = 2, 3, \ldots, N - 1$), and all couplings are pulse shaped. Let us also assume that the Stokes pulse $\Omega_{N,N-1} \equiv \Omega_s$, coupling the last transition, precedes the pump pulse $\Omega_{1,2} \equiv \Omega_p$, coupling the first transition. By setting $\Omega_p = 0$ and $\Omega_s = 0$ in Eq. (69), we find that there are two eigenvalues of $H(t)$ which vanish as $t \to \pm\infty$ [although they are nonzero at finite times because $\det H(t) \neq 0$ in general]. The other eigenvalues tend to the (nonzero) detunings Δ_k and the corresponding adiabatic states tend to the respective bare intermediate states ψ_k. At $\pm\infty$, each of the two adiabatic states, corresponding to the vanishing eigenvalues, is equal to either state ψ_1, or state ψ_N, or a superposition of ψ_1 and ψ_N. Obviously, if an AT state exists, its eigenvalue should be one of these two eigenvalues. Hence, of particular interest are the asymptotic behaviors of the two vanishing eigenvalues and the corresponding adiabatic states at early and late times. We note here that only one eigenvalue vanishes when $\Omega_p \to 0$ and $\Omega_s \ne 0$, which happens at early times, or when $\Omega_p \to 0$ and $\Omega_s \ne 0$, which happens at late times. Hence, at early times ($t \to -\infty$) as soon as the Stokes pulse Ω_s arrives, one of the initially degenerate eigenvalues, ε_l^- (the "large" one), departs from zero, while the other, ε_s^- (the "small" one), remains zero until the pump pulse Ω_p arrives later.

A similar scenario takes place at late times when first one eigenvalue vanishes with the disappearance of the Stokes pulse, and then the other eigenvalue vanishes with the pump pulse. It can be shown that at early times ($t \to -\infty$), the two vanishing adiabatic energies behave as (Vitanov 1998a)

$$\varepsilon_s^- \approx -\frac{\mathcal{D}^{(3,N-2)}}{4\mathcal{D}^{(2,N-2)}}\Omega_p^2, \qquad \varepsilon_l^- \approx -\frac{\mathcal{D}^{(2,N-2)}}{4\mathcal{D}^{(2,N-1)}}\Omega_s^2, \qquad (76)$$

and at late times ($t \to +\infty$) the two vanishing adiabatic energies behave as

$$\varepsilon_s^+ \approx -\frac{\mathcal{D}^{(3,N-2)}}{4\mathcal{D}^{(3,N-1)}}\Omega_s^2, \qquad \varepsilon_l^+ \approx -\frac{\mathcal{D}^{(3,N-1)}}{4\mathcal{D}^{(2,N-1)}}\Omega_p^2. \qquad (77)$$

Here $\mathcal{D}^{(j,k)}$ denotes the determinant of the matrix obtained from H by keeping its columns from jth to kth and its rows from jth to kth,

$$\mathcal{D}^{(j,k)} = \begin{vmatrix} \Delta_j & \frac{1}{2}\Omega_{j,j+1} & 0 & \cdots & 0 \\ \frac{1}{2}\Omega_{j,j+1} & \Delta_{j+1} & \frac{1}{2}\Omega_{j+1,j+2} & \cdots & 0 \\ 0 & \frac{1}{2}\Omega_{j+1,j+2} & \Delta_{j+2} & \cdots & 0 \\ \vdots & \vdots & \vdots & \ddots & \vdots \\ 0 & 0 & 0 & \cdots & \Delta_k \end{vmatrix}. \qquad (78)$$

It is easy to verify that the adiabatic eigenstates corresponding to ε_s^- and ε_l^+ tend to state ψ_1, while those corresponding to ε_l^- and ε_s^+ tend to state ψ_N. Hence, the AT state Φ_T, if it exists, must have an eigenvalue that coincides with ε_s^- as $t \to -\infty$ and with ε_s^+ as $t \to +\infty$. The energies ε_s^- and ε_s^+ do not necessarily correspond to the same eigenvalue and it may happen that ε_s^- is linked to ε_l^+ rather than ε_s^+; then an AT state does not exist. In any case, since the eigenvalues do not cross, the upper (the lower) of the two eigenvalues at $-\infty$ is connected to the upper (the lower) of the two eigenvalues at $+\infty$. Because $|\varepsilon_l^-| \gg |\varepsilon_s^-|$ and $|\varepsilon_l^+| \gg |\varepsilon_s^+|$, the linkage is determined by the signs of the "large" eigenvalues ε_l^- and ε_l^+. If they have the same signs, they will be both above (or below) ε_s^- and ε_s^+ and hence, the desired linkages $\varepsilon_l^- \leftrightarrow \varepsilon_l^+$ and $\varepsilon_s^- \leftrightarrow \varepsilon_s^+$ will take place. If ε_l^- and ε_l^+ have opposite signs, they cannot be connected because such an eigenvalue would cross the one linking

ε_s^- and ε_s^+. Thus, from this analysis and Eqs. (76) and (77), we conclude that the condition for existence of an AT state is

$$\mathcal{D}^{(2,N-2)}\mathcal{D}^{(3,N-1)} > 0. \tag{79}$$

The signs of the $\mathcal{D}$'s have to be determined at the early (or late) times when the Rabi frequencies are much smaller than the detunings (note that each of the $\mathcal{D}$'s has the same sign at early and late times). The above condition (79) remains valid also when one or more intermediate detunings are equal to zero (Vitanov, 1998a).

If all intermediate couplings are pulse shaped (i.e., vanish as $t \to \pm\infty$), we have $\mathcal{D}^{(2,N-2)} \to \Delta_2\Delta_3 \ldots \Delta_{N-2}$, $\mathcal{D}^{(3,N-1)} \to \Delta_3\Delta_4 \ldots \Delta_{N-1}$, and condition (79) reduces to

$$\Delta_2\Delta_{N-1} > 0. \tag{80}$$

Below we provide examples for existence and nonexistence of an AT state (and hence, of STIRAP-like transfer) in the cases of $N = 4$ and 5 states.

- In a four-state chain, an AT state exists when the cumulative detunings of the two intermediate states have the same sign, $\Delta_2\Delta_3 > 0$, whereas it does not exist if $\Delta_2\Delta_3 \leq 0$.
- In the five-state case, condition (79) reads $(\Delta_2\Delta_3 - \Omega_{2,3}^2)(\Delta_3\Delta_4 - \Omega_{3,4}^2) > 0$. Thus if all intermediate detunings are nonzero, an AT state exists only if $\Delta_2\Delta_4 > 0$. If $\Delta_2 = 0$, AT state exists only for $\Delta_3\Delta_4 < 0$. If $\Delta_4 = 0$, AT state exists for $\Delta_2\Delta_3 < 0$. If $\Delta_3 = 0$, an AT state always exists regardless of Δ_2 and Δ_4, which agrees with the result in Section V.A. If $\Delta_2 = \Delta_4 = 0$, an AT state also exists regardless of Δ_3.

Figure 28 displays examples of successful and unsuccessful adiabatic population transfer in four-state (top frames) and five-state (bottom frames) chains. In the on-resonant case (left frames), the final-state population approaches unity for $N = 5$, while it oscillates for $N = 4$. The middle frames show examples for which the AT condition (80) is satisfied; consequently, the transfer efficiency approaches unity for both $N = 4$ and $N = 5$. The right frames show examples when the AT condition (80) is not satisfied; then population transfer fails for both $N = 4$ and $N = 5$.

We point out that while condition (79) provides adiabatic connection between the two end states of the multistate chain, the AT state has in general nonzero components from all states involved, including from the decaying excited states. The even-state components (which correspond to the decaying excited states in the linkage pattern in Fig. 27) vanish only when the odd detunings are zero; then the AT state is a dark state (Section V.A). In the general case, beyond the particular case of chainwise transitions between magnetic sublevels of degenerate two-level systems,

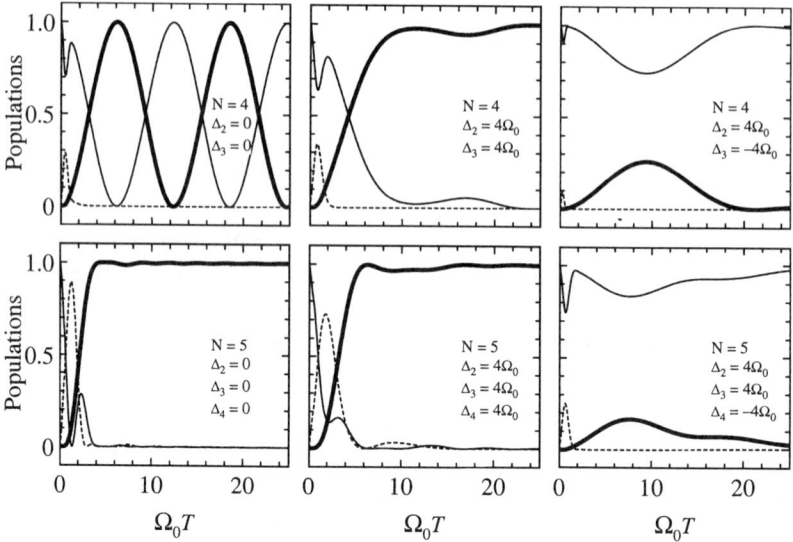

FIG. 28. Examples of success and failure of adiabatic population transfer in four-state (top frames) and five-state (bottom frames) chains. In each frame, we show the numerically calculated final-state population (thick solid curve), the initial-state population (thin solid curve), and the total population in the intermediate states (dashed curve). The dimensionless product $\Omega_0 T$ is proportional to the pulse area and as it increases, adiabaticity improves. The pump and Stokes Rabi frequencies are given by $\Omega_p = \Omega_0 \exp[-(t-\tau)^2/T^2]$, $\Omega_s = \Omega_0 \exp[-(t+\tau)^2/T^2]$, with $\tau = 0.5T$, and all intermediate-pulse Rabi frequencies are equal to $\Omega_0 \exp[-(t/2T)^2]$. The two left frames are for the on-resonance case, when excitation leads to STIRAP-like transfer for $N=5$ states but to Rabi-like oscillations for $N=4$. The middle frames show examples of successful multistate STIRAP for both $N=4$ and $N=5$, whereas the right frames show examples of failure of multistate STIRAP.

the transiently populated intermediate states can decay radiatively during the transfer; then it is important to reduce their populations. (Malinovsky and Tannor, 1997) suggested that these transient populations can be suppressed when the pulses coupling the intermediate transitions are much stronger than the pulses driving the first (pump) and the last (Stokes) transitions. They proposed a pulse sequence, named *straddle STIRAP*, in which all intermediate pulses arrive simultaneously with the Stokes pulse and vanish with the pump pulse, being much stronger than the pump and the Stokes at all times (Malinovsky and Tannor, 1997; Sola et al., 1999).

D. OPTIMIZATION OF MULTISTATE STIRAP: DRESSED-STATE PICTURE

A dressed-state approach (Vitanov et al., 1998) provides a particularly clear picture of multistate STIRAP, valid for both odd and even number of states. It is most useful when the pulses (or the pulse) driving the intermediate transitions arrive before and vanish after the pulses that drive the first (the pump) and the last (the Stokes) transitions, as in Malinovsky and Tannor (1997), and are nearly constant during the time when the pump and the Stokes lasers are present. Then, before the arrival of the pump and Stokes pulses, the $N-2$ intermediate states are coupled into a dressed subsystem, as shown in Fig. 27. By changing the parameters of the dressing pulses (intensities and frequencies), one can manipulate the properties of this dressed subsystem and thus control the population transfer. By tuning the pump and Stokes lasers to one of the dressed eigenstates, Φ_k, the multistate dynamics is essentially reduced to a system of three strongly coupled states: $\psi_1 \leftrightarrow \Phi_k \leftrightarrow \psi_N$; this paves the road for an efficient STIRAP-like population transfer from state ψ_1 to state ψ_N. Furthermore, if the dressing pulses are constant, at least during the time when the pump and Stokes pulses are present, then the couplings between the dressed states vanish. Also, if the dressing pulses are strong, the splittings between the dressed energies are large. This makes the multi-Λ system resemble the single-Λ system in STIRAP and therefore place little population in the intermediate states.

The dressed picture also displays the difference between odd and even number of states in the on-resonance case (all detunings in the original chain equal to zero), when, as we noted above, odd-N chains have a zero eigenvalue and a corresponding trapped state, while even-N chains do not have such an eigenvalue, nor a trapped state. For odd-N chains, one of the dressed states is always on resonance with the pump and Stokes lasers. In contrast, for even-N chains, the pump and Stokes lasers are tuned in the middle between two adjacent dressed eigenvalues and the ensuing interference between different adiabatic paths leads to Rabi-like oscillations. Thus, while the on-resonance choice provides the best results for odd-N chains, the only possibility to achieve STIRAP-like population transfer in an even-N chain is to choose nonzero intermediate-state detunings and ensure that their values fall within an AT region.

Figure 29 shows the final-state population in four- and five-state systems against the cumulative detunings from the intermediate states. In the four-state case, high transfer efficiency (the white zones) is achieved for sufficiently adiabatic evolution only if $\Delta_2 \Delta_3 > 0$, as discussed above. Near the dressed-state resonances (shown by hyperbolas) the transfer efficiency is high even when the evolution is not very adiabatic elsewhere (top left frame). As the pulse areas increase (top right frame), the two regions where $\Delta_2 \Delta_3 > 0$ (first and third quadrants) get filled with white (unity transfer efficiency), while no high transfer is possible in the $\Delta_2 \Delta_3 < 0$ regions (second and fourth quadrants). In the five-state case there are three AT regions in the $\Delta_2 \Delta_4$ plane ($\Delta_2 = \Delta_3$ is assumed in order to display a two-dimensional plot), defined by the AT condition $(\Delta_2^2 - \Omega_{2,3}^2)(\Delta_2 \Delta_4 - \Omega_{3,4}^2) > 0$. In each AT region, there is an intermediate-dressed-state resonance, shown by a thick curve. As in the four-state case, only the AT regions get filled with white as the adiabaticity improves (bottom right frame). Again near the dressed-state resonances the transfer efficiency is high even for poor adiabaticity elsewhere (bottom left frame).

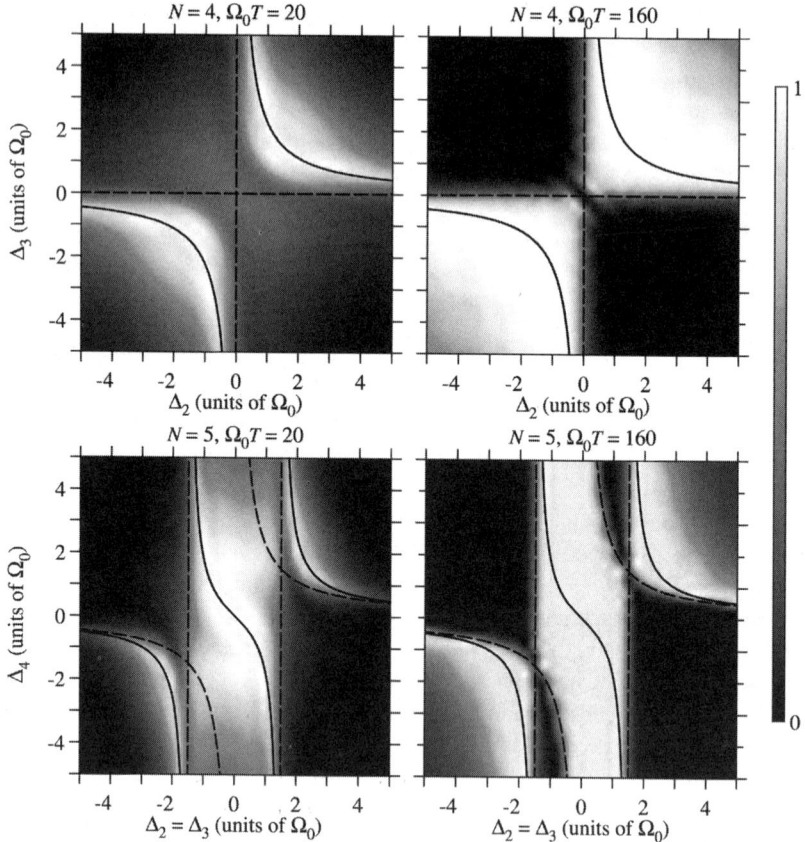

FIG. 29. Numerically calculated transfer efficiency for chain STIRAP. Upper frames: four-state system versus the two intermediate-state detunings Δ_2 and Δ_3. Lower frames: five-state system versus the three intermediate-state detunings $\Delta_2 = \Delta_3$ and Δ_4. In all frames, we have taken $\Omega_p = \Omega_0 \exp[-(t-\tau)^2/T^2]$, $\Omega_s = \Omega_0 \exp[-(t+\tau)^2/T^2]$, with $\tau = 0.5T$, and all intermediate Rabi frequencies constant and equal to $3\Omega_0$. In the left frames, $\Omega_0 T = 20$, while in the right frames, $\Omega_0 T = 160$. The solid curves show the dressed-state resonances and the dashed curves separate the regions where adiabatic transfer state does or does not exist. (From Vitanov, 1998.)

VI. Adiabatic Momentum Transfer

Coherent population transfer between atomic states is always accompanied by transfer of photon momenta to the atoms. Momentum transfer is the basis of atom optics, particularly in the design of its key elements—atom mirrors and beam splitters. An atomic beam splitter separates the single-atom wavefunction into a macroscopic superposition state corresponding to two center-of-mass wave packets propagating in different spatial directions. An atomic mirror, on the other hand, deflects these wave packets so that the matter waves traveling along two paths of an interferometer can be brought together to interfere. This interference will be observed only if these scattering processes are coherent. Because STIRAP enables efficient, robust, and dissipation-free coherent population and momentum transfer, it has been quickly recognized that it is a perfect tool for building practically "ideal" atomic mirrors and beam splitters (Marte et al., 1991).

A. COHERENT MATTER-WAVE MANIPULATION

1. Atomic Mirrors

A particularly suitable system for coherent momentum transfer is the chainwise transition formed between the magnetic sublevels of two degenerate levels with total angular momentum J_g of the ground level and $J_e = J_g$ or $J_g - 1$ of the excited level. An example for such a chain in the case when $J_g = J_e = 2$ is shown in Fig. 30. When such a system is prepared initially in one of the chain-end ground sublevels, e.g., in $M = -J_g$, and is driven adiabatically by two counterintuitively ordered sequential laser pulses with opposite circular polarizations, then, as discussed in Section V, complete population transfer occurs between the two ends of the chain, i.e., from sublevel $M = -J_g$ to sublevel $M = J_g$. In the general case, the chain couples $J_g + 1$ ground sublevels and J_g excited sublevels. Thus, for $J_g = 1 \leftrightarrow J_e = 0$ and $J_g = 1 \leftrightarrow J_e = 1$ transitions, the σ^+ and σ^- laser fields couple three sublevels: two ground sublevels $M = -1$ and $M = +1$ and one excited sublevel $M = 0$. If the two laser beams propagate in opposite directions, each atom will receive, during its journey from the $M = -1$ sublevel to the $M = 1$ sublevel, a total momentum of $2\hbar k$ in the direction of the σ^+ beam: a momentum of $\hbar k$ from absorbing a photon from the σ^+ beam, and another $\hbar k$ in the same direction due to

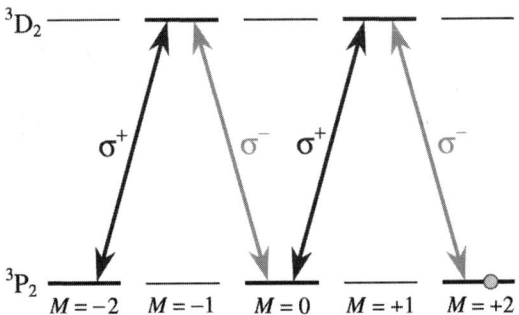

FIG. 30. Linkage pattern between the Zeeman sublevels in the $^3P_2 - {}^3D_2$ transition in Ne* driven by a pair of counterpropagating and displaced σ^+ and σ^- laser beams.

recoil from the stimulated emission of a photon into the σ^- beam. As a result, the atom is deflected in a single well-defined direction, determined by two σ^+ photon momenta.

An example in the case when $J_g = J_e = 2$, demonstrated in a recent experiment (Theuer and Bergmann, 1998), is shown in Fig. 31. A beam of metastable neon atoms, prepared by optical pumping in the $M = 2$ magnetic sublevel of the $^3P_2$ metastable level, crosses two slightly displaced circularly polarized CW laser beams. The two beams are ordered counterintuitively, so the atoms encounter the σ^+ beam (the Stokes) first and then the σ^- beam (the pump). In the adiabatic limit, the population is completely transferred to the $M = -2$ sublevel of the $^3P_2$ level, without residing at any time in the $M = -1$ and $M = 1$ sublevels of the decaying excited level $^3D_2$. Because the two laser beams propagated in opposite directions, each atom received a total momentum of $4\hbar k$ in the direction of the σ^- beam during its journey from the $M = 2$ sublevel to the $M = -2$ sublevel: $2\hbar k$ momentum due

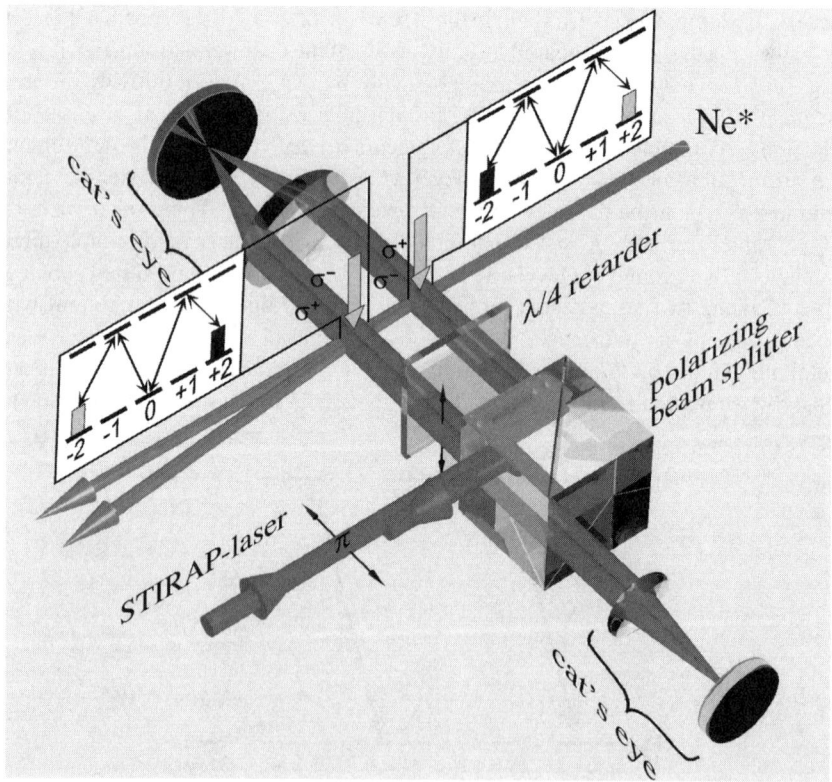

FIG. 31. Experimental setup for the Ne* atomic mirror. (From K. Bergmann, H. Theuer, and B. W. Shore. Coherent population transfer among quantum states of atoms and molecules. *Rev. Mod. Phys.* 1998;70:1003–1025.)

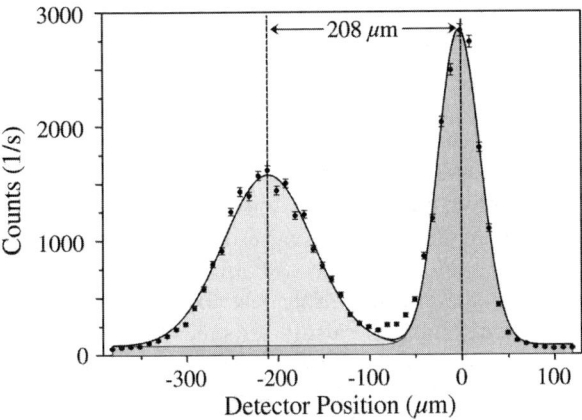

FIG. 32. Experimental results showing deflection of a beam of ^{20}Ne* atoms due to transfer of eight photon momenta after double adiabatic passage from the $M=2$ sublevel to $M=-2$ and then back to $M=2$. The narrower, undeflected original distribution is observed due to the presence of ^{22}Ne isotope atoms which are insensitive to the light. (From H. Theuer and K. Bergmann. Atomic beam deflection by coherent momentum transfer and the dependence on weak magnetic fields. *Eur. Phys. J. D* 1998;2:279–289.)

to absorption of two photons from the σ^- beam (which transfer their momenta to the atom), and another momentum $2\hbar k$ in the same direction due to recoil from the stimulated emission of two photons into the σ^+ beam. The experimental results in Fig. 32 correspond to a double passage from the $M=2$ sublevel to the $M=-2$ sublevel and then back to the $M=2$ sublevel (by using a second interaction zone with a reversed ordering of the σ^+ and σ^- beams), resulting in the transfer of eight photon momenta.

The attraction of using STIRAP for atomic interferometry is based on several features. First, it provides nearly 100% population transfer to a single final state, corresponding to transferring a fixed momentum $2\hbar k$. Thus there is no splitting of the incident wave packet into a superposition corresponding to many momentum peaks $\pm\hbar k$, $\pm 2\hbar k$, $\pm 3\hbar k$, etc., as occurs for deflection of a beam of two-state atoms by a standing light wave. Second, since the process is adiabatic, it is insensitive to changes in the laser properties (intensity and frequency) and the interaction time (i.e., atomic velocity for crossed-beam experiments, or pulse width for pulsed experiments); this is in contrast to transfer by π pulses, which is very sensitive to pulse area and resonance tuning (and, hence, to the laser parameters and the interaction time). Moreover, although the interaction is resonant, the deflection process is unaffected by spontaneous decay, because the excited state is never populated during the transfer. One can therefore avoid the common approach of eliminating spontaneous decay by detuning the laser far off resonance to reduce the excited-state population, at the expense of reducing the effective laser–atom coupling. Finally, using chainwise transitions between degenerate sublevels has the advantage that it requires only a single laser, because the σ^- wave can be derived by reflecting the σ^+ laser light. Having the two laser fields derived from the same laser is very convenient because the two-photon resonance condition, which is crucial for STIRAP, is automatically fulfilled (provided, of course, that there are no residual magnetic fields) and it is immune to laser frequency fluctuations.

STIRAP-based atomic mirrors are superior also to those based on the spontaneous emission force. In the latter, an atom absorbs a photon from a traveling-wave laser beam and then emits it spontaneously. Because the momenta from the absorbed photons are in the direction of the laser beam, while the momenta of the spontaneously emitted photons are distributed randomly (thus averaging to zero), the absorption-emission of N photons will result in a net momentum of $N\hbar k$ in the laser-beam direction. However, the dissipative nature of the emission process leads to a wide distribution of deflection angles. Moreover, it does not preserve any coherence, which is essential in an atom interferometer. STIRAP-based atom mirrors and beam splitters cure these drawbacks.

Coherent momentum transfer by adiabatic passage in similar chains of Zeeman sublevels has been demonstrated in a number of other experiments. Pillet *et al.* (Pillet *et al.*, 1993; Valentin *et al.*, 1994) and Goldner *et al.* (1994a,b) have reported momentum transfer of $8\hbar k$, with about 50% efficiency, resulting from the single-pass adiabatic passage between the $M_F = -4$ and $M_F = 4$ Zeeman sublevels in the hyperfine transition $F_g = 4 \leftrightarrow F_e = 4$ of the cesium D_2 line. Lawall and Prentiss (1994) have demonstrated momentum transfer of $4\hbar k$ with 90% efficiency in the $2^3 S_1 \leftrightarrow 2^3 P_0$ transition of He* with circularly polarized lasers, after double adiabatic passage ($M = -1 \to M = 1 \to M = -1$) between the groundstate sublevels. They demonstrated momentum transfer of $6\hbar k$ with 60% efficiency after a triple pass with linearly polarized lasers.

2. Atomic Beam Splitters

STIRAP can be used not only to transfer population and momentum completely from one state to another in a robust fashion, free of loss or incoherence, but also to create coherent superpositions of atomic states. In momentum space, this corresponds to splitting of the initial momentum distribution into two or more momenta distributions. The most obvious approach to achieve this objective is, starting with population in state ψ_1, to interrupt abruptly the time evolution of the dark state (44b) at a certain intermediate time, when it is the desired superposition of the initial state ψ_1 and the final state ψ_3 (Marte *et al.*, 1991; Weitz *et al.*, 1994a,b), before it has evolved into state ψ_3 as in STIRAP. Then only a fraction of the total population is transferred to ψ_3 and the composition of the superposition depends on the ratio between the pump and Stokes Rabi frequencies at the turn-off time. This fractional STIRAP scheme has been demonstrated experimentally (Weitz *et al.*, 1994a).

A smooth-pulse realization of this scheme that avoids sudden interruption of the pulses has been proposed for chains between degenerate sublevels (Vitanov et al., 1999). In this scheme, starting from the $M = -1$ sublevel, a coherent superposition between states $M = -1$ and $M = 1$ is created by applying first a σ^- polarized pulse (Stokes), followed by a slightly delayed pulse which is elliptically polarized in the same plane as the σ^- pulse. Since elliptical polarization can be represented as a sum of σ^- and σ^+ polarizations, the elliptically polarized pulse couples both the pump and Stokes transitions. As a result, the pulses create a superposition of the $M = -1$ and $M = 1$ sublevels whose composition is controlled by the ellipticity of the latter pulse.

Another possible realization of an atomic beam splitter (Lawall and Prentiss, 1994), suitable for $J_g = 1 \leftrightarrow J_e = 0$ transitions, starts with a coherent superposition of the $M = -1$ and $M = 0$ sublevels. A pair of counterpropagating σ^- and σ^+ pulses transfers the population from $M = -1$ to the $M = 1$ state, which results in $2\hbar k$ momentum deflection with respect to the momentum of the $M = 0$ sublevel, which is unaffected by the laser fields.

3. Atomic Interferometers

Weitz et al. (1994a, 1994b) have built the first atomic interferometer based on STIRAP, by using the transition between the two cesium hyperfine ground states ($6S_{1/2}$, $F = 3$, $M_F = 0$) and ($6S_{1/2}$, $F = 4$, $M_F = 0$) via the excited state ($6P_{1/2}$, $F = 3$ or 4, $M_F = 1$). This atom interferometer had the Bordé four-$\pi/2$ geometry (Bordé, 1989; Riehle et al., 1991) and involved four successive atomic beam splitters. Each of the beam splitters used two σ^+-polarized counterpropagating laser pulses and was based on interrupted STIRAP. Two 40-MHz accoustooptic modulators generated the pulse shapes for adiabatic following. A coherent superposition of two states of different momenta was created by turning the intensities of both pulses to zero in the middle of the transfer. The experiment achieved multiple-pass coherent transfer of more than 140 photon momenta with 95% efficiency per exchanged photon pair.

The transition used in the interferometer is insensitive to any magnetic field, which is essential for precision interferometry. Moreover, the σ^+–σ^+ configuration (the same helicity for the two pulses) increased the interferometer contrast because atoms that were not transfered adiabatically were optically pumped into the $F = 4$, $M_F = 4$ and $F = 3$, $M_F = 3$ states. Furthermore, using the cesium D_1 line (involving the $6P_{1/2}$ state) increased the transfer efficiency in comparison with using the cesium D_2 line (involving the $6P_{3/2}$ state) (Goldner et al., 1994a,b), because the excited-state hyperfine splitting of the $6P_{1/2}$ state is 5.8 times larger and thus off-resonant excitation is significantly lower.

In the experiment (Weitz et al., 1994a,b), a cesium atomic beam, slowed by a chirped laser beam, loaded a magnetooptic trap. Then the trapping magnetic field was shut off and the atoms were further cooled to 4 μK in polarization-gradient

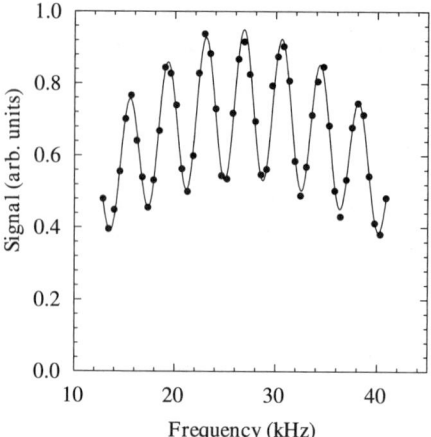

FIG. 33. Interference fringes for an atomic interferometer based on adiabatic passage. The dots are experimental data and the curve is a fit by a cosine function with a Gaussian envelope. (Reprinted with permission from M. Weitz, B. C. Young, and S. Chu. Atomic interferometer based on adiabatic population transfer. *Phys. Rev. Lett.* 1994;73:2563–2566.)

optical molasses. The atoms were then launched in a vertical ballistic trajectory at 2.3 m/s in a moving molasses. The molasses then was shut off. On their way up, the atoms were optically pumped into the $F = 4$, $M = 0$ sublevel and entered a magnetically shielded region with a homogeneous 100-mG magnetic bias field oriented parallel to the Raman beams. While the atoms were in the shielded region, a series of adiabatic pulses of the Raman beams were applied, which constituted the interferometer. The pulse sequence was designed to leave the coherently transferred atoms in the $F = 3$, $M = 0$ sublevel. As the atoms dropped back, a laser beam first removed the residual $F = 4$ population and then the $F = 3$, $M = 0$ population was transferred to $F = 4$, $M = 0$ by a microwave π pulse and measured by recording the fluorescence induced by a probe laser. Figure 33 shows the observed interference fringes.

Burnett and co-workers (Featonby *et al.*, 1996, 1998; Morigi *et al.*, 1996, Godun *et al.*, 1999; Webb *et al.*, 1999) have also demonstrated coherent momentum transfer in trapped and laser-cooled cesium atoms, both with laser beams having circular/circular ($\sigma^+\sigma^-$) and with others having circular/linear ($\sigma^+\pi$) polarizations. They have built a separated-path Ramsey atom interferometer (Featonby *et al.*, 1998) in which the closed loop constituting the Mach–Zehnder-type interferometer was produced by manipulating the atomic internal and external states separately, which provided greater flexibility. This interferometer used a sequence of ground-state microwave interactions and optical adiabatic transfer pulses. The microwaves were used to create a superposition of the ground hyperfine levels, and the adiabatic transfer then selectively manipulated the momentum of the $F = 4$ component of this superposition. The scheme started with the application of a $\pi/2$ microwave pulse to create, starting from the $F = 3$, $M = 0$ sublevel, an equal superposition of the $F = 3$, $M = 0$ and $F = 4$, $M = 0$ sublevels. At the end, a second $\pi/2$ pulse was applied in order to induce Ramsey fringes (which were observed by scanning the phase of the latter $\pi/2$ pulse). Between the microwave pulses, the momentum of the $F = 4$ component was manipulated by multistate STIRAP, using orthogonally propagating light pulses of linear (π) and right circular (σ^+) polarization. The laser fields were resonant with the D_1 $6S_{1/2}$, $F = 4 \leftrightarrow 6P_{1/2}$, $F' = 4$ transition in cesium, and thus only the atoms in the $F = 4$, $M = 0$ sublevel were subjected to adiabatic transfer. Since the $M = 0 \leftrightarrow M' = 0$ transition is forbidden, a sequence of partly overlapping π and σ^+ laser pulses transferred the population from $F = 4$, $M = 0$ to $F = 4$, $M = 4$ in an eight-photon transition. Because of the orthogonal $\sigma^+\pi$ geometry, the net momentum transferred to the atom was $4\sqrt{2}\hbar k$, rather than $8\hbar k$ as with $\sigma^+\sigma^-$ geometry. Once the $M = 4$ was reached, the STIRAP was reversed in time and the population returned to $M = 0$. Thus the first STIRAP was used to split the paths and the second reestablished the spatial overlap between the components of the superposition, $F = 3$, $M = 0$ and $F = 4$, $M = 0$. Although the net momentum transfer of such a process is zero, a small displacement was produced because of the finite momentum of the atom during the interaction. This atom interferometer was used to develop a method for measuring the temperature of an atomic ensemble (Featonby *et al.*, 1998) and a method for measuring the Berry phase (Webb *et al.*, 1999).

Incidentally, an interesting multiple-beam atomic interferometer not using adiabatic transfer has been demonstrated: a beam of cesium atoms has been split into five spatially distinct beams (separated by two-photon momenta), corresponding to the magnetic sublevels $M = -4, -2, 0, 2, 4$, and then recombined (Weitz *et al.*, 1996).

B. COHERENT MANIPULATION OF LASER-COOLED AND TRAPPED ATOMS

The STIRAP technique has been successfully applied in laser cooling experiments to coherently manipulate the atomic wave packets resulting from subrecoil laser cooling by velocity-selective coherent population trapping (VSCPT) (Aspect *et al.*, 1988, 1989; Kasevich and Chu, 1991; 1992; Kasevich *et al.*, 1991; Lawall *et al.*, 1994, 1995, 1996; Chu, 1998; Cohen-Tannoudji, 1998; Phillips, 1998). The momentum distribution of atoms cooled by VSCPT has two peaks, at $+\hbar k$ and $-\hbar k$, both with widths smaller than the photon recoil momentum $\hbar k$. Esslinger *et al.* (1996) have used adiabatic passage to coherently transfer rubidium atoms cooled by VSCPT into a single momentum state, still with a subrecoil momentum spread. Kulin *et al.* (1997) have demonstrated adiabatic transfer of metastable helium atoms into a single wave packet or into two coherent wave packets, while retaining the subrecoil momentum dispersion of the initial wave packets. They have achieved nearly 100% transfer efficiency in one and two dimensions, and 75% in three dimensions, while being able to choose at will the momentum direction and the internal state of the atoms. The three-dimensional manipulation is particularly important because it can be used to produce an ultraslow, spin-polarized atomic beam with subrecoil momentum spread in all directions.

The wave packet manipulation uses the fact that the atomic state after VSCPT— the dark state—has the same structure as the laser field. Hence a slow, adiabatic change in the laser field induces a corresponding change of the trapping state. If the evolution is adiabatic, the atoms remain decoupled from the laser fields during the transfer process and hence are immune to spontaneous emission. The final state may be chosen at will, and in particular, if at the end of such a change a single laser beam remains, the ensuing atomic state will consist of a single wave packet. Hence, this manipulation can be seen as inverted fractional STIRAP (Section VI.A.2).

It is obvious that for such a wave packet manipulation, the coherence of both the two components of the initial momentum distribution and the adiabatic transfer are crucial. Hence, this operation can be also used as a proof of the coherence of the two momentum peaks at $+\hbar k$ and $-\hbar k$ (Esslinger et al., 1996).

C. MEASUREMENT OF WEAK MAGNETIC FIELDS WITH LARMOR VELOCITY FILTER

The potential of STIRAP for inducing atomic beam deflection by coherent momentum transfer has been used to create a technique (called a *Larmor velocity filter*) for measuring very small magnetic fields along the axis of the atomic beam (Theuer and Bergmann, 1998). The scheme, which was demonstrated with metastable neon

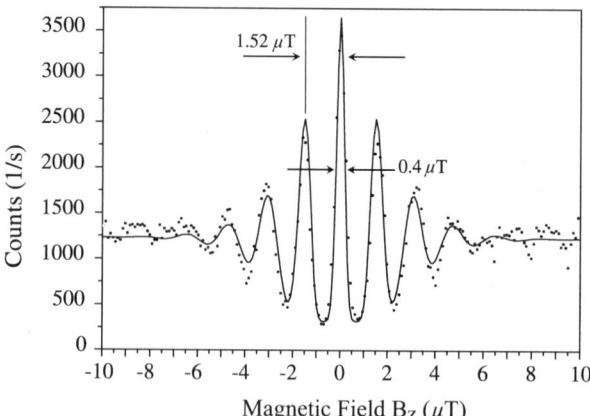

FIG. 34. Variation of the flux of deflected Ne* atoms in the Larmor velocity filter with the magnetic field strength. (From H. Theuer and K. Bergmann. Atomic beam deflection by coherent momentum transfer and the dependence on weak magnetic fields. *Eur. Phys. J. D* 1998;2:279–289.)

atoms, consisted of two STIRAP zones. In the first zone, atoms were prepared in the $M = 2$ sublevel of the $^3P_2$ metastable state and transferred to the $M = -2$ sublevel. They were transferred back to the initial $M = 2$ sublevel in the second zone provided they remained in the $M = -2$ state along the path between the two zones (cf. Fig. 31). The magnetic field in the region between the two transfer zones caused Larmor precession, thereby mixing the magnetic sublevels and affecting the momentum transfer. The resulting narrow-peaked pattern, an example of which is shown in Fig. 34, permitted measurement of weak magnetic fields.

VII. Branched-Chain Excitation

A. Branched Linkage Patterns

Just as the admission of a third energy state offers a wealth of new options for excitation in comparison to a two-state system, so too does the presence of a fourth (or fifth) state. The simplest cases are those in which the additional state merely links onto the end of the three-state chain, thereby forming a four-state chain, as discussed in Section V.

The four-state atom also offers interesting opportunities for competing and interfering paths. If the terminal level of the chain is the initial level, then the four states form a closed loop. The relative phases of the four Rabi frequencies then have special significance: they determine whether there is constructive or destructive interference and whether there are population nodes on some of the states.

Another possibility is that three of the states are linked, by as many as three separate pulsed fields, to a single state. The relative energies may be such that the linkage pattern appears as the letter Y (i.e., two of the levels are highly excited, perhaps dissociating or photoionizing) or as a tripod (i.e., a single state, connected to all the others, lies highest).

The variety of linkage patterns admits a variety of uses for pulsed coherent excitation. In either the Y or the tripod configuration, one might wish to consider a three-state main chain, to which the fourth state provides a branch. The effects of branches, weak or strong, on a main chain have been discussed not only for steady fields (Shore, 1990, chap. 21) but also for pulsed excitation (Kobrak and Rice, 1998a,b,c).

As with three-state excitation, mathematical analysis of four-state behavior is greatly facilitated by the use of dressed states (adiabatic states, if the fields are pulsed). These have been presented and utilized by several groups (Unanyan et al., 1998a, 1999; Kobrak and Rice, 1998a,b,c). Notably, these configurations have two null-eigenvalue states instead of the single one obtained with the three-state system. The occurrence of this degeneracy has some interesting consequences, as will be noted below.

In the remainder of this section we consider the tripod configuration explicitly.

B. THE TRIPOD LINKAGE

1. *Concept*

In the tripod version of STIRAP, proposed by Unanyan et al. (1998a, 1999) and demonstrated by Theuer et al. (1999), the usual three-state STIRAP system is supplied with an additional state ψ_4, coupled to the intermediate state ψ_2 by a third, control laser with Rabi frequency $\Omega_c(t)$. Such a *tripod STIRAP* scheme has two, rather than one, zero-energy dark states. Because they are degenerate, transitions between them take place even in the adiabatic limit. Time evolution will therefore eventually lead to the creation of a coherent superposition of states, rather to a population transfer to a single state. The composition of this superposition depends on, and therefore can be controlled by, the ordering of the pulses, by the time delay between the pump and Stokes pulses, and by the strength of the control pulse.

2. *The Tripod Linkage*

The RWA Hamiltonian describing an on-resonance tripod system has the form

$$\mathsf{H}(t) = \frac{\hbar}{2} \begin{bmatrix} 0 & \Omega_p(t) & 0 & 0 \\ \Omega_p(t) & 0 & \Omega_s(t) & \Omega_c(t) \\ 0 & \Omega_s(t) & 0 & 0 \\ 0 & \Omega_c(t) & 0 & 0 \end{bmatrix}, \tag{81}$$

where the real-valued functions of time $\Omega_p(t)$, $\Omega_s(t)$, and $\Omega_c(t)$ are the Rabi frequencies of the pump, Stokes, and control pulses, respectively. It is easy to verify that the Hamiltonian (81) has the following eigenvalues, two of which are degenerate,

$$\varepsilon_1(t) = \varepsilon_2(t) = 0, \qquad \varepsilon_3(t) = -\varepsilon_4(t) = \tfrac{1}{2}\hbar\Omega(t), \tag{82}$$

where $\Omega \equiv \sqrt{\Omega_p^2 + \Omega_s^2 + \Omega_c^2}$. The corresponding eigenvectors (the adiabatic states) are expressible in terms of two time-dependent angles $\vartheta(t)$ and $\varphi(t)$, defined as

$$\tan \vartheta(t) = \frac{\Omega_p(t)}{\sqrt{\Omega_s^2(t) + \Omega_c^2(t)}}, \qquad \tan \varphi(t) = \frac{\Omega_c(t)}{\Omega_s(t)}. \tag{83}$$

The angle $\vartheta(t)$ is the mixing angle used in standard STIRAP (where $\Omega_c \equiv 0$), and $\varphi(t)$ is an additional mixing angle related to the additional pulse. The adiabatic states corresponding to the two null-valued eigenenergies are (Unanyan et al., 1998a)

$$\Phi_1(t) = \psi_1 \cos\vartheta(t) - \psi_3 \sin\vartheta(t)\cos\varphi(t) - \psi_4 \sin\vartheta(t)\sin\varphi(t), \quad (84a)$$

$$\Phi_2(t) = \psi_3 \sin\varphi(t) - \psi_4 \cos\varphi(t), \quad (84b)$$

while the remaining adiabatic states are

$$\Phi_3(t) = \tfrac{1}{\sqrt{2}}[\psi_1 \sin\vartheta(t) + \psi_2 + \psi_3 \cos\vartheta(t)\cos\varphi(t)$$
$$+\psi_4 \cos\vartheta(t)\sin\varphi(t)], \quad (84c)$$

$$\Phi_4(t) = \tfrac{1}{\sqrt{2}}[\psi_1 \sin\vartheta(t) - \psi_2 + \psi_3 \cos\vartheta(t)\cos\varphi(t)$$
$$+\psi_4 \cos\vartheta(t)\sin\varphi(t)]. \quad (84d)$$

When the $\Omega_c(t)$ pulse is absent, one has the usual three-state atomic system and the adiabatic states turn into the adiabatic states (44) for STIRAP. However, the occurrence of two degenerate null-eigenvalue states here adds complications, and flexibility, that is not present with three-state STIRAP.

The systems of interest for the present discussion are those for which the atomic states ψ_1, ψ_3, and ψ_4 are stable states. Spontaneous emission occurs, if at all, only from state ψ_2. The two degenerate adiabatic states $\Phi_1(t)$ and $\Phi_2(t)$ receive no contribution from state ψ_2. They are therefore immune to loss of coherence and population that may occur due to spontaneous emission from ψ_2—these are dark or trapped states—and hence there is no difficulty in considering long pulses, as needed to ensure adiabatic evolution.

We note that because the zero-eigenvalue adiabatic states $\Phi_1(t)$ and $\Phi_2(t)$ are degenerate, any linear superposition of them will also be a zero-eigenvalue eigenstate of the Hamiltonian (81), and the choice of the two orthogonal zero-eigenvalue eigenstates of (81) is merely a matter of convenience. Furthermore, we point out that there is an obvious symmetry in the linkage of the three states ψ_1, ψ_3, and ψ_4. Our choice of the definitions of the angles $\vartheta(t)$ and $\varphi(t)$ and the dark states $\Phi_1(t)$ and $\Phi_2(t)$ is determined by the special role of state ψ_1 as the initial state.

3. Adiabatic Evolution

Nonadiabatic transitions between any pair of adiabatic states Φ_m and Φ_n are suppressed if the nonadiabatic coupling between these states is small compared to the difference between their energies (Messiah, 1962; Crisp, 1973; Shore, 1990); i.e.,

$$|\langle \dot\Phi_m(t)|\Phi_n(t)\rangle| \ll |\varepsilon_m - \varepsilon_n|.$$

If the adiabatic energies ε_m and ε_n are nondegenerate, this condition can always be satisfied for sufficiently large pulse areas, because the nonadiabatic couplings $|\langle\dot\Phi_m|\Phi_n\rangle|$ are proportional to $\dot\vartheta$ or $\dot\varphi$, which are in turn proportional to the inverse pulse width $1/T$, while the eigenenergy splittings $|\varepsilon_m - \varepsilon_n|$ are proportional to the maximum Rabi frequencies. Therefore, for large pulse areas the adiabatic states Φ_3 and Φ_4 are decoupled from the two dark states Φ_1 and Φ_2 and from each other. Then, unless states Φ_3 and Φ_4 are populated initially, the population dynamics is confined within the Hilbert subspace of the two nondecaying dark states, which allows us to realize coherent processes on time scales exceeding the lifetime of the excited state ψ_2.

Because the two dark states Φ_1 and Φ_2 are degenerate ($\varepsilon_1 = \varepsilon_2 = 0$), the adiabatic condition cannot be satisfied for the $\Phi_1 \leftrightarrow \Phi_2$ transition and hence transitions between Φ_1 and Φ_2 always occur, unless the nonadiabatic coupling between them vanishes identically (e.g., for constant fields). It is these transitions that lead to the controlled creation of a coherent superposition of states. The nonadiabatic coupling between the two trapped states is

$$\langle\dot\Phi_1(t)|\Phi_2(t)\rangle = \dot\varphi(t)\sin\vartheta(t).$$

If initially $\Omega_p = 0$ but one (or both) of the couplings Ω_s and Ω_c is nonzero, we have $\vartheta = 0$, meaning that $\psi_1 = \Phi_1$. Thus the system starts initially in state Φ_1, and if the laser pulses have large enough pulse areas to prevent transitions to the other adiabatic states Φ_3 and Φ_4, the system will end in a superposition of states Φ_1 and Φ_2,

$$\Phi_1(-\infty) \xleftarrow{-\infty} \Psi(t) \xrightarrow{+\infty} \Phi_1(+\infty)\cos\alpha - \Phi_2(+\infty)\sin\alpha. \tag{85}$$

Here the mixing angle α is the "area" of the nonadiabatic coupling,

$$\alpha = \int_{-\infty}^{\infty} \dot\varphi(t)\sin\vartheta(t)\,dt. \tag{86}$$

Because $\vartheta(t)$ and $\varphi(t)$ depend on the relative strengths of the pulses and their relative delays [see Eqs. (83)], the mixing angle α of the created superposition (85) of the dark states Φ_1 and Φ_2 also depends on, and therefore can be controlled by, these laser parameters.

According to Eqs. (84a), (84b), and (83), the asymptotic correspondence between the adiabatic states and the bare states at $\pm\infty$, and hence the bare-state composition of the created superposition (85), can be controlled by suitably choosing the pulse ordering. We provide four particular examples below.

- If the pulses are ordered so that the Stokes pulse starts before and ends after the pump pulse, while the control pulse is delayed with respect to both of

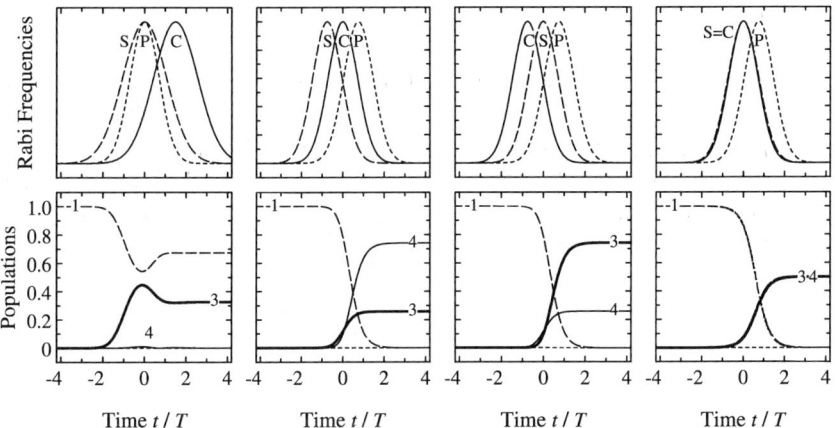

FIG. 35. Various possible orderings for the pump (P), Stokes (S), and control (C) laser pulses in tripod-STIRAP (top), and the corresponding population evolutions (bottom), where the numbers on the curves label the respective states.

them (Fig. 35, col. 1) (Unanyan et al., 1998a), the following asymptotic relations apply: $\vartheta(-\infty) = \vartheta(+\infty) = 0, \varphi(-\infty) = 0, \varphi(+\infty) = \pi/2$. Then

$$\psi_1 \overset{-\infty}{\longleftarrow} \Phi_1(t) \overset{+\infty}{\longrightarrow} \psi_1,$$
$$-\psi_4 \overset{-\infty}{\longleftarrow} \Phi_2(t) \overset{+\infty}{\longrightarrow} \psi_3.$$

Hence, as follows from Eq. (85), the system, which by assumption starts in the bare state ψ_1, will end in a superposition of states ψ_1 and ψ_3,

$$\Psi(+\infty) = \psi_1 \cos\alpha - \psi_3 \sin\alpha. \tag{87}$$

- Alternatively, one can arrange the three pulses to arrive in the ordering Stokes–control–pump (Fig. 35, col. 2) (Theuer et al., 1999). In this case, the following asymptotic relations apply: $\vartheta(-\infty) = 0, \vartheta(+\infty) = \pi/2$, $\varphi(-\infty) = 0, \varphi(+\infty) = \pi/2$. Therefore

$$\psi_1 \overset{-\infty}{\longleftarrow} \Phi_1(t) \overset{+\infty}{\longrightarrow} -\psi_4,$$
$$-\psi_4 \overset{-\infty}{\longleftarrow} \Phi_2(t) \overset{+\infty}{\longrightarrow} \psi_3.$$

Hence the system, starting from the bare state ψ_1, will end in a superposition of states ψ_3 and ψ_4,

$$\Psi(+\infty) = -\psi_3 \sin\alpha - \psi_4 \cos\alpha. \tag{88}$$

- A third possibility is to arrange the three pulses in the ordering control–Stokes–pump (Fig. 35, col. 3) (Theuer et al., 1999). Then the following asymptotic relations apply: $\vartheta(-\infty) = 0$, $\vartheta(+\infty) = \pi/2$, $\varphi(-\infty) = \pi/2$, $\varphi(+\infty) = 0$. Therefore

$$\psi_1 \xleftarrow{-\infty} \Phi_1(t) \xrightarrow{+\infty} -\psi_3,$$

$$\psi_3 \xleftarrow{-\infty} \Phi_2(t) \xrightarrow{+\infty} -\psi_4.$$

Hence the system, starting from the bare state ψ_1, will end in a superposition of states ψ_3 and ψ_4, but with reversed populations compared to the previous case (88),

$$\Psi(+\infty) = -\psi_3 \cos\alpha + \psi_4 \sin\alpha. \tag{89}$$

- Finally, if the pulses are ordered so that the Stokes and control pulses coincide in time and precede the pump pulse (Fig. 35, col. 4) (Unanyan et al., 1998a), the following asymptotic relations apply: $\vartheta(-\infty) = 0$, $\vartheta(+\infty) = \pi/2$, $\varphi(-\infty) = \varphi(+\infty) = \pi/4$. Then

$$\psi_1 \xleftarrow{-\infty} \Phi_1(t) \xrightarrow{+\infty} -\tfrac{1}{\sqrt{2}}(\psi_3 + \psi_4),$$

$$\tfrac{1}{\sqrt{2}}(\psi_3 - \psi_4) \xleftarrow{-\infty} \Phi_2(t) \xrightarrow{+\infty} \tfrac{1}{\sqrt{2}}(\psi_3 - \psi_4).$$

Because in this case $\varphi(t)$ is constant (equal to $\pi/4$), we have $\dot{\varphi}(t) = 0$ and thus the diabatic mixing angle is zero, $\alpha = 0$. Hence the system, starting in state ψ_1, will remain at all times in the dark state $\Phi_1(t)$ and will end in an equal superposition of states ψ_3 and ψ_4,

$$\Psi(+\infty) = -\tfrac{1}{\sqrt{2}}(\psi_3 + \psi_4). \tag{90}$$

It is important to note that the analysis above will also apply when there is one-photon detuning Δ from state ψ_2, as long as any pair of states ψ_1, ψ_3, and ψ_4 are on two-photon resonance. For nonzero Δ, the two dark states Φ_1 and Φ_2 remain eigenstates of the Hamiltonian, but the other eigenstates Φ_3 and Φ_4 are different. This change has no effect on the results, which rely on adiabatic evolution to maintain the state vector as a combination of the dark states Φ_1 and Φ_2.

The relative phase in each of the created superpositions can be altered by changing the relative phases of the laser fields. On the other hand, once a superposition is created, its parameters can be measured independently by the Newton–Young method using Stern–Gerlach analyzers (Newton and Young, 1968) or by coupling the superposition states to a third, excited state and measuring the subsequent fluorescence for different ratios of laser intensities and different relative laser phases (Vitanov et al., 2000; Vitanov, 2000).

In conclusion, branched-chain excitation, such as the tripod scheme described in this section, provides more freedom in manipulating the internal quantum state of atoms and molecules. In more complex systems, one can use the decomposition method of Morris and Shore (1983) to identify possible trapped (dark) states and devise excitation schemes that confine the dynamics within the trapped subspace.

C. EXPERIMENTAL DEMONSTRATION

The first experimental implementation of the tripod scheme was achieved in a beam of metastable neon atoms crossing three suitably arranged laser beams at right angles. The level scheme for this experiment is shown in Fig. 36a. The initially populated state $2p^5 3s\,^3P_0 (M=0)$ (state ψ_1) is coupled by a π-polarized pump laser field (Rabi frequency Ω_p) to an intermediate state $2p^5 3p\,^3P_1 (M=0)$ (state ψ_2), which in turn is coupled via σ^+ (Ω_s) and σ^- (Ω_c) laser fields to two final magnetic sublevels of level $2p^5 3s\,^3P_2$, $M=-1$ (state ψ_3) and $M=+1$ (state ψ_4). The sequence of interaction with the three laser beams is controlled by the spatial displacement of their axes. The σ^+ and σ^- beams propagate in opposite directions, while the axis of the π polarized beam is at right angles to the others (see Fig. 36b).

In the experiment, the beam of metastable neon atoms emerged from a liquid-nitrogen-cooled cold cathode discharge. The mean longitudinal velocity was 600 m/s with FWHM of 200 m/s. The metastable states $2p^5 3s\,^3P_0$ and $2p^5 3s\,^3P_2$ were populated with an efficiency of the order of 10^{-4}. The on-axis beam intensity was increased by a factor of 27 by two-dimensional transverse polarization gradient

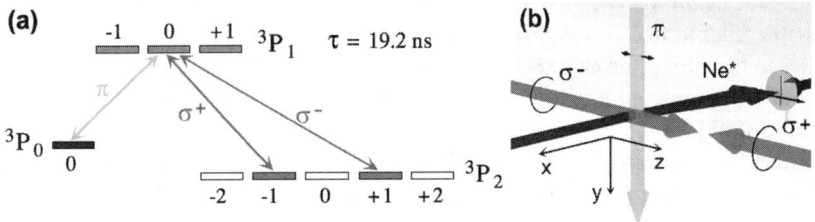

FIG. 36. Linkage pattern in tripod STIRAP (left) and laser beam geometry (right). (From H. Theuer, R. G. Unanyan, C. Habscheid, K. Klein, and K. Bergmann. Novel laser controlled variable matter wave beamsplitter. *Opt. Express* 1999;4:77–83.)

laser cooling. Next, the atoms in state $^3P_2$ were transferred to state $^3P_0$ by optical pumping. An excitation laser at 588 nm drove the transition to the excited level $^3P_1$, which has a lifetime of 18 ns. The atomic beam was highly collimated (1:47000) by two collimation slits, which is equivalent to a transverse velocity component of $\pm$ 1.3 cm/s (0.4 recoil velocities). The magnetic field was reduced to less than 1 μT in the relevant region using the Larmor velocity filter setup (Section VI.C). The transverse atomic beam profile was monitored farther downstream with a channeltron behind a 25-μm slit driven perpendicularly to the atomic beam axis by a stepper motor.

Three independent continuous single-mode dye lasers (Coherent 699) were used in this experiment. The cooling laser operated at 640 nm. The optical pumping laser and the Stokes beams were provided by the same dye laser (588 nm). The third laser generated the 616-nm radiation needed for the pump $^3P_0 \leftrightarrow {}^3P_1$ transition. The Stokes laser passed through a $\lambda/4$ waveplate, interacted with the atomic beam and was back-reflected by a cats-eye retroreflector with an integrated $\lambda/4$ retarder plate (Theuer et al., 1999). The translation of the cats-eye parallel to the atomic beam axis allowed precise adjustment of the spatial displacement of the two Stokes lasers.

The laser beams were arranged in such a way that the atoms encountered the pump laser last, while the timing of the Stokes and control laser beams was varied by displacing their axes. Hence, it was possible to create with this setup various superpositions of the magnetic sublevels $M = -1$ (state ψ_3) and $M = +1$ (state ψ_4) of state $2p^53s^3P_2$, as described by Eqs. (88)–(90). Furthermore, since the σ^+ and σ^- beams propagated in opposite directions, the momentum transfer to the $M = +1$ and $M = -1$ states had opposite signs, resulting in coherent beam splitting.

Figure 37 shows examples of atomic beam profiles recorded for different displacements of the Stokes and control laser beams. Two maxima separated by (122 ± 2) μm are observable. This separation corresponds to a difference in transverse momentum in the direction of Stokes propagation of $2\hbar k_s$. The momentum which was accumulated by an atom during the transfer process was $\hbar(\vec{k}_p \pm \vec{k}_s)$. Since the beam was collimated by slits and was detected behind a narrow slit, which was parallel to the π-polarized beam, only the component of the momentum parallel to the Stokes beam axis was observed. The data in Fig. 37 demonstrate that the splitting ratio could be smoothly controlled by the displacement of the Stokes and control laser beams. When the axes of the two beams coincided, a 50:50 beam splitting was observed, as predicted by Eq. (90). When the σ^+ (Stokes) beam precedes the σ^- (control) beam, the population was transferred predominantly to the $M = +1$ sublevel (state ψ_4), in agreement with Eq. (88). When the σ^- beam preceded the σ^+ beam, the population was transferred predominantly to the $M = -1$ sublevel (state ψ_3), in agreement with Eq. (89).

We note in conclusion that the robustness and relative simplicity of the tripod scheme can be beneficial for various fields in atomic and molecular physics utilizing

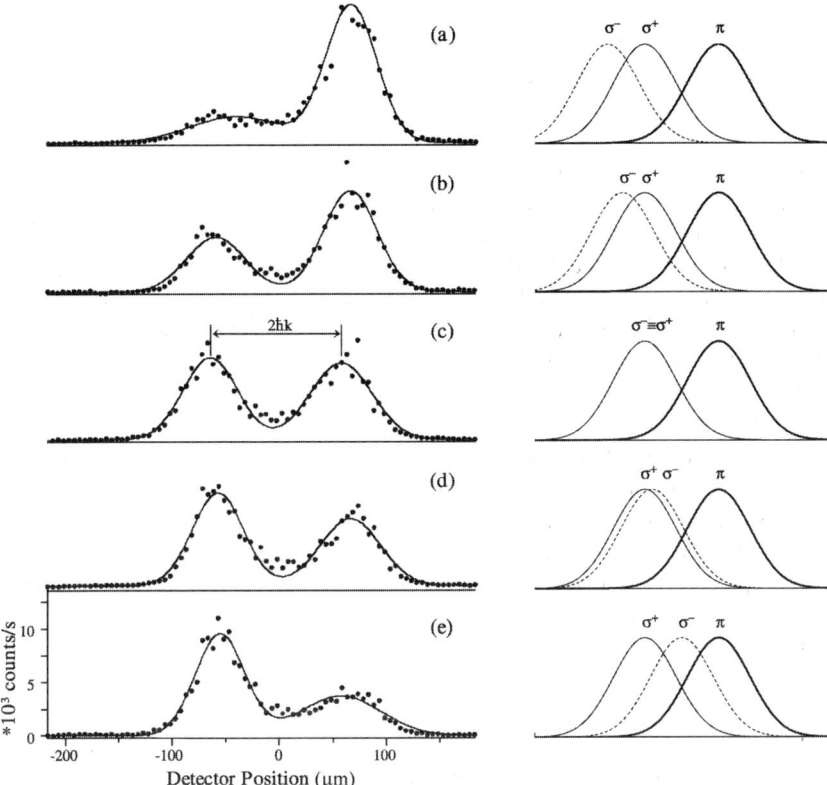

FIG. 37. Momentum distribution in the experiment of Theuer *et al.* (1999).

coherent superpositions of states, for instance, in quantum information [see, e.g., Williams and Clearwater (1997) and Steane (1998)].

VIII. Population Transfer via a Continuum of Intermediate States

Until the widespread use of lasers in atomic and molecular physics, the photoionization continuum was regarded as an incoherent terminus of probability flow, lacking any possibility for participating in a coherent process. Ionization was treated by means of rate equations, typically using the "Fermi golden rule" approximation (Messiah, 1962) to calculate ionization rates. Contemporary physics no longer views continuum as an irreversible drain of population. Laser interactions with a continuum have been found to exhibit coherent features, such as Rabi-like oscillations (Frishman and Shapiro, 1996), nearly complete population transfer into a continuum (i.e., photoionization or photodissociation) (Vardi and Shapiro, 1996) or from a continuum (i.e., photorecombination or photoassociation) (Vardi et al., 1997, 1999; Javanainen and Mackie, 1999; Mackie and Javanainen, 2000). Continuum coherence is essential in autoionization (Fano, 1961; Fano and Cooper, 1968), where it leads to a resonance in photoionization cross section plotted versus frequency; this is caused by destructive interference between two ionization channels, which can lead to complete suppression of photoionization at a specific wavelength. Laser-induced continuum structure (LICS) (Knight, 1984; Knight et al., 1990; Halfmann et al., 1998; Yatsenko et al., 1999a) is another coherence phenomenon, closely related to autoionization, where a strong laser field embeds a discrete state in an otherwise flat, structureless photoionization continuum; this laser-induced resonance can be detected by a second, probe laser field.

A few years ago it was suggested (Carroll and Hioe, 1992, 1993) that a continuum can serve as an intermediary for population transfer between two discrete states in an atom or a molecule by using a sequence of two delayed partially overlapping laser pulses, ordered in the STIRAP fashion: with the Stokes pulse, coupling the initially unpopulated state ψ_2 to the continuum, preceding the pump pulse, coupling the initial state ψ_1 to the continuum. This intriguing scheme—which has yet to be demonstrated experimentally—can be seen as a variant of STIRAP in which the discrete intermediate state is replaced by a continuum of states. The advantage of this scheme for population transfer would be its flexibility because a continuum offers a continuous range of possible combinations to match the pump and Stokes laser frequencies to the two-photon resonance between the initial state and the target state. The Carroll–Hioe analytic model (Carroll and Hioe, 1992, 1993), which involves an infinite quasicontinuum of equidistant discrete states, equally strongly coupled to the two bound states, suggests that complete population transfer is possible, the ionization being completely suppressed. The physical reason for this unexpected conclusion is closely related to LICS created in the continuum by the Stokes laser. Another reason supporting this scheme is that, as in a discrete three-state Λ system, there exists a trapped (dark) state—a coherent superposition of the two bound states—that is immune to ionization. Nakajima et al. (1994) later demonstrated, however, that the completeness of the population transfer in the Carroll–Hioe model derives from the very stringent restrictions of the model which cannot be met in a realistic physical system with a real continuum, in particular with a nonzero Fano parameter (Fano, 1961; Knight, 1990) and Stark shifts. It has subsequently been recognized that although complete population transfer is unrealistic, significant partial transfer may still be feasible (Carroll and Hioe, 1995, 1996; Yatsenko et al., 1997; Paspalakis et al., 1997, 1998; Vitanov and Stenholm, 1997c). It has also been suggested that STIRAP-like process can take place via an autoionizing state (Nakajima and Lambropoulos, 1996; Paspalakis and Knight, 1998).

COHERENT MANIPULATION OF ATOMS AND MOLECULES

Below, we describe the problems in transferring population via a continuum and point out a few possible solutions.

A. LASER-INDUCED CONTINUUM STRUCTURE

1. LICS Equations

The problem of two bound states coupled by two laser fields via a common continuum has been studied by a number of authors in the context of laser-induced continuum structure (LICS) (Knight et al., 1990, and references therein). The most general expression of the wave function of such a system must be written as a superposition of the two bound states and the continuum with time-dependent coefficients (Knight et al., 1990). By substituting this expansion in the Schrödinger equation and taking the Fourier transforms of the resulting differential equations for the probability amplitudes, one can eliminate the continuum adiabatically by substituting the adiabatic solution for its amplitudes in the other two equations. The summation over continuum states leads to a real part and an imaginary part representing a decay (via a pole) into the continuum (ionization or dissociation) and a coupling between the two bound states (via the principal value part). As a result of this elimination the wave function can be written as $\Psi(t) = C_1(t)\psi_1 + C_2(t)\psi_2$, where the probability amplitudes $\mathbf{C} = [C_1, C_2]^T$ satisfy the coupled equations

$$i\hbar \frac{d}{dt}\mathbf{C} = \mathbf{H}\mathbf{C} \tag{91}$$

where

$$\mathbf{H} = \hbar \begin{bmatrix} S_1 - \frac{1}{2}i\Gamma_1 & -\frac{1}{2}\sqrt{\Gamma_1^p \Gamma_2^s}(q+i) \\ -\frac{1}{2}\sqrt{\Gamma_1^p \Gamma_2^s}(q+i) & S_2 - \frac{1}{2}i\Gamma_2 + \delta \end{bmatrix}. \tag{92}$$

We shall assume that the system is initially in state ψ_1, $C_1(-\infty) = 1$, $C_2(-\infty) = 0$, and the quantities of interest are the populations of the discrete states $P_k(t) = |C_k(t)|^2 (k = 1, 2)$ and the ionization probability $P_i(t) = 1 - P_1(t) - P_2(t)$, particularly their values after the excitation ($t \to +\infty$).

The Hamiltonian matrix appearing here is that of two-photon excitation, but with a significant twist: the sum over intermediate states includes a continuum, and the consequent matrix elements therefore have both real and imaginary parts. On the diagonal, the elements appear as dynamic Stark shifts S_n and photoionization loss rates Γ_n. The off-diagonal elements are here parameterized by means of the Fano parameter q, the ratio of real to imaginary parts of the two-photon Rabi frequency (Fano, 1961; Knight et al., 1990; Yatsenko et al., 1997). As with auto ionization, the Fano parameter governs the (asymmetric) shape of the curve of ionization versus frequency [as embodied in the static two-photon detuning $\delta = (E_2 - E_1)/\hbar + \omega_s - \omega_p$]. It also plays an important role in the context of population transfer (Nakajima et al., 1994).

The Stark shifts and the ionization rates receive contributions from both of the fields,

$$\Gamma_n = \Gamma_n^p + \Gamma_n^s, \qquad S_n = S_n^p + S_n^s, \qquad (n = 1, 2).$$

The superscript labels p and s refer to the pump and Stokes lasers, respectively. The ionization widths and the Stark shifts are proportional to the laser fields intensities $I_p(t)$ and $I_s(t)$,

$$\Gamma_n^j(t) = G_n^j I_j(t), \qquad S_n^j(t) = S_n^j I_j(t),$$

with ($n = 1, 2$; $j = p, s$), where the parameters G_n^j and S_n^j depend on the particular atomic states and the laser frequencies.

With the exception of the Fano parameter, which is a dimensionless constant determined by the atomic structure, all variables involved in Eq. (91) can be controlled externally by the laser intensities and are generally time dependent.

2. Coherent and Incoherent Channels

The pump pulse applied on the ψ_1-continuum transition and the Stokes pulse applied on the ψ_2-continuum transition (the solid arrows in Fig. 38a) form a two-photon Raman transition which enables coherent population transfer between states ψ_1 and ψ_2. But the pulses also have other effects: the pump pulse applied on the ψ_2-continuum transition as well as the Stokes pulse applied on the ψ_1-continuum transition (the dashed arrows in Fig. 38a) cause irreversible ionization with rates $\Gamma_1^s(t)$ and $\Gamma_2^p(t)$. These two incoherent ionization channels turn out to be the main problem for population transfer. One of these channels may be eliminated by choosing a sufficiently small laser frequency, as for the Stokes laser in Fig. 38a, which cannot connect state ψ_1 to the continuum and hence $\Gamma_1^s = 0$. However, at least one of the incoherent channels is always present, as is $\Gamma_2^p(t)$ in Fig. 38a which prevents complete population transfer.

3. Fano Profile

The Fano profile emerges in a configuration similar to that in Fig. 38a, when the Stokes laser is strong (dressing laser) and its frequency is held fixed, while the pump laser is weak (probe laser) and its frequency is scanned across the two-photon resonance region. When plotted as a function of the probe-laser frequency ω,

COHERENT MANIPULATION OF ATOMS AND MOLECULES 147

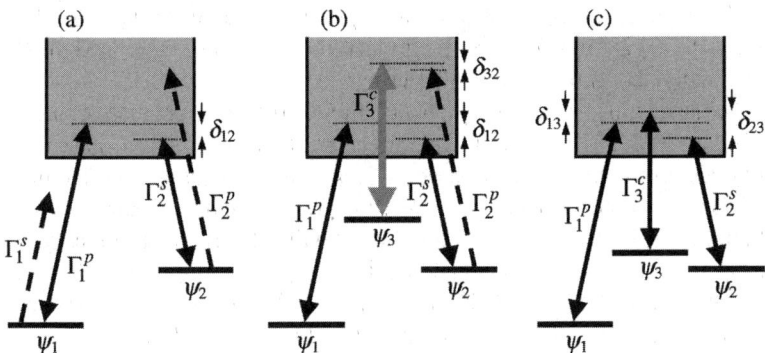

FIG. 38. Linkage patterns for bound states ψ_1 and ψ_2 coupled via a common continuum. (a) Two bound states coupled via a continuum. The solid arrow depict the ionization channels induced by the pump and Stokes lasers that form a Raman-type linkage. The dashed arrows show the irreversible (incoherent) ionization channels. (b) Same as (a) but with a third, compensatory laser (thick gray arrow), embedding a third auxiliary state ψ_3 and used to suppress incoherent ionization from state ψ_2 by a Fano-type resonance. (c) Three bound states coupled via a continuum (tripod-continuum scheme).

the photoionization probability $P_i(x)$ exhibits an asymmetric dependence—the famous Fano profile. It is described by (Fano, 1961; Fano and Cooper, 1968) as

$$P_i(x) = P_i^b + P_i^a \frac{(x+q)^2}{x^2+1} \quad (93)$$

where P_i^b is a background ionization, P_i^a is a scaling parameter for the resonance, and the dimensionless variable $x = (\omega - \omega_0)/\Gamma$ is the detuning from the resonance frequency ω_0 of the probe frequency ω in units of the ionization width Γ. As Eq. (93) shows, the Fano parameter q establishes the profile of the resonance. When $q = 0$, the photo ionization cross section is a symmetric, inverted-bell-shaped function of x and has a minimum at $x = 0$, i.e., at $\omega = \omega_0$, a frequency at which there is destructive interference between two photo ionization channels. For nonzero q, the Fano profile (93) is asymmetric, with a minimum $P_i^{\min} = P_i^b$ at $x = -q$, and maximum $P_i^{\max} = P_i^b + P_i^a(1+q^2)$ at $x = 1/q$. Far from resonance, we have $P_i^\infty = P_i^b + P_i^a$. The Fano parameter q can be determined from experimental data by measuring the values of the ionization cross section at its minimum, its maximum, and far from resonance and taking the ratio

$$\frac{P_i^{\max} - P_i^{\min}}{P_i^\infty - P_i^{\min}} = 1 + q^2.$$

Alternatively, one can measure the distance between the minimum and the maximum, which is $q + 1/q$.

At the trapping frequency $\omega_{trap} = \omega_0 - q\Gamma$, the ionization is suppressed. In the absence of background, $P_i^b = 0$, the ionization probability there is zero, $P_i = 0$, and the material will be completely transparent.

Equation (93) is derived using perturbation theory and assuming strong dressing (Stokes) laser and weak probe (pump) laser, and that both lasers have constant amplitudes. When the probe laser gets stronger, the distinction between dressing and probing becomes inaccurate because the probe field itself begins to affect the continuum. Also, when the two laser fields are pulse shaped, the Fano formula (93) becomes inaccurate. A detailed theoretical and experimental study of the behavior of LICS for pulsed lasers, both for coincident and delayed pulses, has been carried out by Halfmann et al. (1998), Yatsenko et al. (1999a) and Kylstra et al. (1998). Figure 39 shows a typical LICS profile, observed experimentally in metastable helium.

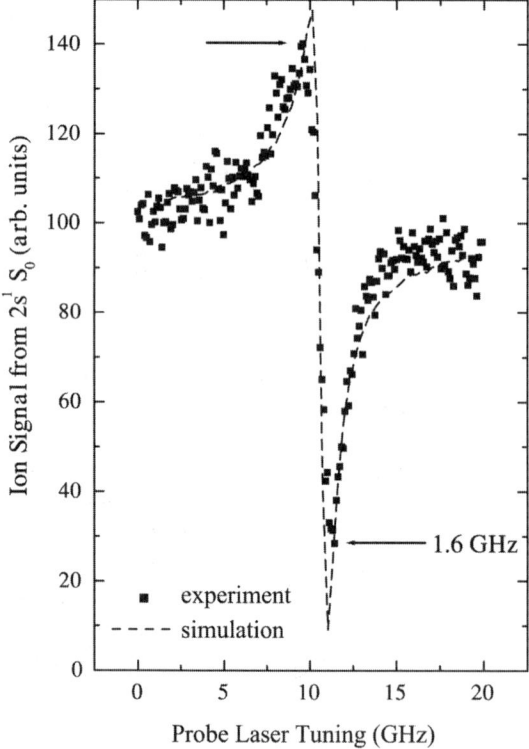

FIG. 39. Observation of laser-induced continuum structure in helium. The observed (almost 70%) reduction in the far-off-resonance ionization signal results from the coherence of the interaction. (From T. Halfmann, L. P. Yatsenko, M. Shapiro, B. W. Shore, and K. Bergmann. Population trapping and laser-induced continuum structure in helium: Experiment and theory. *Phys. Rev. A* 1998;58:R46–R49.)

4. Population Trapping: Dark and Bright States

Coherent excitation in a Raman configuration—with a continuum as well as with a discrete intermediate state—leads to the phenomenon of population trapping. Population trapping occurs when the imaginary part of one of the eigenvalues of the Hamiltonian vanishes; then the corresponding eigenstate is not coupled to the continuum—it is a nondecaying (trapped, dark) state. The trapping condition reads (Knight et al., 1990)

$$\delta_0 = \tfrac{1}{2} q \left[\Gamma_1^p(t) - \Gamma_2^s(t)\right] + S_1(t) - S_2(t). \tag{94}$$

Population trapping is most easily revealed in the basis of the dark and bright states, which are defined as

$$\Phi_d(t) = \psi_1 \cos \vartheta(t) - \psi_2 \sin \vartheta(t), \tag{95a}$$

$$\Phi_b(t) = \psi_1 \sin \vartheta(t) + \psi_2 \cos \vartheta(t), \tag{95b}$$

with

$$\tan \vartheta(t) = \sqrt{\frac{\Gamma_1^p(t)}{\Gamma_2^s(t)}}. \tag{96}$$

The reason for the names "bright" and "dark" is that, as follows from Eq. (91), the total ionization rate is

$$\frac{d}{dt} P_i = \Gamma_1^s P_1 + \Gamma_2^p P_2 + \sqrt{\Gamma_1^p + \Gamma_2^s} P_b. \tag{97}$$

Hence, in the absence of incoherent ionization ($\Gamma_1^s = \Gamma_2^p = 0$), the rate of change of the ionized population is proportional to the population in the bright state $\Phi_b(t)$. Therefore only the population in the bright state is exposed to ionization, whereas the ionization cannot occur from the dark state. If transitions between the dark and bright states are negligible, the population residing initially in the dark state remains trapped there. For constant laser field amplitudes (then the coupling between the dark and bright states vanishes), with the system starting in state ψ_1, the trapped population is

$$P_1 + P_2 = P_d = \frac{\Gamma_2^s}{\Gamma_1^p + \Gamma_2^s}.$$

Hence, for $\Gamma_1^p \ll \Gamma_2^s$, i.e., strong Stokes (dressing) laser and weak pump (probe) laser, almost all population is trapped, provided the trapping condition (94) is satisfied. For $\Gamma_1^p = \Gamma_2^s$, half of the population is trapped, and for $\Gamma_1^p \gg \Gamma_2^s$, the entire population can be ionized.

For pulsed lasers, there is some time-dependent nonzero coupling between the dark and bright states. However, if the evolution can be made adiabatic, this coupling can be reduced to negligible values. In this case, while the population of the dark state Φ_d remains nearly constant, its composition changes in time and the populations of the bare states ψ_1 and ψ_2 change accordingly; this ultimately enables population transfer between them.

When the trapping condition (94) is satisfied and when incoherent ionization is negligible, the dark and bright states are adiabatic states, i.e., eigenstates of the Hamiltonian in Eq. (91). When the trapping condition is not satisfied, the dark and bright states are no longer adiabatic states, i.e., adiabatic evolution does not guarantee staying in the dark state. Then the equations for the amplitudes of the dark and bright states $\mathbf{B} = [B_d, B_b]^T$ read

$$\hbar \frac{d}{dt}\mathbf{B} = -i\mathsf{H}_b \mathbf{B}$$

where

$$\mathsf{H}_b = \hbar \begin{bmatrix} D \sin^2 \vartheta & -\tfrac{1}{2} D \sin 2\vartheta - i\dot{\vartheta} \\ -\tfrac{1}{2} D \sin 2\vartheta + i\dot{\vartheta} & D \cos^2 \vartheta - \tfrac{1}{2}(q+i)\left(\Gamma_1^p + \Gamma_2^s\right) \end{bmatrix},$$

and

$$D(t) = \delta + S_2(t) - S_1(t) - \tfrac{1}{2} q \left[\Gamma_1^p(t) - \Gamma_2^s(t)\right]$$

is the deviation from the trapping condition. Here $\dot{\vartheta}$ is the usual nonadiabatic coupling, which can be overcome by making the evolution sufficiently adiabatic. However, the deviation D from the trapping condition introduces an additional coupling between the dark and bright states, which does not vanish in the adiabatic limit and which drives some population from the dark state into the bright state, where it is subjected to ionization.

B. Population Transfer via a Continuum

1. Population Transfer in the Ideal Case

Let us assume that the trapping condition (94) is satisfied exactly, i.e., $D=0$, and that there is no incoherent ionization, $\Gamma_1^s = \Gamma_2^p = 0$. Then, if the evolution is made adiabatic, the transitions between the dark and bright states will be suppressed completely.

If, as in STIRAP, the Stokes pulse precedes the pump, then the mixing angle $\vartheta(t)$ will change from $\vartheta = 0$ initially to $\vartheta = \pi/2$ at the end; hence the dark state

$\Phi_d(t)$ will coincide with the initial state ψ_1 before the interaction and with state $-\psi_2$ after it, i.e., $\psi_1 \xleftarrow{-\infty} \Phi_d(t) \xrightarrow{+\infty} -\psi_2$. Therefore complete population transfer is possible (in principle) if the evolution is adiabatic, if the trapping condition (94) is maintained, and if there is no incoherent ionization; then no population is lost from the dark state.

In contrast, if the pump pulse precedes the Stokes, then $\vartheta = \pi/2$ initially and $\vartheta = 0$ at the end; hence now the bright state Φ_b coincides with the initial state ψ_1 before the interaction and with state ψ_2 after it, $\psi_1 \xleftarrow{-\infty} \Phi_b(t) \xrightarrow{+\infty} \psi_2$. In this case, adiabatic evolution leads to maximal ionization, rather than to population transfer to state ψ_2, because no population resides in the dark state and thus all population is exposed to ionization.

In reality, neither the trapping condition (94) can be satisfied exactly for pulsed excitation, nor can incoherent ionization be eliminated completely. As a result, complete population transfer is ruled out. Various authors have proposed, however, schemes that reduce the negative effects of the deviation from the trapping condition and the incoherent ionization; we summarize them below.

2. Satisfying the Trapping Condition

If $q = 0$ and there are no Stark shifts, the trapping condition (94) is satisfied on two-photon resonance, $\delta = 0$. For $q \neq 0$ and for nonzero Stark shifts, which is the case in real atoms, the trapping condition (94) becomes time dependent in the general case of time-dependent ionization rates, as for pulse-shaped laser fields. For delayed pump and Stokes pulses, the trapping condition can be satisfied exactly only if the detuning δ is made time dependent and matches exactly the time dependence of the RHS of Eq. (94). This can be achieved, at least in principle, by using chirped laser pulses with a carefully tailored time-dependent frequency (chirp) (Paspalakis *et al.*, 1997; Vitanov and Stenholm, 1997c). Another possibility to enforce the trapping condition on the effective detuning is to make use of controlled Stark shifts induced by a pair of auxiliary far-off-resonant laser pulses with suitable intensities (whose frequencies are small enough not to influence the system otherwise) (Carroll and Hioe, 1995; Yatsenko *et al.*, 1997), which show up as additional time-dependent terms in the two-photon detuning δ.

If the trapping condition (94) is not satisfied exactly, the nonzero D-coupling causes transitions from the dark state to the bright state with subsequent ionization losses. These losses can be reduced by suitably tuning the laser frequencies slightly off two-photon resonance (i.e., for nonzero detuning δ) to a value that minimizes D. We also point out that it is easier to fulfill the trapping condition (94) approximately when the Fano parameter q is small; then the Γ terms have a less detrimental effect.

3. Suppressing Incoherent Ionization

The main difficulty in achieving efficient population transfer is related to the incoherent ionization channels (Yatsenko *et al.*, 1997; Vitanov and Stenholm, 1997c), of which at least one is always present; these lead to inevitable irreversible population losses. It has been suggested (Carroll and Hioe, 1996; Yatsenko *et al.*, 1997) that these losses can be reduced (although not eliminated) by choosing an appropriate region in the continuum where the incoherent-ionization probability is minimal.

Unanyan *et al.* (1998b) have proposed another approach to suppress incoherent ionization from state ψ_2. It makes use of a Fano-type resonance induced by an additional, strong compensatory laser which embeds a third, highly lying bound state ψ_3 into the region in the continuum where the incoherent ionization takes place. This laser, depicted by a thick gray arrow in Fig. 38b, forms with the incoherent channel Γ_2^p a nearly resonant Raman transition between states ψ_2 and ψ_3. It is assumed that the compensatory laser does not affect the system otherwise and, particularly, that its frequency is small enough so that it cannot ionize states ψ_1 and ψ_2 directly, i.e., $\Gamma_1^c \approx \Gamma_2^c \approx 0$. Also, if state ψ_3 is close enough to the ionization threshold, ionization from ψ_3 by the pump and Stokes lasers can be ignored because these lasers point deeply into the continuum where the probability for ionization is small ($\Gamma_3^p \approx \Gamma_3^s \approx 0$). The Hamiltonian describing the interaction between the bound states reads

$$\mathsf{H} = \frac{\hbar}{2} \begin{bmatrix} 2\Delta_{12} - i\Gamma_1^p & -\sqrt{\Gamma_1^p \Gamma_2^s}(q_{12}+i) & 0 \\ -\sqrt{\Gamma_1^p \Gamma_2^s}(q_{12}+i) & -i\Gamma_2^s - i\Gamma_2^p & -\sqrt{\Gamma_2^p \Gamma_3^c}(q_{23}+i) \\ 0 & -\sqrt{\Gamma_2^p \Gamma_3^c}(q_{23}+i) & 2\Delta_{32} - i\Gamma_3^c \end{bmatrix},$$

where $\Delta_{12}(t) = \delta_{12} + S_1(t) - S_2(t)$ and $\Delta_{32}(t) = \delta_{32} + S_3(t) - S_2(t)$, with δ_{12} and δ_{32} being the static detunings for the $\psi_1 \leftrightarrow \psi_2$ and $\psi_2 \leftrightarrow \psi_3$ transitions, respectively. Here q_{jk} is the Fano parameter characterizing the two-photon transition between states ψ_j and ψ_k, while $\Gamma_k(t)$ and $S_k(t)$ are the total ionization rate and the total Stark shift for state ψ_k.

The additional laser creates a new structure in the continuum in the region where the incoherent ionization from state ψ_2 takes place. The idea is to suppress the incoherent ionization channel Γ_2^p by tuning the parameters of the compensatory laser pulse near a Fano-type minimum in the ionization probability. For this to occur, the ionization width Γ_3^c should be sufficiently larger than Γ_2^p. Then $|2\Delta_{32} - i\Gamma_3^c| \gg |\sqrt{\Gamma_2^p \Gamma_3^c}(q_{23}+i)|$ and, hence, state ψ_3 can be eliminated adiabatically. The effective two-state LICS system has a new modified two-photon detuning δ

and incoherent-ionization width γ of state ψ_2,

$$\delta = \Delta_{12} + S_2 - S_1 - \tfrac{1}{2}q_{12}(\Gamma_1^p - \Gamma_2^s) - \mathrm{Re}\frac{\tfrac{1}{4}\Gamma_3^c\Gamma_2^p(q_{23}+i)^2}{\delta_{32} + S_3 - \tfrac{1}{2}i\Gamma_3^c}, \qquad (98a)$$

$$\gamma = \tfrac{1}{2}\Gamma_2^p + \mathrm{Im}\frac{\tfrac{1}{4}\Gamma_3^c\Gamma_2^p(q_{23}+i)^2}{\delta_{32} + S_3 - \tfrac{1}{2}i\Gamma_3^c}. \qquad (98b)$$

By appropriately choosing the parameters of the compensatory laser, one can make both γ and δ vanish; this cancellation takes place when

$$\delta_{21} = S_1 - S_2 + \tfrac{1}{2}q_{12}(\Gamma_1^p - \Gamma_2^s) - \tfrac{1}{2}q_{23}\Gamma_2^p, \qquad (99a)$$

$$\delta_{32} = -S_3 - \tfrac{1}{2}q_{23}\Gamma_3^c. \qquad (99b)$$

Condition (99b) ensures that the incoherent ionization is suppressed, whereas Eq. (99a) is the new trapping condition. Then, as discussed above, the efficiency of the population transfer from state ψ_1 to ψ_2 can approach unity for the counterintuitive pulse order in the adiabatic limit.

This idea for Fano-type suppression of incoherent ionization is illustrated in Fig. 40, where the populations of the two bound states ψ_1 and ψ_2 and the ionization

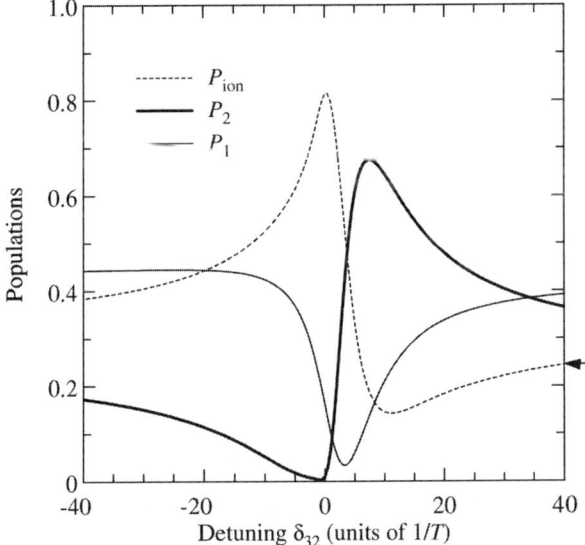

FIG. 40. Numerically calculated populations of the initial state ψ_1 and the final state ψ_2 and the ionization against the two-photon detuning δ_{32} between the target state ψ_2 and the auxiliary state ψ_3. The arrow on the RHS shows P_2 in the absence of compensatory laser. (From R. G. Unanyan, N. V. Vitanov, and S. Stenholm. Suppression of incoherent ionization in population transfer via continuum. Phys. Rev. A 1998;57:462–466.)

probability are plotted versus the detuning δ_{32}, i.e., versus the frequency of the compensatory laser. At a certain value of δ_{32} the population of the target state ψ_2 reaches 0.69, which is considerably higher than the value 0.26 in the absence of a compensatory laser. The P_2 curve, as well as the ionization curve, are reminiscent of the Fano profile of Fig. 39.

In Fig. 41, the target-state population P_2 is plotted versus the peak ionization rate Γ_0 of the three lasers in four cases. The transfer efficiency to state ψ_2 is lowest when there is no compensation of incoherent ionization and the trapping condition (94) is not fulfilled (curve 1). Satisfying the trapping condition alone, still without incoherent-ionization reduction, increases the transfer efficiency (curve 2). When the incoherent ionization is (partly) suppressed by a compensatory laser, without satisfying the trapping condition $\delta = 0$, the transfer efficiency increases further (curve 3). Finally, when the incoherent ionization is (partly) suppressed and the trapping condition $\delta = 0$ is satisfied, the transfer efficiency is highest and can exceed 0.8 in this example. If both the $\delta = 0$ and $\gamma = 0$ conditions were satisfied exactly, the transfer efficiency could approach unity.

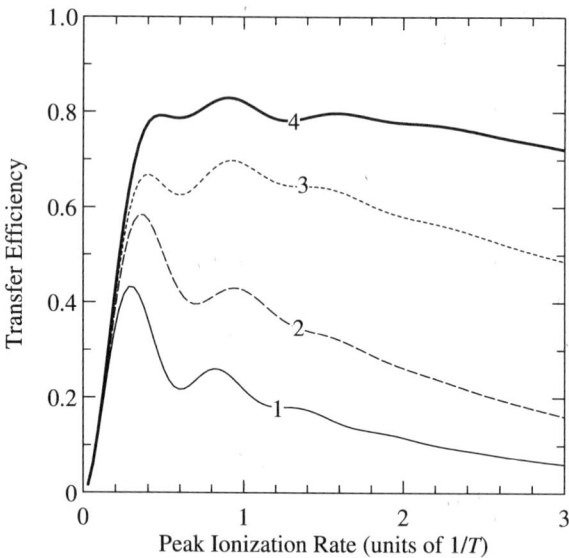

FIG. 41. Numerical calculated population of the target state ψ_2 against the peak ionization rate Γ_0 of the three lasers. The four curves correspond to the following cases: (1) no compensatory laser and $\delta_{12} = 0$; (2) the condition $\delta = 0$ is satisfied, but the condition $\gamma = 0$ is not; (3) the condition $\gamma = 0$ is satisfied approximately, but the condition $\delta = 0$ is not satisfied; (4) the condition $\delta = 0$ is satisfied exactly, and the condition $\gamma = 0$ is satisfied approximately. (From R. G. Unanyan, N. V. Vitanov, and S. Stenholm. Suppression of incoherent ionization in population transfer via continuum. *Phys. Rev. A* 1998;57:462–466.)

C. Tripod Coupling via a Continuum

Recently, the two-state LICS scheme has been extended to a tripod-continuum scheme involving three discrete states coupled to each other by two-photon processes via a common continuum (Unanyan et al., 2000a). All three lasers are tuned approximately in the same region of the continuum, as shown in Fig. 38c. This scheme may also be seen as an extension of the discrete-state tripod scheme discussed in Section VII. The Hamiltonian describing the tripod-continuum system is given by

$$\mathsf{H} = \frac{1}{2} \begin{bmatrix} 2\Delta_{13} - i\Gamma_1 & -\sqrt{\Gamma_1\Gamma_2}(q_{12} + i) & -\sqrt{\Gamma_1\Gamma_3}(q_{13} + i) \\ -\sqrt{\Gamma_1\Gamma_2}(q_{12} + i) & 2\Delta_{23} - i\Gamma_2 & -\sqrt{\Gamma_2\Gamma_3}(q_{23} + i) \\ -\sqrt{\Gamma_1\Gamma_3}(q_{13} + i) & -\sqrt{\Gamma_2\Gamma_3}(q_{23} + i) & -i\Gamma_3 \end{bmatrix},$$

where $\Delta_{13}(t) = \delta_{13} + S_1(t) - S_3(t)$ and $\Delta_{23}(t) = \delta_{23} + S_2(t) - S_3(t)$, with δ_{13} and δ_{23} being the static detunings for the $\psi_1 \leftrightarrow \psi_3$ and $\psi_2 \leftrightarrow \psi_3$ transitions, respectively, q_{jk} are the Fano parameters, and $\Gamma_k(t)$ and $S_k(t)$ are the total ionization rate and the total Stark shift for state ψ_k.

Unlike the two-state Raman coupling via a continuum, there are two trapping conditions for the tripod-continuum system (Unanyan 2000a),

$$\Delta_{13}(t) = \tfrac{1}{2}q_{13}[\Gamma_3(t) - \Gamma_1(t)] + \tfrac{1}{2}(q_{12} - q_{23})\Gamma_2(t), \tag{100a}$$

$$\Delta_{23}(t) = \tfrac{1}{2}q_{23}[\Gamma_3(t) - \Gamma_2(t)] + \tfrac{1}{2}(q_{12} - q_{13})\Gamma_2(t). \tag{100b}$$

If these conditions are satisfied, there are two dark states, as for the discrete-state tripod system (Section VII). Likewise, the presence of two dark states provides greater flexibility in performing coherent population transfer between the bound states, compared to the two-state scheme. Moreover, for large and constant Γ_2, the trapping conditions (100) can be satisfied approximately at the (constant) detunings $\Delta_{13} = \tfrac{1}{2}(q_{12} - q_{23})\Gamma_2$ and $\Delta_{23} = \tfrac{1}{2}(q_{12} - q_{13})\Gamma_2$; hence, it may be easier to satisfy the two trapping conditions (100) for the tripod-continuum system than the single-trapping condition (94) in the two-state-via-continuum scheme.

In the case when one of the discrete states is strongly coupled to the continuum, for example ψ_3, it can be eliminated adiabatically. Then the population dynamics reduces to an effective two-state LICS problem, involving the other two states ψ_1 and ψ_2, described by Eqs. (91), but with modified parameters. In particular, the effective Fano parameter is (Unanyan, 2000a)

$$q = \frac{q_{12} - q_{13} - q_{23}}{q_{13}q_{23}}. \tag{101}$$

Thus, using an auxiliary third laser embedding a third state into the continuum provides the possibility to customize the parameters of a given two-state LICS system.

IX. Extensions and Applications of STIRAP

A. CONTROL OF CHEMICAL REACTIONS

The remarkable properties of STIRAP have already had applications in many diverse areas. The first implementation of STIRAP has been in a crossed-beam reactive scattering experiment (Dittmann et al., 1992). It allowed investigation of the effect of vibrational excitation on the cross section for the chemiluminescent channel in the process $Na_2(v) + Cl \rightarrow NaCl + Na^*$ in crossed particle beams. It was found that the cross sections increased by about 0.75% per vibrational level in the range $3 \simeq v \simeq 19$.

Another example is the detailed study of the reaction $Na_2(v'', j'') + H \rightarrow NaH(v', j') + Na$, where the angular distribution and the population distribution have been determined for the product molecule NaH for a range of selectively populated levels v'' of the reagent molecule Na_2 (Pesl, 1999). STIRAP has been used also to investigate the dependence of the dissociative attachment process $Na_2(v'', j'') + e^- \rightarrow Na + Na^-$ (with electron energies <1 eV) on the vibrational excitation by exciting efficiently and very selectively the Na_2 molecules to a specific vibrationally excited level (Külz et al., 1993, 1995, 1996; Ekers et al., 1999a,b; Keil et al., 1999; Kaufmann et al., 2001). The vibrational excitation to $v'' = 12$ has increased the state-dependent dissociative attachment rate by more than three orders of magnitude.

B. HYPER-RAMAN STIRAP (STIHRAP)

The application of standard STIRAP to molecules, using two single-photon transitions (pump and Stokes), is often impeded by the fact that most molecules require ultraviolet or even vacuum ultraviolet (VUV) pump photons to reach the first electronically excited states. For the Stokes pulse, which connects the excited electronic state to a high vibrational level of the ground electronic state, optical wavelengths are usually sufficient. It is difficult to provide VUV pulses with adequate power and coherence properties. It is natural to consider achieving the pump excitation (and possibly also the Stokes excitation) by a two-photon transition. The corresponding $(2 + 1)$ and $(2 + 2)$ versions of STIRAP have been named hyper-Raman STIRAP (Yatsenko et al., 1998; Guérin and Jauslin, 1998; Guérin et al., 1998, 1999). Although these extensions seem obvious, they turn out to be nontrivial.

The main obstacle in hyper-Raman STIRAP are the dynamic Stark shifts induced by the two-photon coupling. These Stark shifts, which are proportional to the laser intensities, modify the Bohr frequencies of the pump and Stokes transitions and destroy the multiphoton resonance between the initial and final states, which is crucial for the existence of the dark state. It has been found, both numerically and analytically, that high transfer efficiency in such a scheme can still be achieved by a suitable choice of static detunings of the carrier frequencies of the two pulses, which suppress the detrimental effect of the Stark shifts; these detuning ranges have been estimated analytically (Guérin *et al.*, 1998). It is interesting to note that, unlike the purely adiabatic evolution in STIRAP, successful population transfer in hyper-Raman STIRAP occurs as a result of a combination of adiabatic and diabatic time evolution, as in SCRAP (see below). Moreover, unlike STIRAP, the intermediate state does acquire some transient population; again, it can be reduced by suitable static detunings.

It should be pointed out that the Stark shifts are also nonzero in traditional STIRAP, but they are usually negligible compared to the one-photon on-resonance couplings (given by the Autler–Townes splittings). In (2 + 1) STIRAP the fundamental field (ω_p) is very strong and the related Stark shift is usually not small compared to the two-photon coupling ($2\omega_p$).

C. STARK-CHIRPED RAPID ADIABATIC PASSAGE

An interesting alternative of STIHRAP for electronic excitation of molecules is the technique of Stark-chirped rapid adiabatic passage (SCRAP), introduced (Yatsenko *et al.*, 1999b) and demonstrated (Rickes *et al.*, 2000) recently for two-state systems. Like STIRAP, it makes use of two delayed and partially overlapping laser pulses, but unlike STIRAP, here one of the pulses is far off resonance and its objective is to induce dynamic Stark shifts in the coupled levels. This time-dependent Stark shift, combined with an appropriate detuning and a time delay of the pump laser, leads to an effective level crossing and adiabatic population transfer between the two states of the pump transition.

Using laser-induced Stark shifts to modify the transition frequency appears the easiest method to induce a level-crossing transition with laser pulses of nanosecond duration. Techniques for producing frequency-swept pulses are not well developed for nanosecond pulses. Nanosecond laser systems are used for many applications because they provide a very good combination of sufficiently high intensity (and hence large interaction strength) as well as long interaction time. The successful implementation of adiabatic evolution of laser–matter interaction relies on a combination of both parameters. Laser frequency chirping by active phase modulation, although possible in principle, is difficult for nanosecond laser pulses because it requires driving modulators at gigahertz frequencies. On the other hand, the spectral bandwidth of nanosecond pulses is too small for successful application of the techniques based on spatial dispersion that are well developed for femtosecond pulses.

1. Theory

Figure 42 illustrates the idea of SCRAP. One of the pulses—the pump pulse—is slightly detuned off resonance with the transition frequency and moderately strong; it serves to drive the population from the ground to the excited state. The other pulse—the Stark pulse—is far off resonant and strong; it is used merely to modify the atomic transition frequency by inducing Stark shifts in the energies of the two states. Because the Stark shifts $S_1(t)$ and $S_2(t)$ of the ground and excited states are generally different (usually $|S_2(t)| \gg |S_1(t)|$) and each of them is proportional to the intensity of the Stark pulse, the transition frequency will experience a net Stark shift $S(t) = S_2(t) - S_1(t)$.

By choosing an appropriate detuning for the pump pulse, it is always possible to create two diabatic level crossings in the wings of the Stark pulse: one crossing

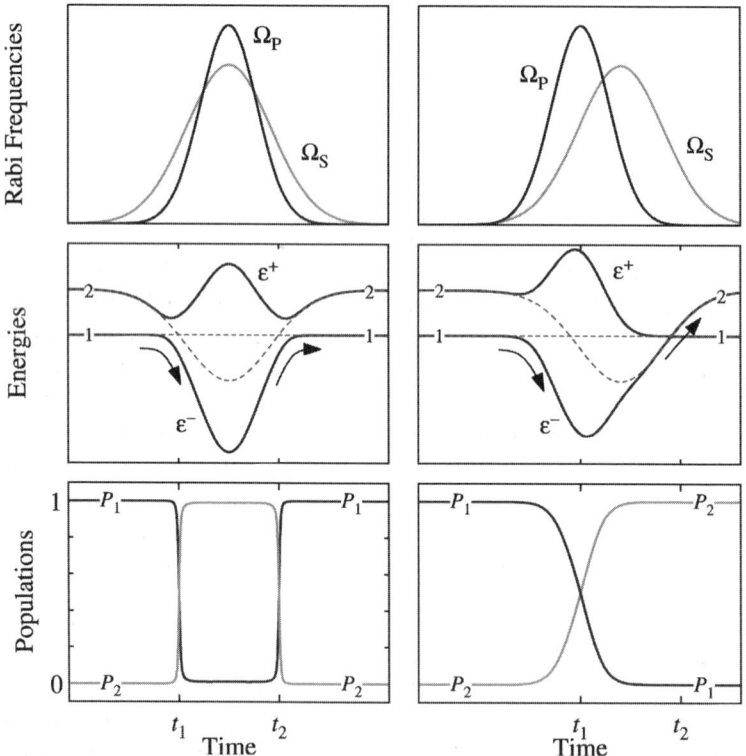

FIG. 42. Time evolution of the Rabi frequencies (top frames), the level energies (middle frames), and the populations (bottom frames) in a two-state system driven by a pump pulse Ω_p and a Stark-shifting pulse Ω_s. Left-hand frames: simultaneous pump and Stark pulses. Right-hand frames: pump pulse before Stark pulse (SCRAP method).

during the growth and another during the decline of the Stark pulse. For successful population transfer, the evolution must be adiabatic at one, and only one, of these crossings. This asymmetry can only occur if the pump and Stark pulses are not applied simultaneously. Rather, the pump pulse must be strong at one and only one of the crossings.

It proves appropriate to set the time delay between the two pulses so that the maximum of the pump pulse occurs at one of the crossings in order to optimize the adiabatic passage there. It is also appropriate that the pump pulse width be smaller than both the Stark pulse width and the delay between the pulses, in order to suppress adiabatic passage at the other crossing. In this adiabatic–diabatic scenario the system will follow the path shown in the middle right frame in Fig. 42: the state vector will adiabatically follow the lower adiabatic state through the first crossing, while during the second crossing it will follow the diabatic state ψ_2 (rather than an adiabatic state) and remain there till the end of the interaction. The net result is complete population transfer from state ψ_1 to state ψ_2. It should be appreciated that the adiabatic and diabatic intervals can occur in either ordering: the pump pulse may either precede or follow the Stark pulse.

The SCRAP technique resembles the early experiment by Loy (1974), who used adiabatic quasistatic pulses of about 5 ms duration to induce Stark shifts. However, he induced two sequential population transfers per pulse—excitation for the leading edge and deexcitation for the trailing edge of each pulse as in the left column of Fig. 42—resulting in no net population transfer. In contrast, the time delay between the pump and Stark pulses in SCRAP ensures that population transfer takes place at just one of the crossings, thus leading to overall population transfer.

It should be obvious from the above description that complete population transfer will only occur within finite ranges of values of the various interaction parameters. For example, in order that there be level crossings, the static detuning Δ_0 must be smaller than the maximum Stark shift S_0 and must have the same sign as S_0. Also, the pump pulse should be strong enough to ensure adiabatic passage at one of the crossings, but weak enough to prevent adiabatic passage at the other. For Gaussian pulse shapes, $\Omega(t) = \Omega_0 \exp(-t^2/T_p^2)$ and $S(t) = S_0 \exp[-(t-\tau)^2/T_s^2]$, the latter requirements lead to the conditions (Rickes et al., 2000)

$$1 \ll \frac{(\Omega_0 T_s)^2}{\Delta_0 \tau} \ll \exp\left(\frac{8\tau^2}{T_p^2}\right). \tag{102}$$

These conditions set upper and lower limits on the peak pump Rabi frequency Ω_0 and the static detuning Δ_0.

The SCRAP technique benefits from the fact that strong fixed-frequency long-wavelength pulsed laser radiation, suitable for Stark-shifting the levels, is often available because it is used to generate (by frequency conversion) the visible or ultraviolet radiation needed for the pump interaction. Moreover, its pulse width is longer than the pump pulse width, which is beneficial for SCRAP.

As with simple adiabatic passage, the SCRAP technique can produce population transfer in an ensemble of atoms having a distribution of Doppler shifts. The peak value of the Stark shift sets the maximum detuning that can be accessed; in turn, this sets the range of Doppler shifts for which population transfer can be produced.

2. Experimental Demonstration

The first experimental demonstration of SCRAP was achieved in metastable helium (Rickes *et al.*, 2000). The initial state $1s2s\,^3S_1$ was coupled to the target state $1s2s\,^3S_1$ by a two-photon transition induced by a 855-nm pump laser pulse with a pulse duration of 3 ns (half-width at 1/e of intensity), as shown in Fig. 43 (left). The Stark shift was induced by a 1064-nm laser pulse with a pulse duration of 4.6 ns, delayed by 7 ns with respect to the pump pulse. Both laser pulses were mildly focused into the atomic beam. Nearly complete population transfer was observed with typical intensities of 20–30 MW/cm^2 for the pump pulse and 200–500 MW/cm^2 for the Stark pulse.

As an example, Fig. 43 (right) displays the transfer efficiency plotted versus the static two-photon detuning $\Delta_0 = \omega_{12} - 2\omega_p$. Nearly complete population transfer was observed within a certain detuning range, as predicted by analytical estimates. For large positive detuning, the adiabatic condition at the first crossing is violated and the transfer efficiency decreases. For small positive detunings (near $\Delta_0 = 0$),

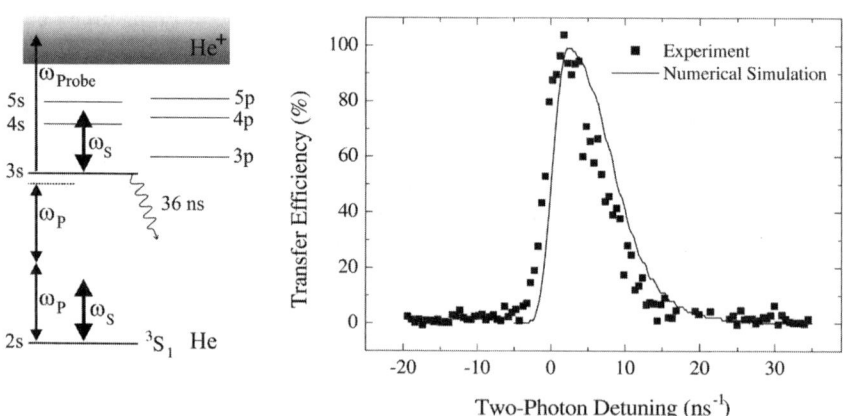

FIG. 43. Left-hand plot: simplified energy-level diagram of helium atom used in the demonstration of SCRAP. Right-hand plot: population transfer efficiency versus the static two-photon detuning Δ_0. (From T. Rickes, L. P. Yatsenko, S. Steuerwald, T. Halfmann, B. W. Shore, N. V. Vitanov, and K. Bergmann. Efficient adiabatic population transfer by two-photon excitation assisted by a laser-induced Stark shift. *J. Chem. Phys.* 2000;113:534–546.)

the diabatic condition at the second crossing is violated and the transfer efficiency decreases. For $\Delta_0 < 0$ and for very large positive Δ_0, there are no level crossings at all and little population is transferred to the excited state.

An interesting extension of SCRAP—potentially very important for molecules—is the application of two sequential SCRAP processes. For example, the first SCRAP can transfer the population from the electronic ground state to an electronically excited state via a two-photon excitation. The second SCRAP then transfers the population to a target state in the electronic ground state, e.g., a highly vibrationally excited state. The second step can take place via a one-photon process [(2 + 1) SCRAP] or by a two-photon process [(2 + 2) SCRAP]. It is easily seen that only one Stark-shifting laser is needed in the (2 + 1) SCRAP scheme. The (2 + 2) SCRAP can be realized even without a separate Stark pulse because the Stokes pulse can induce Stark shifts for the pump transition, and the pump pulse can induce the Stark shift for the Stokes transition (Rickes et al., 2000).

D. Adiabatic Passage by Light-Induced Potentials (APLIP)

Recently, Garraway and Suominen (1998) (see also Kalush and Band, 2000, and Sola et al., 2000) have suggested, on the basis of numerical calculations for sodium dimers, that the STIRAP ideas of counterintuitively ordered laser pulses and adiabatic evolution can be applied to the transfer of a wave packet from one molecular potential to the displaced ground vibrational state of another. This process—termed adiabatic passage by light-induced potentials (APLIP)—seemingly violates the Frank–Condon principle because the overlap between the initial and final wavefunctions is very small (the two wave packets were displaced at a distance seven times larger than their widths). There is, however, no such violation because the time scale of the process is close to, but longer than, the vibrational time scale. APLIP shares many features with STIRAP, such as high efficiency and insensitivity to pulse parameters. However, in contrast to STIRAP, the two-photon resonance condition in APLIP cannot be satisfied (except at a certain time), and the main mechanism for the transfer of the wave packet is through a "valley" which emerges in the time-dependence of the light-induced potential, as shown in Fig. 44 (left). Figure 44 (right) shows how the wave packet gradually disappears from the ground-state potential (lower plot) and appears in the excited-state one (upper plot). While the original proposal assumed transitions between the lowest vibrational states ($v = v' = v'' = 0$), recent calculations (Rodriguez et al., 2000) extended APLIP to excited vibrational states.

E. Photoassociative STIRAP as a Source for Cold Molecules

With the experimental realization of Bose–Einstein condensation of weakly interacting atoms, the physics of cold gases has become a topic of increasing interest. While the formation of condensates of atomic gases is by now an established

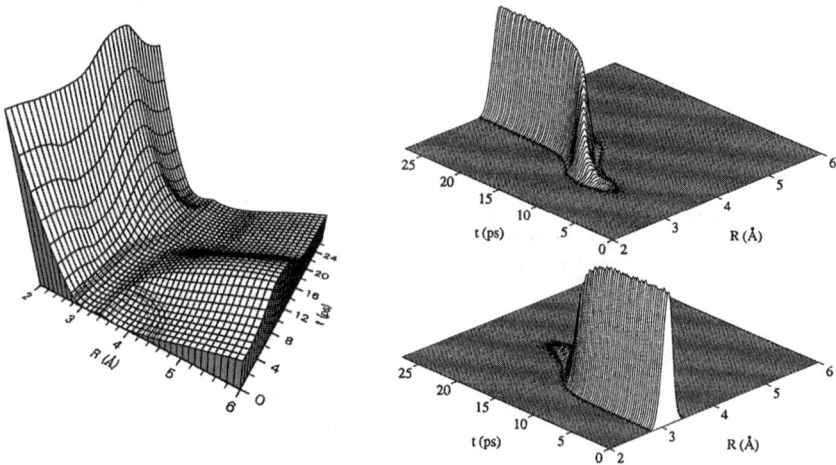

FIG. 44. Left: APLIP potential. The population flows through the "valley" in this potential. Right: time evolution of the ground-state (lower plot) and excited-state (upper plot) populations. (Reprinted with permission from B. M. Garraway and K.-A. Suominen. Adiabatic passage by light-induced potentials in molecules. *Phys. Rev. Lett.* 1998;80:932–935.)

technique in many laboratories, the generation of the molecular counterpart is still an open problem. Standard cooling techniques cannot be applied to molecules due to the lack of closed two-level transitions and the many rovibrational degrees of freedom. Instead, it has been suggested to produce molecular condensates by photoassociation of a condensate of atoms. An essential limitation of this process, however, is the fast stimulated dissociation of the molecular condensate into atoms in noncondensate states (Goral *et al.*, 2000). Due to their state selectivity and directionality, adiabatic transfer techniques such as STIRAP may be used to overcome this problem of "rogue dissociation." Numerical simulations based on a two-mode model, which does not include rogue dissociation, have shown that STIRAP can be used to induce coherent two-color photoassociation of an atomic Bose–Einstein condensate in free–bound–bound transitions and convert it to a molecular on short time scales (Javanainen and Mackie, 1998, 1999; Mackie and Javanainen 1999; Mackie *et al.*, 2000). It was predicted that Bose stimulation can enhance the atomic free–bound dipole matrix element to the extent enabling photoassociative STIRAP.

X. Propagation Phenomena

The preceding sections of this chapter all treated the dynamics of atoms exposed to prescribed fields, emphasizing the transfer of population. As noted in Section II.B, populations $P_n(t)$ can be regarded as the diagonal elements of a density matrix. The off-diagonal elements describe another important property of coherent excitation, the induced dipole moments. The density of dipole moments, whether intrinsic or induced, provides the polarization field **P** for use in the Maxwell equations for the fields (cf. Shore, 1990; chap. 12). More specifically, the acceleration (second time derivative) of the dipole moments contributes to the radiation field. As a pulse of radiation passes by an atom, it induces a dipole moment that modifies the original field, and subsequent atoms along the propagation path will experience this modified field. In this section we discuss some aspects of a self-consistent treatment of atoms responding to a pulsed field, and traveling waves being modified by such atomic response.

The importance of the induced polarization field depends on the number density of atoms, $\mathcal{N}$, the length of the propagation path (measured as the product of the absorption coefficient and the physical distance), and on the strength of the induced dipole moment. For a two-state atom the quantity relevant for coherent propagation modification is the expectation value of instantaneous dipole moment $d_{12}C_1(t)C_2(t)^*$, where d_{ij} is the transition dipole moment. This same quantity alters the pump field in a Raman process; the Stokes field is affected by $d_{23}C_2(t)C_3(t)^*$. (These expressions provide slowly varying dipole moments, for use with the RWA and carrier frequency ω.)

When the atomic density is low and the path length is short, one can disregard any action of the atoms on the field. But Raman adiabatic passage has some very interesting effects for beam propagation. For example, STIRAP can be employed to make an otherwise optically thick ensemble of three-level atoms transparent to a pair of pump and Stokes pulses. This phenomenon, called *electromagnetically induced transparency* (EIT) (Boller *et al.*, 1991; Harris, 1997; Marangos, 1998), has a number of important applications ranging from communication to laser design and nonlinear optical processes (Harris *et al.*, 1990). EIT does not require careful control of pulse shapes or areas. Therefore it is qualitatively different from self-induced transparency and other soliton-like phenomena (cf. Allen and Eberly, 1975), which will not be discussed here.

A. Electromagnetically Induced Transparency (EIT)

1. Complete Decoupling of Light and Matter

As is customary when simplifying the description of propagation effects, we consider an infinite medium through which pass pulses of radiation that travel in the z direction and that are uniform in the transverse x, y plane. We assume that the atoms are distributed uniformly, and that their properties are described by a state vector $\Psi(z, t)$ and probability amplitudes $C_n(z, t)$. For further simplicity we assume that the atoms are stationary (i.e., no Doppler shifts) and that both the pump and the Stokes carriers are resonant with their respective Bohr frequencies. This assumption allows us to remove from each field a rapidly varying carrier $\exp(ik_j z - \omega_j t)$

and to treat the probability amplitudes and the Rabi frequencies $\Omega_j(z, t)$ as slowly varying functions of the coordinates z and t (cf. Shore, 1990, sect. 12.4). The resonance assumption permits us to choose the Rabi frequencies to be real valued; in general, it is not possible to separate phase and amplitude so clearly.

The evolution of the atoms is described by the three-state Schrödinger equation for the amplitudes $C_n(z, t), n = 1, 2, 3$. The needed RWA Hamiltonian is given by Eq. (40), with $\Delta_p = \Delta_s = 0$. Because the decay of the polarization plays an important role for the propagation of the fields, even if it is negligible for each individual atom, a decay out of the excited state ψ_2 with rate Γ is included. This can be done by adding in Eq. (40) an imaginary term $i\hbar\Gamma/2$ to the energy of the excited state. The propagation of the slowly varying Rabi-frequencies is described by one-dimensional first-order wave equations,

$$\left(\frac{\partial}{\partial z} + \frac{1}{c}\frac{\partial}{\partial t}\right)\Omega_p(z, t) = i\frac{\alpha_p}{2}\Gamma C_2(z, t)C_1^*(z, t), \tag{103}$$

$$\left(\frac{\partial}{\partial z} + \frac{1}{c}\frac{\partial}{\partial t}\right)\Omega_s(z, t) = i\frac{\alpha_s}{2}\Gamma C_2(z, t)C_3^*(z, t), \tag{104}$$

where the effect of the atoms on the fields is parameterized by the resonant absorption coefficients

$$\alpha_p = \frac{\omega_p \mathcal{N}|d_{12}|^2}{2\epsilon_0 c\hbar\Gamma}, \qquad \alpha_s = \frac{\omega_s \mathcal{N}|d_{32}|^2}{2\epsilon_0 c\hbar\Gamma}. \tag{105}$$

An important property of the adiabatic dark state of the atomic system is the absence of a dipole moment for either of the two transition $C_2 C_1^* = C_2 C_3^* = 0$. Atoms in that state therefore have no effect on the fields, i.e., light and matter are exactly decoupled. Typically, all atoms are initially in the ground state rather than in the dark state. The question is how to transfer atoms from the ground state into the dark state. This is best done by adiabatic passage. If all atoms start in the ground state ψ_1, approximate decoupling of the pulses requires (1) the instantaneous dark state be connected asymptotically to ψ_1, and (2) deviations from the dark state are small.

The first requirement can be fulfilled by a counterintuitive pulse ordering. The second condition is more involved. Satisfying the adiabaticity criterion of Section IV is not sufficient, because the fields interact with many atoms during propagation. The range of validity of the approximation can be estimated by considering the atomic eigenstates in lowest order of nonadiabatic corrections (cf. Section III.A.8).

$$|C_2| \sim \left|\frac{1}{\Omega T}\right|, \qquad |C_1| \sim \frac{\Omega_s}{\Omega}, \qquad |C_3| \sim \frac{\Omega_p}{\Omega}, \tag{106}$$

where $\Omega = \sqrt{\Omega_p^2 + \Omega_s^2}$. Although $\Omega T \gg 1$ is a sufficient condition to allow neglect of nonadiabatic corrections for individual atoms, one recognizes, by integrating the field equations (103) and (104) in steady state, that laser pulses propagate freely over a distance L only if

$$\frac{\Omega^2 T}{\Gamma} \gg \alpha_S L \tan \vartheta \quad \text{and} \quad \frac{\Omega^2 T}{\Gamma} \gg \alpha_P L \cot \vartheta, \tag{107}$$

where ϑ is the mixing angle defined in Eq. (45). If condition (107) is fulfilled, the interaction of the pulses with the otherwise optically thick medium ($\alpha L \gg 1$) is completely eliminated.

It is worth noting that the adiabaticity condition (107) implies that the spectral width $\Delta \omega \sim T^{-1}$ over which undisturbed propagation is possible is inversely proportional to the density-length product $\alpha_s L$ or $\alpha_p L$ (Lukin et al., 1997).

2. Elimination of Absorption; Slow Light

It is possible to achieve transparency, i.e., an elimination of the absorptive part of the interaction, for even smaller intensities than those required by condition (107). However, the light-induced modification of the refractive effects remains. Since nonadiabatic corrections are relevant here, they need to be taken into account in an approximate way. For this it is convenient to work in the basis of dark Φ_d and bright states Φ_b,

$$\begin{bmatrix} \Phi_d(z,t) \\ \Phi_b(z,t) \end{bmatrix} = \begin{bmatrix} \cos \vartheta(z,t) & -\sin \vartheta(z,t) \\ \sin \vartheta(z,t) & \cos \vartheta(z,t) \end{bmatrix} \begin{bmatrix} \psi_1 \\ \psi_2 \end{bmatrix}. \tag{108}$$

The probability amplitudes in this basis, $\mathbf{C}(t) = [C_d, C_b, C_2]^T$, obey a Schrödinger equation with a Hamiltonian,

$$\tilde{\mathsf{H}} = \hbar \begin{bmatrix} 0 & i\dot{\vartheta} & 0 \\ -i\dot{\vartheta} & 0 & \frac{1}{2}\Omega \\ 0 & \frac{1}{2}\Omega & -\frac{i}{2}\Gamma \end{bmatrix},$$

that is equivalent to a three-state system driven by two pulsed "fields," of strength $\dot{\vartheta}$ and Ω. Under adiabatic conditions $\dot{\vartheta}$ is small and can be treated perturbatively. Solving the Schrödinger equation to leading order of $\eta = |\dot{\vartheta}/\Omega|$, one finds:

$$C_d = 1 + \mathcal{O}(\eta), \quad C_b = \mathcal{O}(\eta), \quad C_2 = \frac{2i}{\Omega}\dot{\vartheta} + \mathcal{O}(\eta^2),$$

or, after transformation back to the bare atomic basis,

$$C_1 = \cos\vartheta \qquad C_2 = \frac{2i}{\Omega}\dot{\vartheta} \qquad C_3 = -\sin\vartheta. \tag{109}$$

An important special case is that of a strong and approximately constant Stokes field, as shown in Fig. 45, i.e., $|\Omega_s| \gg |\Omega_p|$ and

$$\dot{\vartheta} = \frac{\dot{\Omega}_p \Omega_s - \Omega_p \dot{\Omega}_s}{\Omega^2} \approx \frac{\dot{\Omega}_p}{\Omega}. \tag{110}$$

Substitution of (109) and (110) into the propagation equations (103) and (104) yields, in lowest order of η,

$$\left(\frac{\partial}{\partial z} + \frac{1}{v_g}\frac{\partial}{\partial t}\right)\Omega_p(z,t) = 0, \tag{111}$$

$$\left(\frac{\partial}{\partial z} + \frac{1}{c}\frac{\partial}{\partial t}\right)\Omega_s(z,t) = 0, \tag{112}$$

with

$$v_g = \frac{c}{1+n_g} \quad \text{and} \quad n_g = \frac{\alpha_p \Gamma c}{\Omega^2}. \tag{113}$$

Thus the pump pulse propagates without changing form (form stable) at a reduced group velocity v_g, while the strong Stokes fields remains unaffected by the interaction (see Fig. 45). It is important to note that the only effect of the medium is a

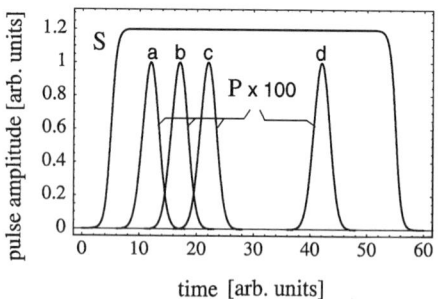

FIG. 45. Slow light propagation: Amplitudes of pump ($\times 100$) and Stokes fields as function of time in a co-moving frame for $n_g z/c = 0$ (a), 5 (b), 10 (c), and 30 (d). Weak pump field propagates with reduced group velocity. Arbitrary space and time units with $c = 1$.

temporal delay of the pump pulse. No energy is lost by absorption or spontaneous emission. At the leading edge of the pump pulse the atoms absorb photons from this pulse and transfer them into the Stokes field (which is much stronger and therefore effectively unchanged) by Raman adiabatic passage. The process is reversed at the tail of the pump pulse, i.e., all energy is returned to it.

In typical optical materials such as glass, the group index n_g is of the order of unity. In a dense medium (large α_p) with EIT the alteration of the propagation velocity can be quite substantial, however. Hau *et al.* (1999) observed a reduction of the group velocity to 17 m/s in a Bose condensate of Na atoms, corresponding to a group index of the order of 10^7. Similarly small values where obtained in a buffer-gas cell of hot Rb (Kash *et al.*, 1999) and Cs atoms (Budker *et al.*, 1999).

One easily verifies that the group-velocity reduction has little effect in the strong-field limit of Eq. (107). In this case the delay time $\tau_d = L/v_g - L/c = n_g L/c$, i.e., the time delay by which the Stokes pulse in the medium lags behind a corresponding pulse in vacuum, is much smaller than the pulse duration T.

The maximum group delay is limited by the finite lifetime of the dark state and higher-order nonadiabatic corrections. The latter lead to a maximum ratio of delay time to pulse length (Harris and Hau, 1999),

$$\left.\frac{\tau_d}{T}\right|_{\max} = \sqrt{\alpha_p L}. \qquad (114)$$

The slowing down of light has a number of important applications. When a pulse enters a medium with a smaller group velocity, it becomes spatially compressed by the ratio of group velocity to the speed of light outside the medium. Thus, information contained in long pulses can be compressed to a small spatial volume. A reduction of the group velocity leads to an enhanced interaction time, important for efficient nonlinear optical processes (Harris and Hau, 1999; Hemmer *et al.*, 1995; Lukin and Imamoglu, 2000). When the light velocity matches the speed of sound, a new type of Brillouin scattering is possible (Matsko *et al.*, 2001). The enhancement of the field gradient, associated with the spatial pulse compression, can lead to very large pondermotive forces on atoms (Harris, 2000) with potential applications for atom optics and cooling.

B. ADIABATONS

It was shown by Grobe *et al.* (1994) that the equations of nonlinear propagation [(103), (104)] are adiabatically integrable for fields of comparable strength and— within the adiabatic approximation—of arbitrary shape. To see this, we adopt the method of Fleischhauer and Manka (1996) and transform the field equations for Ω_p and Ω_s in propagation equations for the total Rabi frequency Ω and the nonadiabatic coupling $\dot{\vartheta}$ assuming equal coupling strength $\alpha_p = \alpha_s = \alpha$:

$$\left(\frac{\partial}{\partial z} + \frac{1}{c}\frac{\partial}{\partial t}\right)\Omega(z,t) = i\frac{\alpha}{2}\Gamma C_2 C_b^*, \tag{115}$$

$$\left(\frac{\partial}{\partial z} + \frac{1}{c}\frac{\partial}{\partial t}\right)\dot{\vartheta}(z,t) = i\frac{\alpha}{2}\Gamma\frac{\partial}{\partial t}\left(\frac{C_2 C_d^*}{\Omega}\right). \tag{116}$$

The transformation of both the atomic states and the fields leads to three dressed states coupled to two new "fields," characterized by Ω and $\dot{\vartheta}$. Under adiabatic conditions $|\dot{\vartheta}| \ll \Omega$, so a weak-field approximation is justified. Substituting the lowest-order adiabatic solutions (109) into these equations, one finds that the total Rabi frequency fulfills the free-space equation

$$\left(\frac{\partial}{\partial z} + \frac{1}{c}\frac{\partial}{\partial t}\right)\Omega(z,t) \approx 0.$$

No photons are lost by absorption and there is only a coherent transfer from one field into the other.

At the same time, $\dot{\vartheta}$ obeys the equation

$$\left(\frac{\partial}{\partial z} + \frac{1}{c}\frac{\partial}{\partial t}\right)\dot{\vartheta}(z,t) = -\alpha\Gamma\frac{\partial}{\partial t}\left(\frac{\dot{\vartheta}}{\Omega^2}\right).$$

This equation is exactly integrable (Grobe et al., 1994). The corresponding solutions, called *adiabatons*, are particularly simple if Ω is approximately constant over the time interval of interest. In that case the pump and Stokes pulses have complementary envelopes and $\dot{\vartheta}$ propagates without changing form, at the group velocity v_g given in Eq. (113). The quasi-form-invariant propagation of an adiabaton is shown in Fig. 46. First experimental evidence of adiabaton propagation was reported by Kasapi et al. (1995) in Pb vapor. Adiabatons in more complicated configurations, such as double-lambda systems and double pairs of pulses, were studied by Cerboneschi and Arimondo (1995).

C. MATCHED PULSES

An interesting feature of the interaction of bichromatic fields with three-level systems was noted by Harris (1993): the dark state corresponding to a pair of pulses with identical envelope (*matched pulses*), i.e., $\Omega_p(z,t) = \Omega_p f(z,t)$ and $\Omega_s(z,t) = \Omega_s f(z,t)$, is time independent. After an appropriate preparation of the medium, matched pulses will remain exactly decoupled from the interaction for all times.

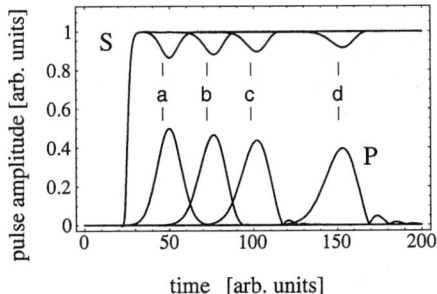

FIG. 46. Propagation of adiabatons. Shown are amplitudes of Pump and Stokes fields as function of time in a co-moving frame for $n_g z/c = 0$ (a), 25 (b), 50 (c), and 100 (d) from numerical solution of propagation equations (Fleischhauer and Manka, 1996). Arbitrary space and time units with $c = 1$.

If a pair of matched pulses is applied to an atomic ensemble in the ground state, they will prepare the atoms by a STIRAP process (Harris and Luo, 1995): during the first few single-photon absorption lengths, the front end of the pump pulse experiences a small loss. In this way a counterintuitive pulse ordering is established. This provides asymptotic connectivity of the dark state to the initial state of the atoms. The leading end of this slightly deformed pair of pulses will then prepare all atoms in the pathway via STIRAP and the pulses can propagate unaffected through the rest of the medium (Harris, 1994; Eberly et al., 1994).

Apart from the preparation at the front end, matched pulses are stable solutions of the propagation problem. They should therefore be formed whenever pulses of arbitrary shape are applied to optically thick three-level media. In fact, it has been shown (Harris, 1993) that pairs of pulses with a strong CW carrier,

$$\Omega_p(z, t) = [1 + f(z, t)]\Omega_p e^{-i\omega_p(t-z/c)}, \tag{117a}$$

$$\Omega_s(z, t) = [1 + g(z, t)]\Omega_s e^{-i\omega_p(t-z/c)}, \tag{117b}$$

tend to adjust their amplitude modulations $f(z, t)$ and $g(z, t)$ in the course of propagation:

$$\left. \frac{f(z, t)}{g(z, t)} \right|_{z \to \infty} = 1.$$

Fast fluctuations of f and g lead to a nonadiabatic coupling of the dark state, established by the CW components, to other states and will be absorbed. Thus pulses of identical envelope are formed. This phenomenon of *pulse matching* also causes a correlation of quantum fluctuations in both fields (Fleischhauer, 1994; Jain, 1994; Agarwal, 1993).

The tendency to generate pulses with identical envelopes is not restricted to fields with a strong CW carrier. The adiabaton solutions are approximately stable over many single-photon absorption lengths. However, as shown by Fleischhauer and Manka (1996), they eventually decay to matched pulses after sufficiently long propagation distances.

D. COHERENCE TRANSFER BETWEEN MATTER AND LIGHT

It was shown in Section A that a strong and constant Stokes field leads to a slowing down of the propagation velocity of the pump pulse, while the shape and photon flux of the pump remained unaffected. The coherent information of the pump pulse is temporarily stored in the medium (in contrast to its energy, which is mostly transferred back and forth to the Stokes field). However, the maximum delay time is rather limited [see Eq. (114)]. It was shown recently (Fleischhauer et al., 2000) that this limit can be overcome. In fact it is possible to control the propagation velocity of the pump pulse, to bring it to a full stop, and to reaccelerate it on demand. This behavior is associated with the existence of quasi-particles called *dark-state polaritons* that are a mixture of atomic and field components,

$$F(z, t) = \cos\theta(t)\,\Omega_p(z, t) - \sin\theta(t)\,\sqrt{\alpha c \Gamma}\; C_3(z, t)C_1^*(z, t).$$

The angle θ (not to be confused with the mixing angle ϑ used earlier) is defined by

$$\tan\theta(z, t) = \frac{\sqrt{\alpha c \Gamma}}{\Omega_s(z, t)}.$$

The dark-state polariton obeys the simple propagation equation

$$\left[\frac{\partial}{\partial t} + c \cos^2\theta(z, t)\frac{\partial}{\partial z}\right] F(z, t) = 0. \tag{118}$$

If Ω_s (and hence θ) is approximately uniform in z, Eq. (118) describes a form-invariant propagation of the quasi-particle with propagation velocity

$$v(t) = c \cos^2\theta(t).$$

When $\theta(t)$ is adiabatically rotated from 0 to $\pi/2$ by externally controlling the amplitude of the Stokes field Ω_s, an initially pure electromagnetic polariton ($F = \Omega_p$) is transformed into a pure atomic polarization ($F = \sqrt{\alpha c \Gamma}\,C_3 C_1^*$). At the same

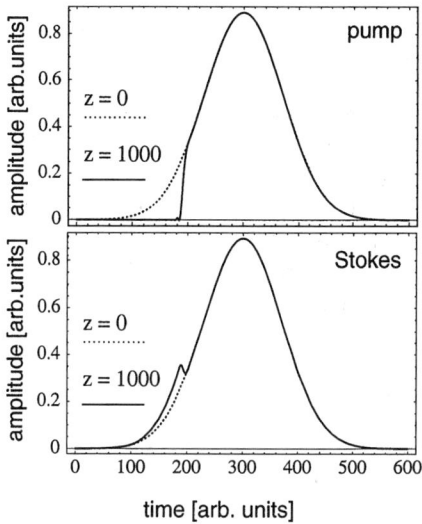

FIG. 47. Top: Amplitude of pump field as a function of time in arbitrary units at medium entrance $z = 0$ (dotted line) and for $z = 1000$ (full line). Bottom: The same for the Stokes field. Distance is measured in units of absorption length $\alpha_P^{-1} = \alpha_S^{-1}$ and peak values are $\Omega_s = \Omega_p = 10\Gamma$.

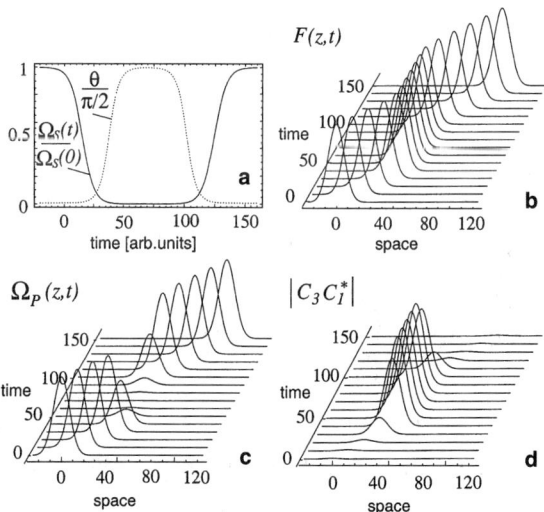

FIG. 48. Stopping and reaccelerating a dark-state polariton. Rabi frequency of control (Stokes) field and mixing angle are shown in (a). Coherent amplitude of dark-state polariton F is plotted in (b), and the amplitudes of pump field Ω_p and atomic coherence $|C_3 C_1^*|$ are shown in (c) and (d), respectively. Space and time have arbitrary units, with $c = 1$. (From M. Fleischhauer and M. D. Lukin. Dark-state polaritons in electromagnetically induced transparency. *Phys. Rev. Lett.* 2000;84:5094–5097.)

time, the propagation velocity is changed from the vacuum speed of light to zero. The pump pulse is thus "stopped," which means that its coherent information is transferred to collective atomic states. During the transfer process the spectrum of the electric field component is narrowed. As a consequence, the limitation (114) does not apply. The atomic polarization can be extracted by reversing the transfer process, i.e., rotating θ back from $\pi/2$ to 0 and re-creating the pump pulse. The deceleration, storage, and reacceleration of a polariton is illustrated in Fig. 48. First experimental demonstrations of light stopping have recently been reported (Phillips *et al.*, 2001; Liu *et al.*, 2001). Figure 49 shows the observed light pulse storage in Rb vapor from Phillips *et al.* (2001).

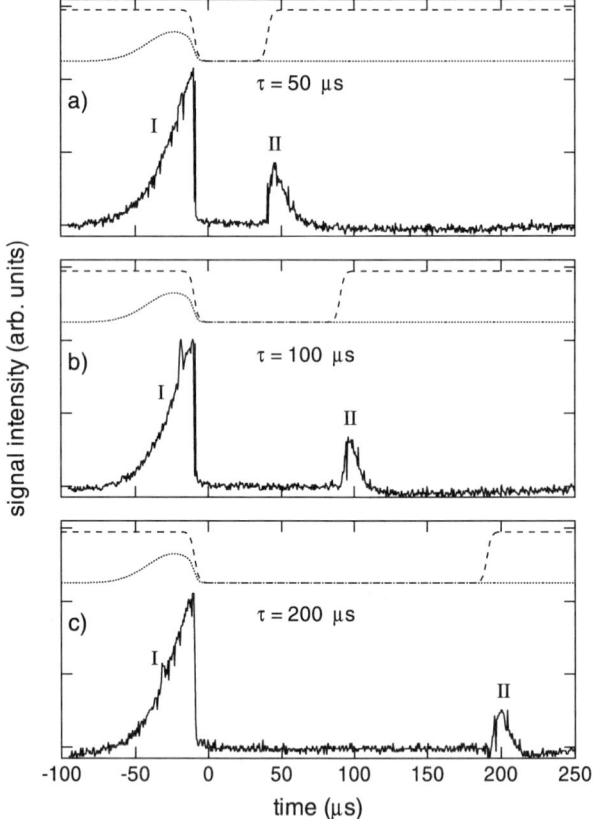

FIG. 49. Light pulse storage in ^{87}Rb cell. Shown above the data in each graph are calculated values of control field (dashed line) and input signal (dotted). (Reprinted with permission from D. F. Phillips, A. Fleischhauer, A. Mair, R. L. Walsworth, and M. D. Lukin. Storage of light in atomic vapor. *Phys. Rev. Lett.* 2001;86:783–786.)

XI. Applications of STIRAP in Quantum Optics and Quantum Information

The growing interest in quantum information science in recent years, and the resulting need for methods that allow controlled and coherent manipulation of quantum states, has led to another field of applications of STIRAP. An important new aspect here is the possibility for coherent control not only of quantum states of atomic and molecular systems, but also of quantum states of the radiation field.

In all aspects of quantum information, *decoherence* (e.g., the loss of coherence through uncontrolled random variations of state vector phases) must be avoided. Here STIRAP, by utilizing radiatively dark states, has definite advantages.

A. Single-Atom Cavity Quantum Electrodynamics

The potential usefulness of STIRAP in studies of cavity quantum electrodynamics (QED) was first pointed out by Parkins et al. (1993, 1995, 1999). They proposed to use STIRAP to create coherent superpositions of photon-number states by strongly coupling an atom to the field within a cavity of volume V. The quantized field of the single-mode resonator provided the Stokes field. Within the RWA, the interaction coupled only triplets of bare eigenstates of the combined atom–field system, viz., $\psi_1^{n+1} \equiv |\psi_1, n+1\rangle$, $\psi_2^n \equiv |\psi_2, n\rangle$, and $\psi_3^n \equiv |\psi_3, n\rangle$, where n denotes the number of photons in the mode. The appropriate Hamiltonian is that given in Eq. (40) with the Rabi frequency of the pump field replaced by the *vacuum Rabi frequency* (in cavity volume V),

$$\Omega_{\text{vac}} = d_{12}/\hbar \sqrt{\hbar \omega_{12}/2\varepsilon_0 V}, \tag{119}$$

the coupling strength between the atom and the resonator mode. The system has an infinite set of adiabatic dark states,

$$\Phi_0^n \equiv \psi_1^{n+1} \cos\theta_n(t) - \psi_3^n \sin\theta_n(t), \qquad n = 0, 1, 2, \ldots,$$

where the mixing angles θ_n are defined as

$$\tan\theta_n(t) = \frac{\Omega_{\text{vac}}\sqrt{n}}{\Omega_p(t)}.$$

Of particular interest is the case $n = 1$. Adiabatic rotation of the dark state maps an atomic excitation ($|\psi_3, n = 0\rangle$) onto an excitation of the resonator mode ($|\psi_1, n = 1\rangle$), and vice versa, in a decoherence-free fashion. Because the RMS Rabi frequency $\Omega = \sqrt{\Omega_{\text{vac}}^2 n + \Omega_s^2}$ is equal to $\Omega_{\text{vac}}\sqrt{n}$ in the asymptotic limit $\theta_n \to \pi/2$, the adiabatic condition in the presence of decay takes the form (Fleischhauer and Manka 1996)

$$\frac{\Omega_{\text{vac}}^2 n}{\Gamma} T \gg 1,$$

where Γ is the decay rate of the excited state ψ_2.

The transfer time T is usually limited by the finite decoherence time of the field state, e.g., due to cavity losses. If κ denotes the photon loss rate of the resonator mode, the decoherence time of a photon number state $|n\rangle$ is of the order $1/(n\kappa)$. Thus adiabaticity requires a strong-coupling regime of the cavity

$$\Omega_{\text{vac}}^2 \gg \kappa \Gamma, \tag{120}$$

a condition which is technically very challenging to satisfy.

Despite these difficulties in the practical realization, adiabatic transfer between atomic and resonator excitations is interesting because of a number of potential applications. For example, successive interactions of atoms with the cavity system can be used to create arbitrary coherent superpositions of photon number states (Parkins et al., 1993). This idea can be extended to two degenerate cavity modes of orthogonal polarizations (Lange and Kimble, 2000).

Furthermore, as first suggested by Law and Eberly (1996, 1998) and Law and Kimble (1997), it is possible to generate a single-photon wave packet, with an envelope determined by $\theta_1(t)$, in the output of the resonator. Single-photon wave packets on demand are important for quantum information processing utilizing photons as information carriers. A first step toward an experimental realization of such a "photon pistol" was recently reported by Hennrich et al. (2000). Here individual atoms prepared in state ψ_3 were dropped from a magneto-optical trap (MOT) and fell through a resonator setup as shown in Fig. 50. Raman adiabatic passage induced by a counterintuitive coupling to the resonator field and a subsequent pump laser beam led to stimulated emission into the resonator mode.

The process of transferring excitation from atoms to a field mode is reversible and allows the opposite process of mapping cavity-mode fields onto atomic ground-state coherences. This mapping provides a possibility for measuring cavity fields (Parkins et al., 1995). The atomic angular momentum state can be measured by the Newton–Young method (Newton and Young, 1968) with a finite number of magnetic dipole measurements using Stern–Gerlach analyzers. Alternatively, the parameters of the atomic superposition can be measured by coupling this degenerate atomic state to an excited state by an elliptically polarized laser pulse and measuring the subsequent fluorescence (Vitanov et al., 2000; Vitanov, 2000).

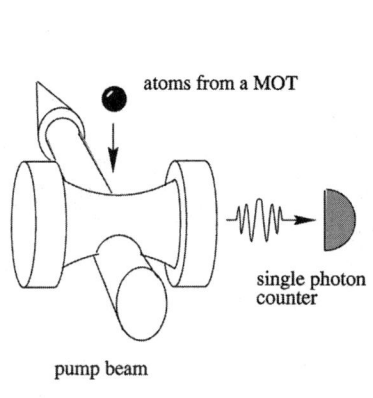
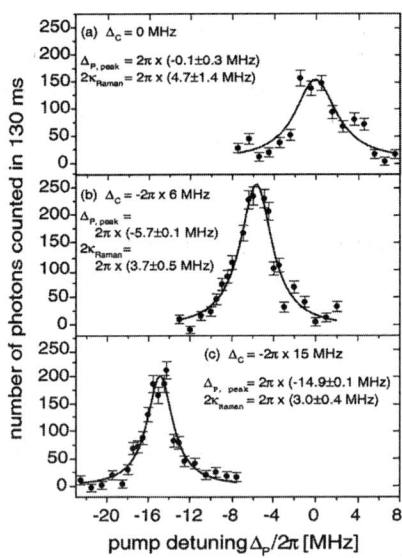

FIG. 50. First evidence of vacuum stimulated Raman scattering into cavity mode via STIRAP. Top: Experimental setup. Bottom: Number of counted photons as function of pump detuning for different probe detunings. Maximum at $\Delta_P = \Delta_C$ proves two-photon nature of the stimulated process. (Reprinted with permission from M. Hennrich, T. Legero, A. Kuhn, and G. Rempe. Vacuum-stimulated Raman scattering based on adiabatic passage in high-finesse optical cavity. *Phys. Rev. Lett.* 2000;85:4872–4875.)

B. Quantum Logic Gates Based on Cavity QED

Quantum information processing requires the coherent manipulation of elementary units, called *quantum bits* (*qubits*), which are equivalent to spin-$\frac{1}{2}$ systems. Besides performing arbitrary rotations within the two-dimensional Hilbert space of individual qubits, H_i, one must be able to perform operations, known as elementary quantum gates, between arbitrary pairs of them. It was shown by Pellizzari *et al.* (1995) that the required two-bit gate can be implemented in a cavity system by employing STIRAP to transfer quantum states between different atoms. The quantum states of two qubits i and j, stored in two different atoms, are first mapped to four different internal states of a single atom. The required gate operation in the 2×2-dimensional Hilbert space $H_i \otimes H_j$ of the ith and jth qubits can then be performed by simply applying the appropriate laser pulses to the second atom. Finally, the qubits are transferred back to the original atoms, where they will generally be in an *entangled* state, i.e., in a quantum state which cannot be factorized.

To illustrate the essence of the qubit transfer from one atom to another via the resonator field, we consider two three-level atoms A and B. The $2 \rightarrow 3$ transitions of both atoms are coupled to the same quantized mode of the resonator. Assume that atom A is initially in a coherent superposition of states 1 and 3 with amplitudes α and β, i.e., $\psi^A = \alpha\psi_1^A + \beta\psi_3^A$. Atom B is assumed to be in state 3, for simplicity, and there are no photons in the cavity. Under these conditions the dynamics of the total system involves only the state

$$\Phi_0^0 = |\psi_3^A, \psi_3^B, 0\rangle, \tag{121}$$

where $|\ldots, 0\rangle$ denotes the photon vacuum. Φ_0^0 is an exact eigenstate of the interaction with zero eigenvalue. In addition there is a resonant, chainwise-coupled system of five states:

$$|\psi_1^A, \psi_3^B, 0\rangle \xleftrightarrow{\Omega_p^A} |\psi_2^A, \psi_3^B, 0\rangle \xleftrightarrow{\Omega_{\text{vac}}} |\psi_3^A, \psi_3^B, 1\rangle,$$

$$\xleftrightarrow{\Omega_{\text{vac}}} |\psi_3^A, \psi_2^B, 0\rangle \xleftrightarrow{\Omega_p^B} |\psi_3^A, \psi_1^B, 0\rangle. \tag{122}$$

The Hamiltonian of this system has the adiabatic dark state (i.e., a state without component of state ψ_2)

$$\Phi_0^1 = \frac{1}{\mathcal{M}} \left[\Omega_p^A(t)\Omega_{\text{vac}} |\psi_3^A, \psi_1^B, 0\rangle \right.$$
$$\left. - \Omega_p^A(t)\Omega_p^B(t)|\psi_3^A, \psi_3^B, 1\rangle + \Omega_p^B(t)\psi_1|\psi_1^A, \psi_3^B, 0\rangle \right], \tag{123}$$

with $\mathcal{M}$ being a normalization constant. Any superposition of the zero-eigenvalue states Φ_0^0 and Φ_0^1 is also a dark state. In particular, if at $t \to -\infty$, the pump field of atom A vanishes, i.e., $\Omega_p^A = 0$, but that of atom B is nonzero, i.e., $\Omega_p^B \neq 0$, then the initial statevector of the system is a superposition of dark states,

$$\Psi(t = -\infty) = \left(\alpha|\psi_1^A\rangle + \beta|\psi_3^A\rangle\right)|\psi_3^B\rangle|0\rangle$$
$$= \alpha|\psi_1^A, \psi_3^B, 0\rangle + \beta|\psi_3^A, \psi_3^B, 0\rangle \tag{124a}$$
$$= \alpha\Phi_0^1(-\infty) + \beta\Phi_0^0.$$

Adiabatic reduction of Ω_p^B to zero while increasing Ω_p^A rotates the superposition of dark states to

$$\Psi(t = +\infty) = \alpha\Phi_0^1(\infty) + \beta\Phi_0^0$$
$$= \alpha|\psi_3^A, \psi_1^B, 0\rangle + \beta|\psi_3^A, \psi_3^B, 0\rangle \tag{124b}$$
$$= |\psi_3^A\rangle\left(\alpha|\psi_1^B\rangle + \beta|\psi_3^B\rangle\right).$$

As a result, the qubit state of atom A is mapped onto atom B—an example of *state swapping*. Implementation of an actual quantum gate requires a slightly more complicated system, in which atom B has two degenerate states "$\pm$" in each energy level 1, 2, and 3. The two Λ systems are coupled by pump and Stokes fields in a parallel way. This allows a mapping of the qubit states of atoms A and B onto the four states $\psi_{1\pm}$ and $\psi_{3\pm}$ of atom B. After this transfer the two-bit operations can be performed by manipulating atom B alone.

As can be seen from Eq. (123), the single-photon state of the cavity mode is excited during the transfer process. To avoid decoherence, the transfer time T needs to be short compared to the decay time of the cavity. On the other hand, the transfer must be sufficiently slow to guarantee adiabaticity. Both requirements can only be fulfilled in a strongly coupled regime (120).

C. Quantum Networking

An important aspect of quantum information processing is the ability to transmit the wavefunction of qubits over mesoscopic or macroscopic distances. A scheme for an ideal quantum-state transfer between atoms at spatially separated nodes of a network was proposed by Cirac *et al.* (1997), using a Raman transfer of quantum states from an atom to a resonator Stokes field, which is coupled to a transmission line such as an optical fiber. Unlike the case described in the previous subsection, this scheme requires that the two single-photon frequencies be far off resonance, so that the excited state can be adiabatically eliminated. The Raman process generates, in the fiber, a photon wave packet that is directed to a second, strongly coupling resonator containing another atom. When the pump field of the first resonator generates time-symmetric photon pulses, a retarded and time-reversed Stokes field in the second resonator leads to a complete transfer of the photon wave packet to the corresponding atom—without any reflection from the input mirror (Cirac *et al.*, 1997). The possibility for suppressing the reflection for arbitrary photon wave packets (quantum impedance matching) was discussed by Fleischhauer *et al.*, 2000; Lukin *et al.*, 2000).

It was shown by Pellizzari (1997) that state swapping can also be performed by means of adiabatic passage via photonic dark states. In this case the two cavities remain in the vacuum states during the whole operation and the effect of resonator losses is diminished. However, the modes of the optical fiber connecting the two cavities are excited during the transfer process. An interesting extension of the adiabatic-passage idea, given by van Enk *et al.*, (1999), avoids any excitation of fiber modes and thus suppresses the effect of fiber losses.

D. Many-Atom Systems

The disadvantages of single-atom cavity-QED schemes are the need for a strongly coupled cavity and the requirement for extremely close control of individual atoms. As shown by Lukin *et al.* (2000), both drawbacks can be overcome if collective excitations of an ensemble of atoms are used as qubit states instead of energy levels of individual atoms. If N three-level atoms are placed inside a resonator, with the $1 \to 2$ transition resonantly coupled by a quantized mode of the resonator and the $2 \to 3$ transition by a resonant classical Stokes field, the interaction couples only symmetric superpositions:

$$\Psi_1 = \psi_1^1 \psi_1^2 \ldots \psi_1^N, \tag{125a}$$

$$\Psi_3 = \frac{1}{\sqrt{N}} \sum_{j=1}^{N} \psi_1^1 \ldots \psi_3^j \ldots \psi_1^N, \tag{125b}$$

$$\Psi_3^2 = \frac{1}{\sqrt{2N(N-1)}} \sum_{i \neq j=1}^{N} \psi_1^1 \ldots \psi_3^i \ldots \psi_3^j \ldots \psi_1^N. \tag{125c}$$

etc.

In particular, if initially the field is in a state with at most one photon, the relevant eigenstates of the bare system are the total ground state $|\Psi_1, 0\rangle$, which is not affected by the interaction at all; the ground state with one photon in the field $|\Psi_3, 1\rangle$; and the singly excited states $|\Psi_2, 0\rangle$ and $|\Psi_3, 0\rangle$. For the case of two excitations, the coupling involves three more states, etc.

The interaction of the N-atom system with the quantized radiation mode has a family of dark states, i.e., states with no component ψ_2 and zero adiabatic eigenvalue (Parkins et al., 1993; Lukin and Imamoglu, 2000; Fleischhauer et al., 2000). The simplest one is

$$\Phi_0^1 = \cos\theta(t)|\Psi_1, 1\rangle - \sin\theta(t)|\Psi_3, 0\rangle$$

with $\tan\theta(t) = \Omega_{\text{vac}}\sqrt{N}/\Omega(t)$, and in general one has

$$\Phi_0^n = \sum_{k=0}^{n} \sqrt{\frac{n!}{k!(n-k)!}} (-\sin\theta)^k (\cos\theta)^{n-k} |\Psi_3^k, n-k\rangle.$$

The existence of dark states provides a very elegant way to transfer an arbitrary quantum state of the single-mode field to collective atomic excitations. Adiabatic rotation of the mixing angle θ from 0 to $\pi/2$ leads to a complete and reversible transfer of the photonic excitation to a collective atomic excitation, if the total number of excitations n is less than the number of atoms. If $\theta = 0 \to \pi/2$ one has, for all $n \leq N$,

$$\Phi_0^n = |\Psi_1\rangle|n\rangle \longrightarrow |\Psi_3^n\rangle|0\rangle.$$

By contrast with the single-atom case (120), the necessary condition for the adiabatic following of the systems state vector with the dark state is now

$$\frac{\Omega_{vac}^2 N}{\Gamma} T \gg n.$$

If the number of atoms N is much larger than the number of photons n, this condition can be much more easily fulfilled and does not require strong coupling resonators.

XII. Summary and Outlook

Many contemporary fields in atomic, molecular, and optical physics require preparation of samples in which almost all of the population resides in a preselected target state. Although a variety of methods have been proposed and tried over the years, many of the most successful ones are based on controlling adiabatic time evolution of the quantum system.

The theoretical description of such time dependence is most easily presented with the aid of adiabatic states: if the evolution is adiabatic, then at all times the state vector remains aligned with one of these states. The progress of the changing state vector can be followed by viewing a plot of adiabatic eigenvalues and noting the crossings of diabatic energies.

The experimenter has various guides for the applicability of adiabatic passage—the adiabatic conditions. Typically these require that excitation pulses be strong and smooth and that the interaction act for a "long" time compared with a characteristic atomic time scale. Often this is expressed as the requirement that a time-integrated Rabi frequency be much larger than unity. What typically sets adiabatic techniques apart from other pulsed-excitation techniques is the relative insensitivity of the transfer efficiency to interaction parameters; adiabatic techniques are not sensitive to pulse area, for example.

In simplest form, adiabatic passage can completely invert the population of a two-state system. Numerous extensions have been devised and put to practical use; we have mentioned some of these. When a quantum system of three (or more) states is subjected to two (or more) coherent pulses, to permit adiabatic evolution of a stimulated Raman process, then remarkable and novel effects become possible. These rest on the presence of a "population trapping" quantum superposition state, a concept first appreciated some 30 years ago and described here in some detail, and the use of "counterintuitive" pulse sequences.

Earliest applications of the technique of stimulated Raman adiabatic passage (STIRAP) aimed to produce complete selective excitation of individual rovibrational states of molecules or magnetic sublevels of atoms. We have discussed the basic STIRAP and numerous extensions. Unlike techniques involving only two quantum states, such procedures can produce population transfer into stable or metastable states. Beyond such simple population transfer uses, we have noted the growing interest in the use of STIRAP and related techniques to the creation of stable superpositions (controlled partial population transfer, with predetermined phases) for a number of applications.

More recently, the response of the fields themselves to such coherent atomic evolution has been considered, and remarkable propagation effects have been demonstrated. For example, inducing an otherwise opaque vapor to allow nearly lossless transmission (electromagnetically induced transparency, or EIT), dramatically slowing pulses of light, and even "stopping photons" for controllable time intervals. These techniques have application, among other things, to the emerging research area of quantum information.

As laser technology continues to improve, and experimenters acquire lasers with ever higher intensity and purer spectral content, one can expect to see imaginative new applications of adiabatic passage to the task of transferring populations between quantum states, of creating novel superposition states, and of modifying pulse propagation in new and novel ways.

XIII. Acknowledgments

This work has been supported by the European Union Research and Training network COCOMO, contract HPRN-CT-1999-00129, by NATO grant 1507-826991, and by Deutsche Forschungsgemeinschaft. NVV has received partial support from the Academy of Finland, project 43336. BWS acknowledges the support of Laserzentrum, University of Kaiserslautern. His work in Germany has been supported, in part, by a Research Award from the Alexander von Humboldt Foundation. His work at Livermore was supported in part under the auspices of the U.S. Department of Energy at Lawrence Livermore National Laboratory under contract W-7405-Eng-48.

XIV. References

Agarwal, G. S. (1993). Coherent population trapping states of a system interacting with quantized fields and the production of the photon statistics matched fields. *Phys. Rev. Lett.* **71**, 1351–1354.
Allen, L., and Eberly, J. H. (1975). "Optical Resonance and Two-level atoms." Dover, New York.
Alzetta, G., Gozzini, A., Moi, L., and Orriols, G. (1976). An experimental method for the observation of RF transitions and laser beat resonances in oriented Na vapour. *Nuovo Cim. B* **36**, 5–20.
Alzetta, G., Moi, L., and Orriols, G. (1979). Nonabsorption hyperfine resonances in a sodium vapour irradiated by a multimode dye-laser. *Nuovo Cim. B* **52**, 2333–2346.

Arimondo, E. (1996). Coherent population trapping in laser spectroscopy. *In* "Progress in Optics XXXV" (E. Wolf, Ed.), pp. 259–356. North-Holland, Amsterdam.
Arimondo, E., and Orriols, G. (1976). Nonabsorbing atomic coherences by coherent two-photon transitions in a three-level optical pumping. *Lett. Nuovo Cim.* **17**, 333–338.
Aspect, A., Armindo, E., Kaiser, R., Vanteenkiste, N., and Cohen-Tannoudji, C. (1988). Laser cooling below the one-photon recoil by velocity-selective coherent population trapping. *Phys. Rev. Lett.* **61**, 826–829.
Aspect, A., Armindo, E., Kaiser, R., Vanteenkiste, N., and Cohen-Tannoudji, C. (1989). Laser cooling below the one-photon recoil energy by velocity-selective coherent population trapping: theoretical analysis. *J. Opt. Soc. Am. B* **6**, 2112–2124.
Autler, S. H., and Townes, C. H. (1955). Stark effect in rapidly varying fields. *Phys. Rev* **100**, 703.
Band, Y. B., and Julienne, P. S. (1991a). Density matrix calculation of population transfer between vibrational levels of Na_2 by stimulated Raman scattering with temporally shifted laser beams. *J. Chem. Phys.* **94**, 5291–5298.
Band, Y. B., and Julienne, P. S. (1991b). Population transfer by multiple stimulated Raman scattering. *J. Chem. Phys.* **95**, 5681–5685.
Band, Y. B., and Julienne, P. S. (1992). Complete alignment and orientation of atoms and molecules by stimulated Raman scattering with temporally shifted lasers. *J. Chem. Phys.* **96**, 3339–3341.
Band, Y. B., and Magnes, O. (1994). Chirped adiabatic passage with temporally delayed pulses. *Phys. Rev. A* **50**, 584–594.
Becker, M., Gaubatz, U., Jones, P. L., and Bergmann, K. (1987). Efficient and selective population of high vibrational levels of Na_2 via a stimulated Raman process. *J. Chem. Phys.* **87**, 5064–5076.
Bergmann, K. (1988). State selection via optical methods. *In* "Atomic and Molecular Beam Methods" (G. Scoles, Ed.), pp. 293–344. Oxford University Press, Oxford.
Bergmann, K., and Demtröder, W. (1971). Inelastic collision cross section for excited molecules: Rotational energy transfer within the $B^1\Pi_u$-state of Na_2 induced by collision with He. *Z. Phys.* **243**, 1–13.
Bergmann, K., and Shore, B. W. (1995). Coherent population transfer. *In* "Molecular Dynamics and Spectroscopy by Stimulated Emission Pumping" (H. L. Dai and R. W. Field, Ed.), pp. 315–73. World Scientific, Singapore.
Bergmann, K., Engelhardt, R., Hetter, U., Hering, P., and Witt, J. (1978). State resolved differential cross sections for rotational transitions in Na_2+Ne(He) collisions. *Phys. Rev. Lett.* **40**, 1446–1450.
Bergmann, K., Theuer, H., and Shore, B. W. (1998). Coherent population transfer among quantum states of atoms and molecules. *Rev. Mod. Phys.* **70**, 1003–1025.
Boller, K. J., Imamoglu, A., and Harris, S. E. (1991). Observation of electromagnetically induced transparency. *Phys. Rev. Lett.* **66**, 2593–2596.
Bordé, C. J. (1989). Atomic interferometry with internal state labeling. *Phys. Lett. A* **140**, 10.
Budker, D., Kimball, D. F., Rochester, S. M., and Yashchuk, V. V. (1999). Nonlinear magneto-optics and reduced group velocity of light in atomic vapor with slow ground state relaxation. *Phys. Rev. Lett.* **83**, 1767–1770.
Carroll, C. E., and Hioe, F. T. (1992). Coherent population transfer via the continuum. *Phys. Rev. Lett.* **68**, 3523–3526.
Carroll, C. E., and Hioe, F. T. (1993). Selective excitation via the continuum and suppression of ionization. *Phys. Rev. A* **47**, 571–580.
Carroll, C. E., and Hioe, F. T. (1995). Excitation using two lasers: effects of continuum-continuum transitions. *Phys. Lett. A* **199**, 145–150.
Carroll, C. E., and Hioe, F. T. (1996). Selective excitation and structure in the continuum. *Phys. Rev. A* **54**, 5147–5151.
Cerboneschi, E., and Arimondo, E. (1995). Transparency and dressing for optical pulse pairs through a double-lambda absorbing medium. *Phys. Rev. A* **52**, R1823–R1826.

Chu, S. (1998). Nobel lecture: The manipulation of neutral particles. *Rev. Mod. Phys.* **70**, 685–706.
Cirac, J. I., Zoller, P., Kimble, H. J., and Mabuchi, H. (1997). Quantum state transfer and entanglement distribution among distant nodes in a quantum network. *Phys. Rev. Lett.* **78**, 3221–3224.
Cohen-Tannoudji, C. N. (1998). Nobel lecture: Manipulating atoms with photons. *Rev. Mod. Phys.* **70**, 707–720.
Coulston, G. W., and Bergmann, K. (1992). Population transfer by stimulated Raman scattering with delayed pulses: Analytical results for multilevel systems. *J. Chem. Phys.* **96**, 3467–3495.
Crim, F. F. (1984). Selective excitation studies of unimolecular reaction dynamics. *Annu. Rev. Phys. Chem.* **35**, 657–691.
Crisp, M. D. (1973). Adiabatic-following approximation. *Phys. Rev. A* **8**, 2128–2135.
Dai, H. C., and Field, R. W. (1995). "Molecular Dynamics and Spectroscopy by Stimulated Emission Pumping." World Scientific, Singapore.
Danileiko, M. V., Romanenko, V. I., and Yatsenko, L. P. (1994). Landau–Zener transitions and population transfer in a three-level system driven by two delayed laser pulses. *Opt. Commun.* **109**, 462–466.
Davis, J. P., and Pechukas, P. (1976). Nonadiabatic transitions induced by a time-dependent Hamiltonian in the semiclassical/adiabatic limit: The two-state case. *J. Chem. Phys.* **64**, 3129–3137.
Demkov, Y. N., and Kunike, M. (1969). Hypergeometric models of the two-state approximation in the theory of atomic collisions. *Vestn. Leningr. Univ. Fiz. Khim.* **16**, 39–45 (in Russian).
Demtröder, W., McClintock, M., and Zare, R. N. (1969). Spectroscopy of Na_2 using laser-induced fluorescence. *J. Chem. Phys.* **51**, 5495–5509.
Dittmann, P., Pesl, F. P., Martin, J., Coulston, G. W., He, G. Z., and Bergmann, K. (1992). The effect of vibrational excitation ($3 \leq v' \leq 19$) on the reaction $Na_2 (v') + Cl \rightarrow NaCl + Na^*$. *J. Chem. Phys.* **97**, 9472–9475.
Drese, K., and Holthaus, M. (1998). Perturbative and nonperturbative processes in adiabatic population transfer. *Eur. Phys. J. D* **3**, 73–86.
Dykhne, A. M. (1962). Adiabatic perturbation of discrete spectrum states. *Sov. Phys. JETP* **14**, 941.
Eberly, J. H., Pons, M. L., and Haq, H. R. (1994). Dressed-field pulses in an absorbing medium. *Phys. Rev. Lett.* **72**, 56–59.
Einstein, A. (1917). Zur Quantentheorie de Strahlung. *Phys. Z.* **18**, 121.
Ekers, A., Kaufmann, O., Bergmann, K., Alnis, J., and Klavins, J. (1999a). Vibrational dependence of the Na_3^+-formation by associate ionization of $Na_3(P_{3/2}) + Na_2$ ($A\ ^1\Sigma_u^+, v''$). *Chem. Phys. Lett.* **304**, 69–72.
Ekers, A., Kaufmann, O., Keil, M., and Bergmann, K. (1999b). Associative ionization in collision of electronically excited Na atoms with vibrationally excited $Na_2(v'')$ molecules. *Eur. Phys. J. D* **7**, 65–71.
Elk, M. (1995). Adiabatic transition histories of population transfer in the Lambda system. *Phys. Rev. A* **52**, 4017–4022.
Esslinger, T., Sander, F., Weidemüller, M., Hemmerich, A., and Hänsch, T. W. (1996). Subrecoil laser cooling with adiabatic transfer. *Phys. Rev. Lett.* **76**, 2432–2435.
Fano, U. (1961). Effects of configuraton interaction on intensities and phase shifts. *Phys. Rev.* **124**, 1866–1878.
Fano, U., and Cooper, J. W. (1968). Spectral distribution of atomic oscillator strengths. *Rev. Mod. Phys.* **41**, 441.
Featonby, P. D., Summy, G. S., Martin, J. L., Wu, H., Zetie, K. P., Foot, C. J., and Burnett, K. (1996). Adiabatic transfer for atomic interferometry. *Phys. Rev. A* **53**, 373–380.
Featonby, P. D., Summy, G. S., Webb, C. L., Godun, R. M., Oberthaler, M. K., Wilson, A. C., Foot, C. J., and Burnett, K. (1998). Separated-path Ramsey atom interferometer. *Phys. Rev. Lett.* **81**, 495–499.
Fewell, M. P., Shore, B. W., and Bergmann, K. (1997). Coherent population transfer among three states: Full algebraic solutions and the relevance of non adiabatic processes to transfer by delayed pulses. *Austral. J. Phys.* **50**, 281–304.

Fleischhauer, M., Keitel, C. H., Narducci, L. M., Scully, M. O., Zhu, S.-Y., and Zubairy, M. S. (1992). Lasing without inversion: Interference of radiatively broadened resonances in dressed atomic systems. *Opt. Commun.* **94**, 599.

Fleischhauer, M. (1994). Correlation of high-frequency phase fluctuations in electromagnetically induced transparency. *Phys. Rev. Lett.* **72**, 989–992.

Fleischhauer, M., and Lukin, M. D. (2000). Dark-state polaritons in electromagnetically induced transparency. *Phys. Rev. Lett.* **84**, 5094–5097.

Fleischhauer, M., and Manka, A. S. (1996). Propagation of laser pulses and coherent population transfer in dissipative three-level systems: An adiabatic dressed-state picture. *Phys. Rev. A* **54**, 794–803.

Fleischhauer, M., Yelin, S. F., and Lukin, M. D. (2000). How to trap photons? Storing single-photon quantum states in collective atomic excitations. *Opt. Commun.* **179**, 395–410.

Frishman, E., and Shapiro, M. (1996). Reversibility of bound-to-continuum transitions induced by a strong short laser pulse and the semiclassical uniform approximation. *Phys. Rev. A* **54**, 3310–3321.

Garraway, B. M., and Suominen, K.-A. (1998). Adiabatic passage by light-induced potentials in molecules. *Phys. Rev. Lett.* **80**, 932–935.

Gaubatz, U., Rudecki, P., Becker, M., Schiemann, S., Külz, M., and Bergmann, K. (1988). Population switching between vibrational levels in molecular beams. *Chem. Phys. Lett.* **149**, 463–468.

Gaubatz, U., Rudecki, P., Schiemann, S., and Bergmann, K. (1990). Population transfer between molecular vibrational levels by stimulated Raman scattering with partially overlapping laser fields. A new concept and experimental results. *J. Chem. Phys.* **92**, 5363–5376.

Glushko, B., and Kryzhanovsky, B. (1992). Radiative and collisional damping effects on efficient population transfer in a three-level system driven by two delayed laser pulses. *Phys. Rev. A* **46**, 2823–2830.

Godun, R. M., Webb, C. L., Oberthaler, M. K., Summy, G. S., and Burnett, K. (1999). Efficiencies of adiabatic transfer in a multistate system. *Phys. Rev. A* **59**, 3775–3781.

Goldner, L. S., Gerz, C., Spreeuw, R. J. C., Rolston, S. L., Westbrook, C. I., Phillips, W. D., Marte, P., and Zoller, P. (1994a). Momentum transfer in laser-cooled cesium by adiabatic passage in a light field. *Phys. Rev. Lett.* **72**, 997–1000.

Goldner, L. S., Gerz, C., Spreeuw, R. J. C., Rolston, S. L., Westbrook, C. I., Phillips, W. D., Marte, P., and Zoller, P. (1994b). Coherent transfer of photon momentum by adiabatic following in a dark state. *Quantum Opt.* **6**, 387–389.

Goral, K., Gajda, M., and Rzazewski, K. (2000). Multi-mode dynamics of a coupled ultracold atomic-molecular system. *Phys. Rev. Lett.* **86**, 1397–1401.

Gottwald, E., Mattheus, A., Schinke, R., and Bergmann, K. (1986). Angularly resolved vibrational excitation in Na_2-He collisions. *J. Chem. Phys.* **84**, 756–763.

Gottwald, E., Mattheus, A., Schinke, R., and Bergmann, K. (1987). Resolution of supernumerary rotational rainbows in Na_2-He-Ne, Ar collisions. *J. Chem. Phys.* **86**, 2685–2688.

Gray, H. R., Whitley, R. M., and Stroud, C. R. (1978). Coherent trapping of atomic populations. *Opt. Lett.* **3**, 218–220.

Grobe, R., Hioe, F. T., and Eberly, J. H. (1994). Formation of shape-preserving pulses in a nonlinear adiabatically integrable system. *Phys. Rev. Lett.* **73**, 3183–3186.

Guérin, S., and Jauslin, H. R. (1998). Two-laser multiphoton adiabatic passage in the frame of the Floquet theory. Applications to (1+1) and (2+1) STIRAP. *Eur. Phys. J. D* **2**, 99–113.

Guérin, S., Yatsenko, L. P., Halfmann, T., Shore, B. W., and Bergmann, K. (1998). Stimulated hyper-Raman adiabatic passage. II. Static compensation of dynamic Stark shifts. *Phys. Rev. A* **58**, 4691–4704.

Guérin, S., Jauslin, H. R., Unanyan, R. G., and Yatsenko, L. P. (1999). Floquet perturbative analysis for STIRAP beyond the rotating wave approximation. *Optics Express* **4**, 84–90.

Halfmann, T., and Bergmann, K. (1996). Coherent population transfer and dark resonances in SO_2. *J. Chem. Phys.* **104**, 7068–7072.

Halfmann, T., Yatsenko, L. P., Shapiro, M., Shore, B. W., and Bergmann, K. (1998). Population trapping and laser-induced continuum structure in helium: Experiment and theory. *Phys. Rev. A* **58**, R46–R49.
Hamilton, C. H., Kinsey, J. L., and Field, R. W. (1986). Stimulated emission pumping: New methods in spectroscopy and molecular dynamics. *Annu. Rev. Phys. Chem.* **37**, 493–524.
Harris, S. E., Field, J. E., and Imamoglu, A. (1990). Nonlinear optical processes using electromagnetically induced transparency. *Phys. Rev. Lett.* **64**, 1107–1110.
Harris, S. E. (1993). Electromagnetically induced transparency with matched pulses. *Phys. Rev. Lett.* **70**, 552–555.
Harris, S. E. (1994). Normal modes for electromagnetically induced transparency. *Phys. Rev. Lett* **72**, 52–55.
Harris, S. E. (1997). Electromagnetically induced transparency. *Phys. Today* **50**, 36–42.
Harris, S. E. (2000). Ponderomotive forces with slow light. *Phys. Rev. Lett.* **85**, 4032–5.
Harris, S. E., and Hau, L. V. (1999). Nonlinear optics at low light levels. *Phys. Rev. Lett.* **82**, 4611–4614.
Harris, S. E., and Luo, Z. F. (1995). Preparation energy for electromagnetically induced transparency. *Phys. Rev. A* **52**, R928–R931.
Hau, L. V., Harris, S. E., Dutton, Z., and Behroozi, C. H. (1999). Light speed reduction to 17 metres per second in an ultracold atomic gas. *Nature* **397**, 594–598.
He, G.-Z., Kuhn, A., Schiemann, S., and Bergmann, K. (1990). Population transfer by stimulated Raman scattering with delayed pulses and by the stimulated-emission pumping method: a comparative study. *J. Opt. Soc. Am. B* **7**, 1960–1969.
Hefter, U., Jones, P., Bergmann, K., and Schinke, R. (1981). Resolution of supernumerary rotational rainbows in state-to-state Na_2-Ne scattering. *Phys. Rev. Lett.* **46**, 915–918.
Hefter, U., Eichert, J., and Bergmann, K. (1985). An optically pumped iodine supersonic beam laser. *Opt. Commun.* **52**, 330–535.
Hefter, U., Ziegler, G., Mattheus, A., Fischer, A., and Bergmann, K. (1986). Preparation and detection of molecular alignment with high state selectivity by saturated optical pumping in beams. *J. Chem. Phys.* **85**, 286–302.
Hemmer, P. R., Katz, D. P., Donoghue, J., Cronin-Golomb, M., Shahriar, M. S., and Kumar, P. (1995). Efficient low-intensity optical phase conjugation based on coherent population trapping in sodium. *Opt. Lett.* **20**, 982–984.
Hennrich, M., Legero, T., Kuhn, A., and Rempe, G. (2000). Vacuum-stimulated Raman scattering based on adiabatic passage in a high-finesse optical cavity. *Phys. Rev. Lett.* **85**, 4872–4875.
Hioe, F. T. (1984). Solution of Bloch equations involving amplitude and frequency modulations. *Phys. Rev. A* **30**, 2100–2103.
Hioe, F. T., and Carroll, C. E. (1988). Coherent population trapping in N-level quantum systems. *Phys. Rev. A* **37**, 3000–3005.
Imamoglu, A., and Harris, S. E. (1989). Lasers without inversion: interference of dressed lifetime-broadened states. *Opt. Lett.* **14**, 1344–1346.
Jain, M. (1994). Excess noise correlation using population-trapped atoms. *Phys. Rev. A* **50**, 1899–1902.
Javanainen, J., and Mackie, M. (1998). Probability of photoassociation from a quasicontinuum approach. *Phys. Rev. A* **58**, R789–R792.
Javanainen, J., and Mackie, M. (1999). Coherent photoassociation of a Bose-Einstein condensate. *Phys. Rev. A* **59**, R3186–R3189.
Jones, P. L., Gaubatz, U., Hefter, U., Wellegehausen, B., and Bergmann, K. (1983). An optically pumped sodium-dimer supersonic beam laser. *Appl. Phys. Lett.* **42**, 222–224.
Kallush, S., and Band, Y. B. (2000). Short-pulse chirped adiabatic population transfer in diatomic molecules. *Phys. Rev. A* **61**, 041401.
Kasapi, A., Jain, M., Yin, G. Y., and Harris, S. E. (1995). Electromagnetically induced transparency: Propagation dynamics. *Phys. Rev. Lett.* **74**, 2447–2450.

Kasevich, M., and Chu, S. (1991). Atomic interferometry using stimulated Raman transitions. *Phys. Rev. Lett.* **67**, 181–184.
Kasevich, M., and Chu, S. (1992). Laser cooling below a photon recoil with three-level atoms. *Phys. Rev. Lett.* **69**, 1741–1744.
Kasevich, M., Weiss, D. S., Riis, E., Moler, K., Kasapi, S., and Chu, S. (1991). Atomic velocity selection using stimulated Raman transitions. *Phys. Rev. Lett.* **66**, 2297–2300.
Kash, M. M., Sautenkov, V. A., Zibrov, A. S., Hollberg, L., Welch, G. R., Lukin, M. D., Rostovtsev, Y., Fry, E. S., and Scully, M. O. (1999). Ultraslow group velocity and enhanced nonlinear optical effects in a coherently driven hot atomic gas. *Phys. Rev. Lett.* **82**, 5229–5232.
Kaufmann, O., Ekers, A., Gebauer-Rochholz, Ch., Mettendorf, K. U., Keil, M., and Bergmann, K. (2001). Dissociative charge transfer from highly excited Na Rydberg atoms to vibrationally excited Na_2 molecules. *Int. J. Mass Spectr.* in press.
Keil, M., Kolling, T., Bergmann, K., and Meyer, W. (1999). Dissociative attachment of low energy electrons to vibrationally excited Na_2-molecules using a novel photoelectron source. *Eur. Phys. J. D* **7**, 55–64.
Kittrell, C., Abramson, E., Kinsey, J. L., McDonald, S. A., Reisner, D. E., Field, R. W., and Katayama, D. H. (1981). Selective vibrational excitation by stimulated emission pumping. *J. Chem. Phys.* **75**, 2056–2059.
Knight, P. L. (1984). Laser-induced continuum structure. *Commun. Atomic Mol. Phys.* **15**, 193–214.
Knight, P. L., Lauder, M. A., and Dalton, B. J. (1990). Laser-induced continuum structure. *Phys. Rep.* **190**, 1–61.
Kobrak, M. N., and Rice, S. A. (1998a). Coherent population transfer via a resonant intermediate state: The breakdown of adiabatic passage. *Phys. Rev. A* **57**, 1158–1163.
Kobrak, M. N., and Rice, S. A. (1998b). Selective photochemistry via adiabatic passage: An extension of stimulated Raman adiabatic passage for degenerate final states. *Phys. Rev. A* **58**, 2885–2894.
Kobrak, M. N., and Rice, S. A. (1998c). Equivalence of the Kobrak- Rice photoselective adiabatic passage and the Brumer-Shapiro strong field methods for control of product formation in a reaction. *J. Chem. Phys.* **109**, 1–10.
Kuhn, A., Coulston, G., He, G. Z., Schiemann, S., Bergmann, K., and Warren, W. S. (1992). Population transfer by stimulated Raman scattering with delayed pulses using spectrally broad light. *J. Chem. Phys.* **96**, 4215–4223.
Kuhn, A., Steuerwald, S., and Bergmann, K. (1998). Coherent population transfer in NO with pulsed lasers: The consequences of hyperfine structure, doppler broadening and electromagnetically induced absorption. *Eur. Phys. J. D* **1**, 57–70.
Kuklinski, J. R., Gaubatz, U., Hioe, F. T., and Bergmann, K. (1989). Adiabatic population transfer in a three-level system driven by delayed laser pulses. *Phys. Rev. A* **40**, 6741–6744.
Kulin, S., Saubamea, B., Peik, E., Lawall, J., Hijmans, T. W., Leduc, M., and Cohen-Tannoudji, C. (1997). Coherent manipulation of atomic wave packets by adiabatic transfer. *Phys. Rev. Lett.* **78**, 4185–4188.
Kurzel, R. B., and Steinfeld, J. I. (1970). Energy-transfer processes in monochromatically excited iodine molecules. III. Quenching and multiquantum transfer from $v' = 43$. *J. Chem. Phys.* **53**, 3293–3303.
Külz, M., Kortyna, A., Keil, M., Schellhaass, B., and Bergmann, K. (1993). Vibrational dependence of negative-ion formation by dissociative attachment of low-energy electrons. *Phys. Rev. A* **48**, 4015–4018.
Külz, M., Kortyna, A., Keil, M., Schellhaass, B., and Bergmann, K. (1995). On the vibrational dependence of electron impact ionization of diatomic molecules. *Z. Phys. D* **33**, 109–117.
Külz, M., Keil, M., Kortyna, A., Schellhaass, B., Hauck, J., Bergmann, K., Meyer, W., and Weyh, D. (1996). Dissociative attachment of low-energy electrons to state-selected diatomic molecules. *Phys. Rev. A* **53**, 3324–3334.

Kylstra, N. J., Paspalakis, E., and Knight, P. L. (1998). Laser-induced continuum structure in helium: ab initio non-perturbative calculations. *J. Phys. B* **31**, L719–L728.

Laine, T. A., and Stenholm, S. (1996). Adiabatic processes in three-level systems. *Phys. Rev. A* **53**, 2501–2512.

Landau, L. D. (1932). Zur Theorie der Energie übertragung II. *Phys. Z. Sowjetunion* **2**, 46.

Lange, W., and Kimble, H. J. (2000). Dynamic generation of maximally entangled photon multiplets by adiabatic passage. *Phys. Rev. A* **61**, 063817.

Law, C. K., and Eberly, J. H. (1996). Arbitrary control of a quantum electromagnetic field. *Phys. Rev. Lett.* **76**, 1055–1058.

Law, C. K., and Eberly, J. H. (1998). Synthesis of arbitrary superposition of Zeeman states in an atom. *Opt. Express* **2**, 368–371.

Law, C. K., and Kimble, H. J. (1997). Deterministic generation of a bit-stream of single-photon pulses. *J. Mod. Opt.* **44**, 2067–2074.

Lawall, J., and Prentiss, M. (1994). Demonstration of a novel atomic beam splitter. *Phys. Rev. Lett.* **72**, 993–996.

Lawall, J., Bardou, F., Saubamea, B., Shimizu, K., Leduc, M., Aspect, A., and Cohen-Tannoudji, C. (1994). Two-dimensional subrecoil laser cooling. *Phys. Rev. Lett.* **73**, 1915–1918.

Lawall, J., Kulin, S., Saubaméa, B., Bigelow, N., Leduc, M., and Cohen-Tannoudji, C. (1995). Three-dimensional laser cooling of helium beyond the single-photon recoil limit. *Phys. Rev. Lett.* **75**, 4194–4197.

Lawall, J., Kulin, S., Saubamea, B., Bigelow, N., Leduc, M., and Cohen-Tannoudji, C. (1996). Subrecoil laser cooling into a single wavepacket by velocity-selective coherent population trapping followed by adiabatic passage. *Laser Phys.* **6**, 153–158.

Lindinger, A., Verbeek, M., and Rubahn, H.-G. (1997). Adiabatic population transfer by acoustooptically modulated laser beams. *Z. Phys. D* **39**, 93–100.

Liu, C., Dutton, Z., Pehroozi, C. H., and Hau, L. V. (2001). Observation of coherent optical information storage in an atomic medium using halted light pulses. *Nature* **409**, 490–493.

Lounis, B., and Cohen-Tannoudji, C. (1992). Coherent population trapping and Fano profiles. *J. Phys. II (France)* **2**, 579.

Loy, M. M. T. (1974). Observation of population inversion by optical adiabatic rapid passage. *Phys. Rev. Lett.* **32**, 814–817.

Lukin, M. D., and Imamoglu, A. (2000). Nonlinear optics and quantum entanglement of ultraslow single photons. *Phys. Rev. Lett.* **84**, 1419–1422.

Lukin, M. D., Fleischhauer, M., Zibrov, A. S., Robinson, H. G., Velichansky, V. L., Hollberg, L., and Scully, M. O. (1997). Spectroscopy in dense coherent media: Line narrowing and interference effects. *Phys. Rev. Lett.* **79**, 2959–2962.

Lukin, M. D., Yelin, S. F., and Fleischhauer, M. (2000). Entanglement of atomic ensembles by trapping correlated photon states. *Phys. Rev. Lett.* **84**, 4232–4235.

Mackie, M., and Javanainen, J. (1999). Quasicontinuum modeling of photoassociation. *Phys. Rev. A* **60**, 3174–3187.

Mackie, M., Kowalski, R., and Javanainen, J. (2000). Bose-stimulated Raman adiabatic passage in photoassociation. *Phys. Rev. Lett.* **84**, 3803–3806.

Malinovsky, V. S., and Tannor, D. J. (1997). Simple and robust extension of the stimulated Raman adiabatic passage technique to N-level systems. *Phys. Rev. A* **56**, 4929–4937.

Marangos, J. P. (1998). Electromagnetically induced transparency. *J. Mod. Opt.* **45**, 471–503.

Marte, P., Zoller, P., and Hall, J. L. (1991). Coherent atomic mirrors and beam splitters by adiabatic passage in multilevel systems. *Phys. Rev. A* **44**, 4118–4121.

Martin, J., Shore, B. W., and Bergmann, K. (1995). Coherent population transfer in multilevel systems with magnetic sublevels. II. Algebraic analysis. *Phys. Rev. A* **52**, 583–593.

Martin, J., Shore, B. W., and Bergmann, K. (1996). Coherent population transfer in multilevel systems with magnetic sublevels. III. Experimental results. *Phys. Rev. A* **54**, 1556–1569.

Matsko, A., Rostovstsev, Y. V., Fleischhauer, M., and Scully, M. O. (2001). Anomalous Brillouin scattering via ultra-slow light. *Phys. Rev. Lett.,* in press.

Mattheus, A., Fischer, A., Ziegler, G., Gottwald, E. E., and Bergmann, K. (1986). Experimental proof of $\Delta m \ll J$ propensity rule in rotationally inelastic differential scattering. *Phys. Rev. Lett.* **56**, 712–715.

Meier, W., Ahlers, G., and Zacharias, H. (1986). State selective population of H_2 ($v' = 1$, $J' = 1$) and D_2 ($v' = 1$, $J' = 2$) and rotational relaxation in collisions with H_2, D_2, and He. *J. Chem. Phys.* **85**, 2599–2608.

Messiah, A. (1962). "Quantum Mechanics." North-Holland, New York.

Milner, V., and Prior, Y. (1998). Multilevel dark states: Coherent population trapping with elliptically polarized incoherent light. *Phys. Rev. Lett.* **80**, 940–943.

Morigi, G., Featonby, P., Summy, G., and Foot, C. (1996). Calculation of the efficiencies and phase shifts associated with an adiabatic transfer atom interferometer. *Quant. Semiclass. Opt.* **8**, 641–653.

Morris, J. R., and Shore, B. W. (1983). Reduction of degenerate two-level excitation to independent two-state systems. *Phys. Rev. A* **27**, 906–912.

Nakajima, T. (1999). Population transfer in N-level systems assisted by dressing fields. *Phys. Rev. A* **59**, 559–568.

Nakajima, T., and Lambropolous, P. (1996). Population transfer through an autoionizing state by pulse delay. *Z. Phys. D* **36**, 17–22.

Nakajima, T., Elk, M., Zhang, J., and Lambropoulos, P. (1994). Population transfer through the continuum. *Phys. Rev. A* **50**, R913–R916.

Newton, R. G., and Young, B.-L. (1968). Measurability of the spin density matrix. *Ann. Phys. (New York)* **49**, 393–402.

Oreg, J., Hioe, F. T., and Eberly, J. H. (1984). Adiabatic following in multilevel systems. *Phys. Rev. A* **29**, 690–697.

Oreg, J., Bergmann, K., Shore, B. W., and Rosenwaks, S. (1992). Population transfer with delayed pulses in four-state systems. *Phys. Rev. A* **45**, 4888–4896.

Orr, B. J., Haub, J. G., and Haines, R. (1984). Time-resolved infrared-ultraviolet double-resonance studies of rotational relaxation in D_2CO. *Chem. Phys. Lett.* **107**, 168.

Parkins, A. S., and Kimble, H. J. (1999). Quantum state transfer between motion and light. *J. Opt. B* **1**, 496–504.

Parkins, A. S., Marte, P., Zoller, P., and Kimble, H. J. (1993). Synthesis of arbitrary quantum states via adiabatic transfer of Zeeman coherence. *Phys. Rev. Lett.* **71**, 3095–3098.

Parkins, A. S., Marte, P., Zoller, P., Carnal, O., and Kimble, H. J. (1995). Quantum-state mapping between multilevel atoms and cavity light fields. *Phys. Rev. A* **51**, 1578–1796.

Paspalakis, E., and Knight, P. L. (1998). Population transfer via an autoionizing state with temporally delayed chirped laser pulses. *J. Phys. B* **31**, 2753–2767.

Paspalakis, E., Protopapas, M., and Knight, P. L. (1997). Population transfer through the continuum with temporally delayed chirped laser pulses. *Opt. Commun.* **142**, 34–40.

Paspalakis, E., Protopapas, M., and Knight, P. L. (1998). Time-dependent pulse and frequency effects in population trapping via the continuum. *J. Phys. B* **31**, 775–794.

Pellizzari, T. (1997). Quantum networking with optical fibres. *Phys. Rev. Lett.* **79**, 5242–5245.

Pellizzari, T., Gardiner, S. A., Cirac, J. I., and Zoller, P. (1995). Decoherence, continuous observation, and quantum computing: A cavity QED model. *Phys. Rev. Lett.* **75**, 3788–3791.

Pesl, F. (1999). Ph.D. thesis. University of Kaiserslautern, Kaiserslautern, Germany.

Phillips, W. D. (1998). Nobel lecture: Laser cooling and trapping of neutral atoms. *Rev. Mod. Phys.* **70**, 721–742.

Phillips, D. F., Fleischhauer, A., Mair, A., Walsworth, R. L., and Lukin, M. D. (2001). Storage of light in atomic vapor. *Phys. Rev. Lett.* **86**, 783–786.

Pillet, P., Valentin, C., Yuan, R.-L., and Yu, J. (1993). Adiabatic population transfer in a multilevel system. *Phys. Rev. A* **48**, 845–848.

Rickes, T., Yatsenko, L. P., Steuerwald, S., Halfmann, T., Shore, B. W., Vitanov, N. V., and Bergmann, K. (2000). Efficient adiabatic population transfer by two-photon excitation assisted by a laser-induced Stark shift. *J. Chem. Phys.* **113**, 534–546.

Riehle, F., Kisters, T., Witte, A., Helmcke, J., and Bord, C. J. (1991). Optical Ramsey spectroscopy in a rotating frame: Sagnac effect in a matter-wave interferometer. *Phys. Rev. Lett.* **67**, 177–180.

Rodriguez, M., Garraway, B. M., and Suominen, K.-A. (2000). Tailoring of vibrational state populations with light-induced potentials in molecules. *Phys. Rev. A* **62**, 53413.

Romanenko, V. I., and Yatsenko, L. P. (1997). Adiabatic population transfer in the three-level lambda-system: Two-photon lineshape. *Opt. Commun.* **140**, 231–236.

Rubahn, H.-G., and Bergmann, K. (1990). Effect of laser-induced vibrational bond stretching in atom-diatom collisions. *Annu. Rev. Phys. Chem.* **41**, 735–773.

Rubahn, H.-G., Konz, E., Schiemann, S., and Bergmann, K. (1991). Alignment of electronic angular momentum by stimulated Raman scattering with delayed pulses. *Z. Phys. D* **22**, 401–406.

Schiemann, S., Kuhn, A., Steuerwald, S., and Bergmann, K. (1993). Efficient coherent population transfer in NO molecules using pulsed lasers. *Phys. Rev. Lett.* **71**, 3637–3640.

Series, G. W. (1978). A semiclassical approach to radiation problems. *Phys. Rep.* **43**, 1–41.

Shore, B. W. (1990). "The Theory of Coherent Atomic Excitation." Wiley, New York.

Shore, B. W. (1995). Examples of counter-intuitive physics (atomic and molecular excitation). *Contemp. Phys.* **36**, 15–28.

Shore, B. W., Bergmann, K., Oreg, J., and Rosenwaks, S. (1991). Multilevel adiabatic population transfer. *Phys. Rev. A* **44**, 7442–7447.

Shore, B. W., Bergmann, K., Kuhn, A., Schiemann, S., Oreg, J., and Eberly, J. H. (1992a). Laser-induced population transfer in multistate systems: A comparative study. *Phys. Rev. A* **45**, 5297–5300.

Shore, B. W., Bergmann, K., and Oreg, J. (1992b). Coherent population transfer: stimulated Raman adiabatic passage and the Landau-Zener picture. *Z. Phys. D* **23**, 33–39.

Shore, B. W., Martin, J., Fewell, M. P., and Bergmann, K. (1995). Coherent population transfer in multilevel systems with magnetic sublevels. I. Numerical studies. *Phys. Rev. A* **52**, 566–582.

Smith, A. V. (1992). Numerical studies of adiabatic population inversion in multilevel systems. *J. Opt. Soc. Am. B* **9**, 1543–1551.

Solá, I. R., Malinovsky, V. S., and Tannor, D. J. (1999). Optimal pulse sequences for population transfer in multilevel systems. *Phys. Rev. A* **60**, 3081–3090.

Solá, I. R., Santamaría, J., and Malinovsky, V. S. (2000). Efficiency and robustness of adiabatic passage by light-induced potentials. *Phys. Rev. A* **61**, 043413.

Steane, A. (1998). Quantum computing. *Rep. Prog. Phys.* **61**, 117–173.

Suominen, K.-A., and Garraway, B. M. (1992). Population transfer in a level-crossing model with two time scales. *Phys. Rev. A* **45**, 374–386.

Süptitz, W., Duncan, B. C., and Gould, P. L. (1997). Efficient 5d excitation of trapped Rb atoms using pulses of diode-laser light in the counterintuitive order. *J. Opt. Soc. Am. B* **14**, 1001–1008.

Theuer, H., and Bergmann, K. (1998). Atomic beam deflection by coherent momentum transfer and the dependence on weak magnetic fields. *Eur. Phys. J. D* **2**, 279–289.

Theuer, H., Unanyan, R. G., Habscheid, C., Klein, K., and Bergmann, K. (1999). Novel laser controlled variable matter wave beamsplitter. *Opt. Express* **4**, 77–83.

Unanyan, R., Fleischhauer, M., Shore, B. W., and Bergmann, K. (1998a). Robust creation and phase-sensitive probing of superposition states via stimulated Raman adiabatic passage (STIRAP) with degenerate dark states. *Opt. Commun.* **155**, 144–154.

Unanyan, R. G., Vitanov, N. V., and Stenholm, S. (1998b). Suppression of incoherent ionization in population transfer via continuum. *Phys. Rev. A* **57**, 462–466.

Unanyan, R. G., Shore, B. W., and Bergmann, K. (1999). Laser-driven population transfer in four-level atoms: Consequences of non-Abelian geometrical adiabatic phase factors. *Phys. Rev. A* **59**, 2910–2919.

Unanyan, R. G., Vitanov, N. V., Shore, B. W., and Bergmann, K. (2000a). Coherent properties of a tripod system coupled via a continuum. *Phys. Rev. A* **61**, 043408.
Unanyan, R., Guerin, S., Shore, B. W., and Bergmann, K. (2000b). Efficient population transfer by delayed pulses despite coupling ambiguity. *Eur. Phys. J. D* **8**, 443–449.
Unanyan, R. G., Guérin, S., and Jauslin, H. R. (2000c). Coherent population trapping under bichromatic fields. *Phys. Rev. A* **62**, 043407.
Valentin, C., Yu, J., and Pillet, P. (1994). Adiabatic transfer In $J \to J$ and $J \to J - 1$ transitions. *J. Phys. II (France)* **4**, 1925–1937.
van Enk, S. J., Kimble, H. J., Cirac, J. I., and Zoller, P. (1999). Quantum communication with dark photons. *Phys. Rev. A* **59**, 2659–2664.
Vardi, A., and Shapiro, M. (1996). Two-photon dissociation/ionization beyond the adiabatic approximation. *J. Chem. Phys.* **104**, 5490–5496.
Vardi, A., Abrashkevich, D., Frishman, E., and Shapiro, M. (1997). Theory of radiative recombination with strong laser pulses and the formation of ultracold molecules via stimulated photorecombination of cold atoms. *J. Chem. Phys.* **107**, 6166–6174.
Vardi, A., Shapiro, M., and Bergmann, K. (1999). Complete population transfer to and from a continuum and the radiative association of cold Na atoms to produce translationally cold Na_2 molecules in specific vib-rotational states. *Opt. Express* **4**, 91–106.
Vitanov, N. V. (1995). Complete population return in a two-state system driven by a smooth asymmetric pulse. *J. Phys. B* **28**, L19–L22.
Vitanov, N. V. (1998a). Adiabatic population transfer by delayed laser pulses in multistate systems. *Phys. Rev. A* **58**, 2295–2309.
Vitanov, N. V. (1998b). Analytic model of a three-state system driven by two laser pulses on two-photon resonance. *J. Phys. B* **31**, 709–725.
Vitanov, N. V. (1999). Pulse-order invariance of the initial-state population in multistate chains driven by delayed laser pulses. *Phys. Rev. A* **60**, 3308–3310.
Vitanov, N. V. (2000). Measuring a coherent superposition of multiple states. *J. Phys. B* **33**, 2333–2346.
Vitanov, N. V. (2001). Two-photon line width in stimulated Raman adiabatic passage. To be published.
Vitanov, N. V., and Knight, P. L. (1995). Coherent excitation by asymmetric pulses. *J. Phys. B* **28**, 1905–1920.
Vitanov, N. V., and Stenholm, S. (1996). Non-adiabatic effects in population transfer in three-level systems. *Opt. Commun.* **127**, 215–222.
Vitanov, N. V., and Stenholm, S. (1997a). Properties of stimulated Raman adiabatic passage with intermediate-level detuning. *Opt. Commun.* **135**, 394–405.
Vitanov, N. V., and Stenholm, S. (1997b). Analytic properties and effective two-level problems in stimulated Raman adiabatic passage. *Phys. Rev. A* **55**, 648–660.
Vitanov, N. V., and Stenholm, S. (1997c). Population transfer by delayed pulses via continuum states. *Phys. Rev. A* **56**, 741–747.
Vitanov, N. V., and Stenholm, S. (1997d). Population transfer via a decaying state. *Phys. Rev. A* **56**, 1463–1471.
Vitanov, N. V., and Stenholm, S. (1999). Adiabatic population transfer via multiple intermediate states. *Phys. Rev. A* **60**, 3820–3832.
Vitanov, N. V., Shore, B. W., and Bergmann, K. (1998). Adiabatic population transfer in multistate chains via dressed intermediate states. *Eur. Phys. J. D* **4**, 15–29.
Vitanov, N. V., Suominen, K.-A., and Shore, B. W. (1999). Creation of coherent atomic superpositions by fractional stimulated Raman adiabatic passage. *J. Phys. B* **32**, 4535–4546.
Vitanov, N. V., Shore, B. W., Unanyan, R. G., and Bergmann, K. (2000). Measuring a coherent superposition. *Opt. Commun.* **179**, 73–83.
Vitanov, N. V., Halfmann, T., Shore, B. W., and Bergmann, K. (2001). Laser-induced population transfer by adiabatic passage techniques. *Annu. Rev. Phys. Chem.* **52**, 763–809.

Webb, C. L., Godun, R. M., Summy, G. S., Oberthaler, M. K., Featonby, P. D., Foot, C. J., and Burnett, K. (1999). Measurement of Berry's phase using an atom interferometer. *Phys. Rev. A* **60**, R1783–R1786.

Weitz, M., Young, B. C., and Chu, S. (1994a). Atomic interferometer based on adiabatic population transfer. *Phys. Rev. Lett.* **73**, 2563–2566.

Weitz, M., Young, B. C., and Chu, S. (1994b). Atom manipulation based on delayed laser pulses in three- and four-level systems: Light shifts and transfer efficiencies. *Phys. Rev. A* **50**, 2438–2444.

Weitz, M., Heupel, T., and Hänsch, T. W. (1996). Multiple beam atomic interferometer. *Phys. Rev. Lett.* **77**, 2356–2359.

Williams, C. P., and Clearwater, S. H. (1997). "Explorations in Quantum Computing." Springer-Verlag, Berlin.

Yatsenko, L. P., Unanyan, R. G., Bergmann, K., Halfmann, T., and Shore, B. W. (1997). Population transfer through the continuum using laser-controlled Stark shifts. *Opt. Commun.* **135**, 406–412.

Yatsenko, L. P., Guerin, S., Halfmann, T., Boehmer, K., Shore, B. W., and Bergmann, K. (1998). Stimulated hyper-Raman adiabatic passage. I. The basic problem and examples. *Phys. Rev. A* **58**, 4683–4690.

Yatsenko, L. P., Halfmann, T., Shore, B. W., and Bergmann, K. (1999a). Photoionization suppression by continuum coherence: Experiment and theory. *Phys. Rev. A* **59**, 2926–2947.

Yatsenko, L. P., Shore, B. W., Halfmann, T., Bergmann, K., and Vardi, A. (1999b). Source of metastable H(2s) atoms using the Stark chirped rapid-adiabatic-passage technique. *Phys. Rev. A* **60**, R4237–R4240.

Zener, C. (1932). Non-adiabatic crossing of energy levels. *Proc. R. Soc. London Ser. A* **137**, 696–969.

Ziegler, G., Rädle, M., Pütz, O., Jung, K., Ehrhardt, H., and Bergmann, K. (1987). Rotational rainbows in electron scattering. *Phys. Rev. Lett.* **58**, 2642–2645.

Ziegler, G., Kumar, S. V. K., Dittmann, P., and Bergmann, K. (1988). Effect of vibrational bond stretching in rotationally inelastic electron-molecule scattering. *Z. Phys. D* **10**, 247–252.

SLOW, ULTRASLOW, STORED, AND FROZEN LIGHT

ANDREY B. MATSKO,[1] OLGA KOCHAROVSKAYA,[1,2] YURI ROSTOVTSEV,[1] GEORGE R. WELCH,[1] ALEXANDER S. ZIBROV,[1,4] MARLAN O. SCULLY,[1,3]

[1]*Department of Physics and Institute for Quantum Studies, Texas A&M University, College Station, Texas 77843*
[2]*Institute of Applied Physics, RAS, 603600 Nizhny Novgorod, Russia*
[3]*Max-Planck-Institut für Quantenoptik, D-85748 Garching, Germany*
[4]*P. N. Lebedev Institute of Physics, RAS, Moscow, 117 924, Russia*

I. Introduction	191
II. Group Velocity: Kinematics	193
III. Slow Light	196
A. High Refractive Index and Slow Light	196
B. Slow Light in Heterostructures	199
C. Slow Light in Two-Level Systems, SIT	202
D. Slow Light in Λ Systems	204
E. Early Slow Light Experiments Involving Atomic Coherence	209
IV. Ultraslow Light	209
A. Slow Light in a Cold Gas	209
B. Slow Light in a Hot Gas	211
C. Slow Light and Nonlinear Magnetooptic Rotation	212
D. Slow Light in Solids	213
V. Storing and Retrieving Quantum Information	214
VI. The Ultimate Slow Light: Frozen Light	219
VII. Applications	225
A. Magnetometry	225
1. General Concept	225
2. Phaseonium Magnetometers	225
3. Magnetometers Based on Nonlinear Faraday Effect	226
B. Intracavity Electromagnetically Induced Transparency	227
C. Slow Light and Phonons	229
D. From Black Hole to Gyro	234
VIII. Conclusion	236
IX. References	237

I. Introduction

In their classic treatment of the propagation of light in dispersive media, Sommerfeld and Brillouin [20, 28] introduced five different kinds of wave velocities.

- The phase velocity, which is the speed at which the zero crossings of the carrier wave move
- The group velocity, at which the peak of a wave packet moves
- The energy velocity, at which the energy is transported by the wave
- The signal velocity, at which the half-maximum wave amplitude moves
- The front velocity, at which the first appearance of a discontinuity moves

All five velocities can differ from each other. In linear passive dispersive media, the group, energy, signal, and front velocities coincide and are usually less than the phase velocity. However, recent experimental demonstrations [23, 77, 98] that the group velocity of light can be reduced by 10–100 million compared with its phase velocity have intrigued the layman and fueled many studies and discussions that are exciting in their own right. These remarkable results are based on usage of very steep frequency dispersion in the vicinity of narrow resonance of electromagnetically induced transparency (EIT) [8, 70, 130].

The goal of this chapter is to address frequent questions asked about this topic. For example: What is special about this phenomenon? Why is slow light interesting for quantum and nonlinear optics? How can pulses of light moving so slowly as to take months to cross the Atlantic Ocean increase the speed of optical communications?

It is well known that any reactive medium leads to a delay of electromagnetic pulses and a system as simple as an infinite chain of RLC circuits can significantly reduce the speed of an electromagnetic pulse resonant within the circuit. But from the RLC analogy it might seem that to have slow light one does not need any optical nonlinearity.

An interesting aspect of ultraslow light is that this essentially "linear" phenomenon appears in coherently driven atomic media with extremely nonlinear optical behavior [68, 97] where the usual optical laws concerning dispersion and absorption are no longer valid. Surprisingly, media prepared for the conditions giving rise to slow light provide new regimes of nonlinear interaction with highly increased efficiency even for very weak light fields. These media hold promise for high-precision spectroscopy and magnetometry [18, 43, 71, 73, 123, 169]. Moreover, the ability to manipulate single photon states is extremely important for telecommunications of the future.

In this chapter we review the theory, experimental implementations, and applications of slow light in coherent media. The layout of the article is as follows.

Section II presents a brief review of the concept of the group velocity and early slow light experiments. We mention the importance of the two kinds of dispersion: frequency and spatial, for achieving ultraslow group velocity.

In Section III we derive expressions for group velocity in two- and three-level atomic systems, and then discuss extreme features of coherently prepared media such as the enhanced dispersion at resonance where the actual absorption is vanishingly small. The experimental results of various authors on the measurement of the dispersion and group velocities in coherent media are given as well.

Section IV is devoted entirely to the experiments in which ultraslow group velocities in Bose–Einstein condensate (BEC), hot atomic gases, and solids were obtained.

Section V discusses the teleportation, storage, and retrieving of the quantum information in a medium sustaining "dark polaritons."

To what limit light may be slowed down is discussed in Section VI. In particular, attention is paid to the atomic motion. Spatial dispersion due to atomic motion makes it possible to "freeze" the light, $v_g = 0$, or even to make its group velocity opposite to the wavevector, $v_g < 0$ [see Eq. (5)]. This idea, based on the fact that the group velocity can be made less than the mean thermal speed of atoms in a gas, results in a kind of optical treadmill, where a pulse of light sits still, swimming in a cell of moving atoms.

Applications of highly dispersive coherent media and slow light are presented in Section VII.

II. Group Velocity: Kinematics

We recall the concept of group velocity by considering the propagation of two waves of the same amplitude E_1 and E_2, where $E_i = E_0 \cos(k_i z - v_i t)$ with $i = 1, 2$. Superposition of the two waves gives rise to the modulation shown in Fig. 1,

$$\begin{aligned} E &= E_0[\cos(k_1 z - v_1 t) + \cos(k_2 z - v_2 t)] \\ &= 2E_0 \cos(\Delta k z - \Delta v t) \cos(kz - vt), \end{aligned} \quad (1)$$

where $\Delta k = (k_1 - k_2)/2$, $\Delta v = (v_1 - v_2)/2$, $k = (k_1 + k_2)/2$, $v = (v_1 + v_2)/2$. The two waves create an interference pattern consisting of a rapid oscillation

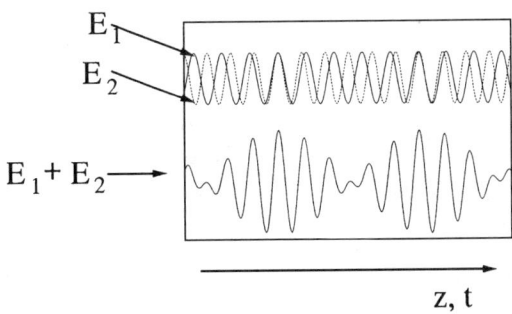

FIG. 1. Interference of two monochromatic waves with different frequencies results in a wave that is modulated in time and space.

propagating with the so-called phase velocity,

$$v_p = \frac{v}{k}, \tag{2}$$

and a slowly varying envelope propagating with the so-called group velocity,

$$v_g = \frac{\Delta v}{\Delta k}. \tag{3}$$

Usually we describe the propagation of wave packets consisting of more than two superimposed harmonics by transforming the ratio $\Delta v/\Delta k$ into dv/dk [86, 110].

We now formally consider the propagation of light in a medium possessing both temporal and spatial dispersion of the index of refraction, $n(v, k) = \sqrt{1 + \chi(v, k)}$, where $\chi(v, k)$ is the susceptibility of the medium. Differentiating the dispersion equation $kc = vn(v, k)$ gives

$$c = v_g \left(n + v \frac{\partial n}{\partial v} \right) + v \frac{\partial n}{\partial k}, \tag{4}$$

where we use the definition $v_g \equiv dv/dk$.

Generally, the susceptibility is a complex quantity. The real part (usually denoted χ') is responsible for the refractive properties of the medium, while the imaginary part (χ'') leads to light absorption. The generalized index of refraction $n = \sqrt{\chi + 1}$ is also a complex quantity $n \equiv n' + in''$. We consider real-valued group velocity and redefine it as $v_g \equiv \text{Re}(dv/dk)$ under the condition that imaginary part of dv/dk is negligible. Otherwise group velocity loses its simple kinematic meaning and strong absorption governs or prevents propagation of the light pulse through the medium. The latter is the reason why the resonant interaction of light with a two-level medium never results in an ultraslow light propagation.

According to Eq. (4), the group velocity of light contains two contributions,

$$v_g \equiv \text{Re} \frac{dv}{dk} = \text{Re} \frac{c - v[\partial n(v, k)/\partial k]}{n(v, k) + v[\partial n(v, k)/\partial v]} = \tilde{v}_g - v_s. \tag{5}$$

This is the basic equation of this chapter, as it shows how to find the group velocity from the index of refraction of the medium. The "only" problem left is to find the actual form of the function $n(v, k)$ for a particular real physical substance, which is not an easy task.

The kinematic meaning of Eq. (5) becomes clear if one turns to the Maxwell equation for the field amplitude E,

$$\left(\frac{\partial^2}{\partial z^2} - \frac{1}{c^2} \frac{\partial^2}{\partial t^2} \right) E = \mu_0 \frac{\partial^2}{\partial t^2} P, \tag{6}$$

where $P(z, t)$ is the macroscopic polarization of the medium. The polarization is related to the electromagnetic field $E(z, t)$ as follows [86].

$$P(z,t) = \epsilon_0 \int \chi(t - t', z - z') E(z', t') \, dt' \, dz', \tag{7}$$

where $\chi(t - t', z - z')$ depends on the properties of the medium. In general, χ is also a function of the electromagnetic field itself. In that case, Eq. (6) becomes nonlinear.

The susceptibility $\chi(\nu, k)$ is the Fourier transformation of $\chi(t - t', z - z')$,

$$\chi(t - t', z - z') = \int d\nu \, dk \, \chi(\nu, k) \exp[ik(z - z') - i\nu(t - t')], \tag{8}$$

and the macroscopic polarization (7) can be written in terms of the the susceptibility as

$$P(z,t) = \epsilon_0 \int dk \, d\nu \, \chi(\nu, k) E(k, \nu) \exp(ikz - i\nu t). \tag{9}$$

We consider the propagation of pulses whose duration is long compared to one optical cycle. We then use the slowly varying amplitude approximation, and the field amplitude E and the polarization P can be written as $E(z,t) = \mathcal{E}(z,t) \exp(ik_0 z - i\nu_0 t) + \text{c.c.}$, $P(z,t) = \mathcal{P}(z,t) \exp(ik_0 z - i\nu_0 t) + \text{c.c.}$, where the functions $\mathcal{E}(z,t)$ and $\mathcal{P}(z,t)$ have spatial and temporal scales much shorter than the inverses of the carrier wavevector k_0 and optical frequency ν_0 ($|\Delta k| \ll k_0$, $|\Delta \nu| \ll \nu_0$). It is convenient to decompose $\chi(\nu, k)$ in a series as

$$\chi(\nu, k) = \chi(\nu_0, k_0) + (\nu - \nu_0) \frac{\partial \chi}{\partial \omega} + (k - k_0) \frac{\partial \chi}{\partial k}$$

and obtain

$$P(z,t) = \epsilon_0 \left[\chi(\nu_0, k_0) \mathcal{E} + i \frac{\partial \chi(\nu_0, k_0)}{\partial \nu} \frac{\partial \mathcal{E}}{\partial t} - i \frac{\partial \chi(\nu_0, k_0)}{\partial k} \frac{\partial \mathcal{E}}{\partial z} \right] \exp(ik_0 z - i\nu_0 t).$$

Simplifying the right- and left-hand sides of the Maxwell equation (6),

$$\mu_0 \frac{\partial^2}{\partial t^2} P \simeq -\frac{\nu_0^2}{c^2} P$$

$$\left(\frac{\partial^2}{\partial z^2} - \frac{1}{c^2} \frac{\partial^2}{\partial t^2} \right) E \simeq 2ik \left(\frac{\partial \mathcal{E}}{\partial z} + \frac{1}{c} \frac{\partial \mathcal{E}}{\partial t} \right) \exp(ik_0 z - i\nu_0 t)$$

we obtain for the slowly varying amplitudes

$$\frac{\partial \mathcal{E}}{\partial z} + \frac{1}{c}\frac{\partial \mathcal{E}}{\partial t} = i\frac{\nu_0}{2c}\left[\chi(\nu_0, k_0)\mathcal{E} - i\frac{\partial \chi(\nu_0, k_0)}{\partial \nu}\frac{\partial \mathcal{E}}{\partial t} + i\frac{\partial \chi(\nu_0, k_0)}{\partial k}\frac{\partial \mathcal{E}}{\partial z}\right].$$

Subsequent rearrangement of terms gives

$$\left(c - \frac{\nu_0}{2}\frac{\partial \chi}{\partial k}\right)\frac{\partial \mathcal{E}}{\partial z} + \left(1 + \frac{\nu_0}{2}\frac{\partial \chi}{\partial \nu}\right)\frac{\partial \mathcal{E}}{\partial t} = \frac{i\nu_0}{2}\chi(\nu, k)\mathcal{E},$$

which implies

$$\left(v_g \frac{\partial}{\partial z} + \frac{\partial}{\partial t}\right)\mathcal{E} = \frac{i\nu_0}{c}\chi(\nu, k)\left(1 + \frac{\nu_0}{2c}\frac{\partial \chi}{\partial \nu}\right)^{-1}\mathcal{E}, \quad (10)$$

where v_g is given by Eq. (5). Under the limit $\chi(\nu, k) \to 0$, the general solution of Eq. (10) can be presented as two waves of arbitrary shape $f(t \pm z/v_g)$ propagating with group velocity v_g in opposite directions.

It follows from Eq. (5) that there are three different ways to reduce group velocity. One way is with very large refractive index $n \gg 1$, which would also result in the decrease of the phase velocity $v_p = c/n$. The second option is to use large frequency dispersion $\partial n/\partial \nu \gg 1$. Both these options influence the first term $\tilde{v}_g$ in Eq. (5), where $\tilde{v}_g = \text{Re}[c/(n + \nu\partial n/\partial \nu)]$. This method has been discussed in many recent papers [18, 43, 68, 71, 73, 77, 97, 98, 123, 169]. The third method that could be employed is to use large spatial dispersion $\partial n/\partial k \gg 1$, i.e., nonlocal response of the medium to a probe field. That is, we would modify v_s in Eq. (5) where $v_s = \text{Re}[(\nu \, \partial n/\partial k)/(n + \nu\partial n/\partial \nu)]$. In the following we discuss all three of these approaches.

III. Slow Light

A. HIGH REFRACTIVE INDEX AND SLOW LIGHT

In transparent semiconductors and dielectric materials the index of refraction can be as large as 10. An important question is whether or not it is possible to make use of the enhanced index of refraction in resonant media. In a two-level medium, the index of refraction near resonance can be very high. However, this high index is accompanied by large resonant absorption, which makes it appear to be useless. In coherent atomic media the situation might be different.

The basic physics underlying the effects described here is the strong modification of the susceptibility that occurs when atoms are prepared in a coherent superposition of states. In particular, one of the most striking phenomena associated with this quantum coherence is the ability to produce a large resonant index of refraction with vanishing absorption. Preparation of matter in such a state (which has been given the nickname "phaseonium" [179]) may provide us with a new type of optical material of interest both in its own right, and in many applications to fundamental and applied physics. The potential application of index-enhanced media are laser particle acceleration, optical microscopy, atomic tests of fundamental interactions, and precision magnetometry.

In a phaseonium gas without population in the excited levels, the absorption cancellation always coincides with vanishing index of refraction. However, with incoherent pumping of a small fraction of atoms to the excited state, the absorption vanishes at a frequency slightly off-resonance, where the real part of the susceptibility has a nonzero value. This gives rise to the possibility of high refractive index in a nonabsorbing medium [39, 40, 54, 158, 176, 178, 206].

When the phenomenon of index of refraction enhancement via quantum interference and coherence was first recognized [176], it was shown that high index of refraction with small absorption can be achieved by operating near an atomic resonance between an excited state $|a\rangle$ and a coherently prepared ground-state doublet $|b\rangle$ and $|b'\rangle$, with an appropriately chosen atom field detuning and level probabilities $\rho_{\alpha\alpha}$ ($\alpha = a, b, c$). Such conditions can be established, for example, by coherent pulse excitation, as in [142].

Other ways to excite the coherence to get high refractive index were discussed by Fleischhauer *et al.* [40], who suggested the possibility of using a microwave field applied to a ground-state doublet or two strong optical coherent fields coupled to the ground-state sublevels via an auxiliary level.

It turns out that the atomic coherence, or interference of pathways, which leads to the high index of refraction can be established even by incoherent processes. For example, the radiative decay of two closely spaced levels to common levels may lead to atomic coherence if the selection rules are such that the corresponding transitions from the excited sublevels couple to the same vacuum modes with the same strength. Unfortunately, the requirement to realize this interference effect is very restrictive. The two closely spaced levels must have the same total angular momentum F, the same projection of the total angular momentum m_F, the same electron angular momentum J, and the same orbital angular momentum L [39].

An effect similar to the interference of radiative decay processes can be achieved, however, by interference of incoherent pump processes, which does not imply such restrictive conditions [39]. In this case the interference can be produced even between magnetic sublevels. Moreover, in contrast to the previous case, the level spacing need only be smaller than the spectral width of the pump field.

Most treatments of enhancement of the index of refraction take into account only the linear response of the medium. Nonlinear analysis of index enhancement via quantum coherence and interference has been done in [158]. Their results show a limit for the continuous-wave (CW) field amplitude for a given index of refraction and a given atomic density that approaches the region of the field amplitudes necessary for example, for laser acceleration. Strong self-focusing or self-defocusing may show up for such high field intensities. Moreover, the properties of phaseonium may become unstable due to intensity fluctuations of the electromagnetic fields.

Proof-of-principle experiments demonstrating resonant enhancement of the index of refraction without absorption were reported in [206]. The experiment was performed in a warm vapor of ^{87}Rb. The resonant index of refraction was determined via phase-shift measurements using a Mach–Zehnder interferometer. Phase shifts up to 7π at the point of complete transparency were observed, which corresponds to resonant change of the refractive index $\Delta n \simeq 10^{-4}$. Such a phase shift is not normally observable in the vicinity of resonance because of the absorption.

Currently, the largest enhancement of the index of refraction reported is still much less than unity, but activity is continuing. Many other schemes have been proposed to enhance refractive index. Examples are four-level systems in which double-dark resonance is achieved [124, 125], an asymmetric double quantum well [163], and index enhancement due to local field effects [129].

For low density, the index of refraction scales linearly with density. However, increase of the density of atomic vapor and application of more powerful drive lasers to suppress absorption is not a simple task. Increase of the density is usually accompanied by increase of collective incoherent processes, such as radiation trapping and collisional broadening.

Reabsorption of spontaneously emitted photons can become important for optically thick media. This process, called radiation trapping, has been studied extensively in astrophysics, plasma physics, and atomic spectroscopy [143]. Radiation trapping has been predicted and demonstrated to have a destructive effect on the orientation produced by optical pumping [6, 42, 66, 156, 173]. Because the spontaneously emitted photons are dephased and depolarized with respect to the coherent fields creating the atomic polarization, the effect of radiation trapping can be described as an external incoherent pumping of the atomic transitions [6, 42]. Under the conditions of electromagnetically induced transparency (EIT), there are not many atoms undergoing spontaneous emission. However, these spontaneous photons destroy the atomic coherence in the same way as incoherent pumping. This effect can change the results of "phaseonium" experiments significantly [134].

Nevertheless, recent experiments on the specular selective reflection of rubidium at atom density 10^{15} cm^{-3} show that some coherence is preserved in a thin boarder layer. In the experiments, the drive field is tuned to the $5P_{3/2} \to 5D_{5/2}$ transition, and the probe field is tuned $5S_{1/2} \to 5P_{3/2}$, and the reflection of the probe beam at the atom–glass interface is measured. Under these conditions the relative reflection due to coherence increases up to 8%, meaning that the refractive index of the media is increased by $\Delta n \sim 0.1$ [209] (see Fig. 2.)

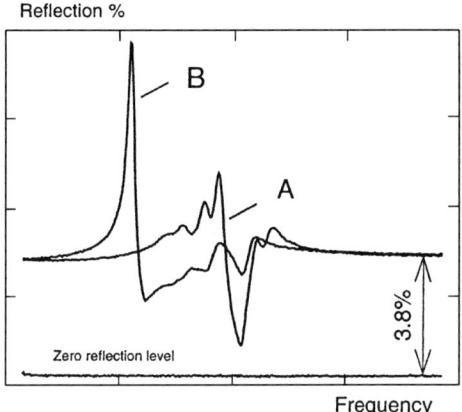

FIG. 2. Selective reflection of the probe beam (A) without and (B) with driving field.

To conclude this subsection, we would like to stress that the problem of creation of a medium with high index of refraction and small absorption remains unsolved. This is an interesting and challenging problem for future work.

B. SLOW LIGHT IN HETEROSTRUCTURES

The largest index of refraction obtainable without strong absorption is still of the order of unity, meaning one cannot reach ultraslow phase velocity. However, the group velocity associated with normal dispersion still can be quite small. The simplest way to realize small group velocity in practice is to construct a chain of linear resonant devices, for example, RLC circuits or use of a Fabry–Perot etalon. Its transparency is

$$\rho = \frac{T \exp(ikL)}{1 + R \exp(2ikL)}, \qquad (11)$$

where k is the wavevector of the light and R, T are the reflectance and the transparency of the mirrors. In Fig. 3, the real and imaginary parts of the transmission are shown. Anomoulos dispersion occurs at almost all frequencies except near the transparency peaks, where there is a region of steep normal dispersion. This steep dispersion leads to a decrease of the group velocity. Such an etalon having extreamly high finesse would provide very long time delay and is applicable to gravitational wave detectors [51]. The time-delay properties of a Fabry–Perot interferometer were revisited in, for example, [6a].

Further development led from the Fabry–Perot etalon to photonic band-gap materials (photonic heterostructures) which have been predicted theoretically and demonstrated experimentally [16, 82, 89, 90, 103, 151–154,188, 194, 195, 202,

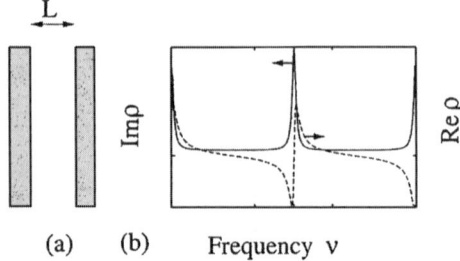

FIG. 3. (a) Scheme of Fabry–Perot etalon. (b) Real (solid line) and imaginary (dashed line) parts of transparency of the Fabry–Perot etalon versus frequency of light ν.

203]. Recent years have seen rapid progress in the development of new optical media with engineered dispersion based on photonic heterostructures. The existence of a frequency gap where the propagation of electromagnetic waves is forbidden can lead to unusual quantum optical phenomena such as inhibition and enhancement of spontaneous emission [94, 109, 131, 202, 203, 205], Anderson localization of light [89, 150], photon-atom bound state [91, 92, 95, 135], gap solitons [29, 30, 35, 93, 104, 139, 161, 166], and anomalous index of refraction [33]. Many applications of photonic crystals utilizing these unique optical properties have been proposed [154, 155]. For example, inhibition of spontaneous emission can be used to substantially enhance semiconductor laser operation.

To see how slow light appears in such structures, we consider the 1D structure as shown in Fig. 4a. Its dispersion [141] is

$$\cos kL = \cos(k_1 L_1 + k_2 L_2) - \frac{(n_1 - n_2)^2}{2 n_1 n_2} \sin(k_1 L_1) \sin(k_2 L_2), \quad (12)$$

where n_1, n_2 are indexes of refraction for the materials forming the photonic crystals (see Fig. 4), L_1 and L_2 are the thicknesses of the slices, and $k_1 = \nu/cn_1$ and $k_2 = \nu/cn_1$ are the wavevectors for each material. The forbidden gap is shown

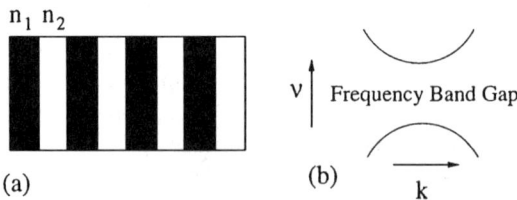

FIG. 4. (a) 1D photonic band gap crystal. (b) Dispersion of band-gap material, $\nu = \nu(k)$.

in Fig. 4b. Analytical expressions for the dispersion of finite photonic structures and the physical effects connected with finite length of the crystals have been studied [12, 182].

It is shown in many papers [11, 34, 35, 93, 182, 193] that the group velocity in photonic heterostructures can be very slow [see Eq. (12) and Fig. 4b]. This research has led to many interesting applications such as a frequency and spatial filtering [83] and a photonic crystal laser mediated by polaritons [149]. Dispersion and group velocities in 2D periodic systems with dissipative metallic components has been considered in [140]. Propagation of ultrashort pulses in photonic crystals has been analyzed in [167]. The dispersion of "heavy" photons at coupled cavity waveguide band edges in 3D photonic crystals has been found in [11]. The appearance of band gaps in optical lattices, formed by contrapropagating electron magnetic waves, have been studied, and their group velocity has been calculated in [193].

An important example of a photonic heterostructure is a fiber Bragg grating. In this structure, considerable slowing of light is feasible in the vicinity of the photonic band gap. Nonlinear phenomena in Bragg gratings have been studied in [36, 162, 183].

To avoid problems associated with high group velocity dispersion in fibers, the use of two cascaded gratings had been suggested and experimentally demonstrated. In these systems, light slows down by a factor of 1.5. Further reduction of the group velocity would require very narrow spacing between the stop bands of two gratings and thus would be extremely difficult to control. If, however, one uses a "superstructure" grating, where two gratings are superimposed rather than cascaded, the situation is different. The spacing between the Bragg wavelengths of the two superimposed gratings will contain a narrow transmission band. Superstructure, or Moiré fiber gratings were first proposed as narrow-transmission filters and are being used in tunable lasers. More recently, their dispersive properties in reflection have attracted attention, and nonlinear propagation and localization effects in some superstructure gratings have been observed [154, 155].

It is instructive to see the way these "superstructures" modify the dispersion in comparison with simple heterostructures. The effective index of refraction for "superstructures" can be written as

$$n(z) = n + \delta n \cos \frac{2\pi z}{\Lambda_s} \cos \frac{2\pi z}{\Lambda}, \qquad (13)$$

where Λ is the Bragg period and Λ_s is the superstructure, or Moiré period. The corresponding dispersion curves are plotted in Fig. 5. The dispersion band between broad stop bands gives rise to slow group velocity. Khurgin [99] analyzed the structure and found that group velocity of light can be 1000 times slower than in vacuum.

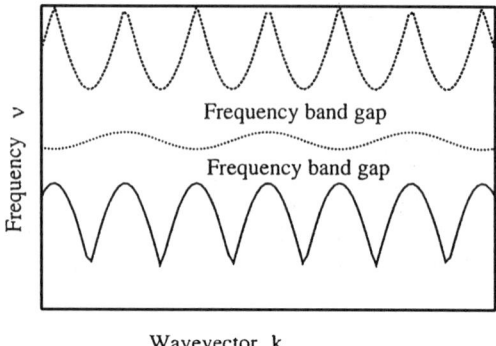

FIG. 5. Dispersion of EM waves in "superstructure," Moiré fiber gratings.

C. SLOW LIGHT IN TWO-LEVEL SYSTEMS, SIT

Systems such as two-level atomic and molecular media and semiconductors also provide the opportunity to decrease group velocity. Because of large absorption, slow light can be observed either for far detuned fields, where the absorption is small, or for very strong fields (such that the Rabi frequency of the field exceeds the homogeneous linewidth), when the absorption decreases because of nonlinear saturation of the medium. For example, in experiments with light interaction with a intense sodium beam in 1977, Mossberg observed a group velocity of $c/60$ [146].

A convenient way to describe the linear propagation of electromagnetic waves in a medium is to study the dispersion relations of the medium and to search for normal modes of the medium-field compound or "polaritons." In an initial-value problem, the temporal evolution of the field is considered with a given real value of the wave vector k. A solution of the dispersion equation in this case has the form $\nu = \nu(k)$. For a homogeneously broadened two-level medium there exist two solutions of the dispersion relation; for three-level media there are three solutions, as shown in Fig. 6. It is these solutions that are called polaritons. One of the polaritons has slow group velocity $\text{Re}(d\nu/dk)$, but its propagation is impossible due to the nonzero value of $\text{Im } \nu$. Taking into account spatial dispersion (in particular, Doppler broadening) modifies the dispersion branches, but does not change the major conclusion about the impossibility of an ultraslow polariton in two-level media. On the wings of the dispersion curve, where the absorption is relatively small, the group velocity can be reduced, keeping the absorption at reasonable level.

The first examples of slow light propagation in resonant two-level systems were demonstrated for self-induced transparency (SIT) solitons [78, 113, 128, 137, 138]. The lowest experimentally observed group velocity for SIT at the time of this writing is about $c/10^4$ [9, 52, 184].

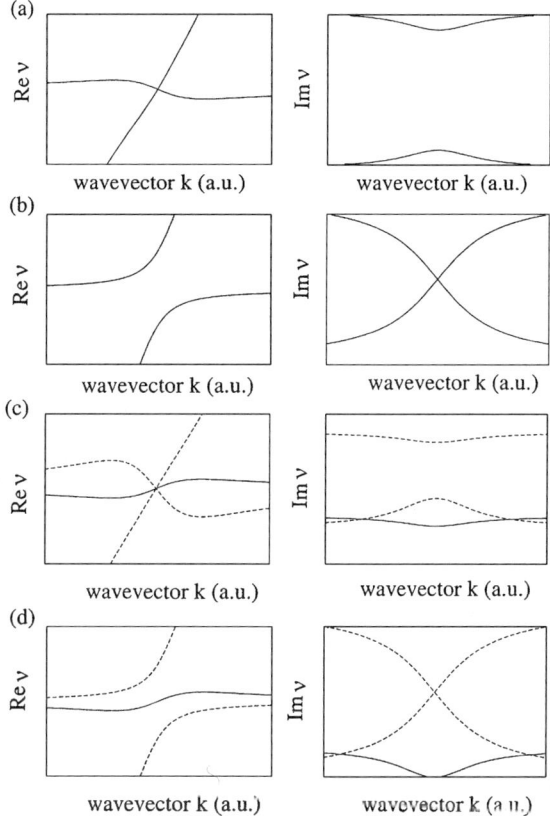

FIG. 6. Real and imaginary parts of frequency dispersion, $\nu = \nu(k)$, for two-level (a, b) and three-level systems (c, d). (a) and (c) correspond to low density ($3\lambda^3 N/16\pi^2 \ll 1$), (b) and (d) correspond to high density ($3\lambda^3 N/16\pi^2 \gg 1$). For three-level atoms, the solid curve corresponds to a slow polariton having low decay rate. Dashed curves are other polaritonic branches of electromagnetic waves with high decay rates.

In the limit of zero absorption, which is valid when the pulse duration is much shorter than the inverse homogeneous linewidth, it is possible to write the solution of the propagation equations for an electromagnetic field in a two-level medium in the form of a 2π soliton [138],

$$\Omega_{\text{SIT}} = A \cosh^{-1}[A(\xi - \tau)] \exp(i \Delta \xi + i\phi), \quad (14)$$

where A is the pulse amplitude (Rabi frequency), τ is the timing, Δ is the detuning of the carrier frequency from the atomic transition, $\xi = t - z/v_g$ is the retarded time, and v_g is the pulse group velocity. The phase factor ϕ depends linearly on the propagation distance, and is uniquely determined by the value of detuning through the dispersion relation,

$$\phi = \frac{\Delta}{A^2 + \Delta^2} \kappa z. \tag{15}$$

Equation (14) implies a specific relationship between the pulse velocity and amplitude,

$$\frac{1}{v_g} = \frac{1}{c} + \frac{\kappa/2}{A^2 + \Delta^2}, \tag{16}$$

where $\kappa = (3/4\pi)N\lambda^2\gamma_r$, λ is the wavelength, N is the atomic density, and γ_r is the natural decay rate.

For the case of resonance $\Delta = 0$ and minimum possible pulse amplitude $A \simeq \gamma_r$, the group velocity for a SIT soliton in a two-level system is $v_{g\min} \approx 2\gamma_r^2/\kappa$. The value of the group velocity can be very small. It is interesting to mention that the minimum soliton velocity exceeds the minimum group velocity in Λ system by a factor of γ_r/γ_{bc}, where γ_{bc} is the decay of the coherence in Λ system [compare (16) and (26)]. It was shown that processes like EIT can also occur in a two-level atomic system and can lead to strong self-action and slow-light effects that are not hampered by material absorption [5a]. For example, Park and Boyd discussed SIT in a two-level atomic system in the presence of an additional control laser field; they showed that the presence of the control field allows SIT to occur under a much broader range of conditions and leads to dramatically reduced values of the group velocity of the SIT soliton [5b].

D. SLOW LIGHT IN Λ SYSTEMS

In multilevel systems, coherent excitation of more than one transition can dramatically alter the group velocity. Previous direct and indirect measurements of increasingly low group velocities in coherently prepared media have ranged from $c/13$ [201] to $c/165$ [97], $c/3000$ [170], about $c/10^6$ [121], and $c/(2 \times 10^7)$ (17 m/s observed for a pulse of light in a Bose condensate of ultracold sodium atoms) [77].

Recently it was shown that by a proper choice of experimental parameters such as atomic density and optical intensity, large group delay (slow group velocity $c/3 \times 10^6$) of light can be observed in a cell of hot (360 K) ^{87}Rb atoms [98]. On the other hand, the dynamics of resonant light propagation in rubidium vapor in a cell with anti-relaxation wall coating was investigated by change the polarization of the input light and measurement the time dependence of the polarization after the cell. Delays up to 13 ms were observed, corresponding to a $c/(4 \times 10^7)$ (8 m/s) group velocity [23]. Recent studies also show that it is possible to achieve slow group velocity of the order of $c/(6 \times 10^6)$ in solids [192].

A three-level system involving one upper level with allowed transitions to two lower levels is called a Λ system. In the conventional usage, the two upper levels are coupled to the lower levels via two lasers. One of the lasers is strong and is called the *coupling* or *driving laser*, and the other is weak and is called the probe laser.

We can write density matrix equations for the coherences of the Λ-type atoms moving with velocity v as

$$\dot{\rho}_{ab} = -\Gamma_{ab}\sigma_{ab} + i(\rho_{aa} - \rho_{bb})\Omega_p - i\Omega_d\rho_{cb}, \quad (17)$$

$$\dot{\rho}_{ca} = -\Gamma_{ca}\sigma_{ab} + i(\rho_{cc} - \rho_{aa})\Omega_d + i\Omega_p\rho_{cb}, \quad (18)$$

$$\dot{\rho}_{cb} = -\Gamma_{bc}\rho_{cb} + i(\rho_{ca}\Omega_p - \Omega_d\rho_{ab}), \quad (19)$$

where $\Gamma_{ab} = \gamma + i\Delta_p$, $\Gamma_{ca} = \gamma - i\Delta_d$, $\Gamma_{cb} = \gamma_{bc} + i\delta$, Δ_d and Δ_p are the detunings of the probe and the drive fields, respectively, and $\delta = \Delta_p - \Delta_s$. We simplify these equations by assuming the case of a strong driving field $|\Omega_d|^2 \gg \gamma\gamma_{bc}$ and a weak probe, so that in the first approximation atomic population is in state $|b\rangle$. For this case, the density matrix equations for Λ-system shown in Fig. 7 are reduced to

$$\dot{\rho}_{ab} = -\Gamma_{ab}\sigma_{ab} - i\Omega_p - i\Omega_d\rho_{cb}, \quad (20)$$

$$\dot{\rho}_{cb} = -\gamma_{cb}\rho_{cb} - i\Omega_d\rho_{ab}. \quad (21)$$

Finally, we obtain

$$\chi = \frac{i\eta\gamma_r\Gamma_{bc}}{\Gamma_{bc}(\gamma + i\Delta_p) + \Omega^2}. \quad (22)$$

Ultraslow light is linked to electromagnetically induced transparency. The transmission of a weak probe pulse in a medium displaying EIT relies on the two-photon coherence ρ_{cb}, which is induced by the joint action of the probe field and a strong

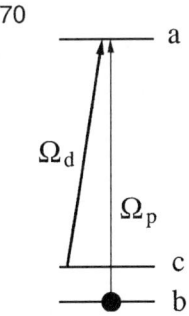

FIG. 7. Energy levels of a Λ system displaying EIT. Ω_p is the Rabi frequency of the probe field, and Ω_d is the Rabi frequency of the driving field.

driving field. EIT offers a wide variety of applications ranging from lasers without population inversion (LWI) (for the earliest papers on EIT/LWI, see [14, 38, 65, 88, 112]; for reviews on EIT/LWI, see [8, 70, 101, 130, 144, 179]) to new trends in nonlinear optics [58, 67, 80, 87, 185, 207].

In a Λ system, EIT appears due to coherent population trapping (CPT). This was first observed in experiments establishing Zeeman coherence in sodium atoms [7, 56, 204]. In these experiments, explained in terms of a three-level Λ-type level scheme, a laser field was used to create superpositions of the ground-state sublevels. One of these superpositions, referred to as the "bright" state, can interact with the laser field, while the other superposition does not and is referred to as the "dark" state [8]. Under conditions of two-photon resonance ($v_d - v_p = \omega_{bc}$), when the driving field is sufficiently strong ($\Omega^2 \gg \gamma \gamma_{bc}$), all the population in the system is eventually optically pumped into the dark state, and resonant absorption of the electromagnetic field [given by the imaginary part of χ in Eq. (22)] almost disappears. The width of this dark resonance is determined by power broadening ($\sim \Omega^2/\gamma$). The dark resonance is accompanied by steep dispersion that leads to ultraslow group velocity of light [8, 70, 130], i.e., to production of a slow EIT polariton. Substitution of χ into the dispersion equation and resolution of this equation in the form $v = v(k)$ leads to three branches $v_i(k)$. Other words, there are three normal modes propagating in the medium. Two branches look similar to those obtained for the two-level system (Figs. 6a and 6b) while the new one, possessing sharp dispersion and vanishing absorption, is the slow EIT polariton (Figs. 6c and 6d).

Let us calculate group velocity in three-level Λ configuration. We consider dilute systems where the susceptibility is small, $|\chi(v, k)| \ll 1$, but $v_g \ll c$, which is the case for all EIT experiments carried out so far. According to Eq. (5), the group velocity in the medium is given by

$$v_g = \frac{c}{1 + (v/2)(d\chi'/dv)},$$

where the derivative is evaluated at the carrier frequency.

We may now consider atomic motion. For atoms moving with velocity v, we replace Δ_p in Eq. (22) with $\Delta_p + k_p v$, assuming $|\Omega_d|^2 \gg (\Delta \omega_D)^2 \gamma_{bc}/\gamma$, where $\Delta \omega_D$ is the Doppler width. Averaging over the velocity distribution gives

$$\chi(v_p) = \int_{-\infty}^{\infty} \frac{i\eta \gamma_r \Gamma_{bc}}{\Gamma_{bc}[\gamma + i(\Delta_p + k_p v)] + \Omega^2} f(v) \, dv. \tag{23}$$

In this expression v_p is the probe laser frequency, $\eta = (3\lambda^3 N)/(8\pi^2)$, where λ is the probe wavelength and N is the atomic density, $\Gamma_{bc} = \gamma_{bc} + i[\delta + (k_p - k_d)v]$, where γ_r is the radiative decay rate of level a to level b, γ_{bc} is the coherence decay rate of the two lower levels (governed here by the time of flight through the laser beams), γ is the total homogeneous half-width of the drive and probe transitions (including radiative decay and collisions), $\Delta_p = \omega_{ab} - \nu_p$ and $\Delta_d = \omega_{ac} - \nu_d$ are the one-photon detunings of the probe and drive lasers, and $\delta = \Delta_p - \Delta_d$ is the two-photon detuning, Ω is the Rabi frequency of the drive transition, k_p and k_d are wave numbers of the probe and driving fields, respectively. One can obtain a simple analytic expression corresponding to Eq. (23) by approximating the thermal distribution $f(v)$ by a Lorentzian, $f(v) = (\Delta\omega_D/\pi)/[(\Delta\omega_D)^2 + (kv)^2]$, where $\Delta\omega_D$ is the Doppler half-width of the thermal distribution and v is the projection of the atomic velocity along the laser beams. The result is

$$\chi(\nu_p) = \eta\gamma_r \frac{i\gamma_{bc} - \delta}{(\gamma + \Delta\omega_D + i\Delta_p)(\gamma_{bc} + i\delta) + \Omega^2}, \quad (24)$$

where we have taken $k = k_p = k_d$. The typical experimental dependence of transmission and absorption for EIT is shown Fig. 8.

Equation (24) leads to propagation with absorption coefficient $\alpha = (k/2)\chi''(\nu_p)$ and group velocity $v_g = c/(1 + n_g)$. Therefore, one obtains

$$\alpha = \frac{3}{8\pi}N\lambda^2 \frac{\gamma_r\gamma_{bc}}{\gamma_{bc}(\gamma + \Delta\omega_D) + \Omega^2}, \quad (25)$$

$$n_g = \frac{3}{8\pi}N\lambda^2 \frac{\gamma_r\Omega^2 c}{[\gamma_{bc}(\gamma + \Delta\omega_D) + \Omega^2]^2}. \quad (26)$$

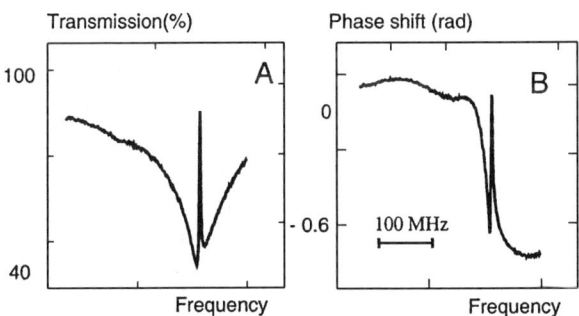

FIG. 8. Transmission (A) and dispersion (B) spectra of coherently prepared atomic [87]Rb vapor. The probe and drive fields are applied to the $5S_{1/2}, F = 1 \to 5P_{1/2}, F' = 2$ and $5S_{1/2}, F = 2 \to 5P_{1/2}, F' = 2$ transitions, respectively. The shape of the curves is the result of two processes: atomic coherence, which gives rise to absorption decreasing (the narrow central resonance); and optical pumping between ground-state levels F = 1 and F = 2, which leads to absorption increasing (the cone-shaped resonance around the central part).

After propagation through a dense coherent ensemble of length L the intensity of the pulse is attenuated by $\exp(-2\alpha L)$, whereas its envelope is delayed compared to free-space propagation by $T_g = n_g L/c$.

To relate the results of direct measurements of slow group velocity of light to earlier studies of EIT-based spectroscopy and prior group delay measurements, we note that the group delay is essentially the reciprocal of the so-called dispersive width associated with the EIT resonance $\Delta\omega_{\text{dis}} = \pi/(2T_g)$. This width was defined in [121] as the detuning from the line center at which the phase of the probe laser shifts by $\pi/2$. The significance of this quantity is that it determines the ultimate resolution of interferometric measurements using EIT. When the group delay is large, the dispersive width is correspondingly small. This is the basis for high-precision spectroscopy in dense coherent media. Let us stress here that for large densities of atomic vapors, we should take into account the density narrowing [121].

Clearly, an essential difference between hot [23, 98] and cold [77] atom experiments concerns Doppler broadening. Equation (26) shows that for a hot gas, the experimental regime where $\Omega^2 \gg \gamma_{bc}(\gamma + \Delta\omega_D)$, the effect of Doppler averaging is not important. The point is that in many current experiments (EIT, LWI, high-resolution dense medium spectroscopy, and ultraslow group velocities) the results are two-photon Doppler free for co-propagating drive and probe fields. That is, as shown in Eq. (23), when $k_p \approx k_d$, only the single-photon denominator [the square-bracketed expression in Eq. (23)] depends on atomic velocity. This has a negligible effect near two-photon resonance, provided that the Rabi frequency of the driving field is sufficiently large. This analysis, experimental demonstrations, and numerical calculations allow us to conclude that for strong driving fields, the effects of Doppler averaging are not of central importance to the group velocity. For the recent experiments [98], in which $\gamma_{bc}/(2\pi) \approx 10^3$ Hz and $(\gamma + \Delta\omega_D)/(2\pi) \approx 4 \times 10^8$ Hz, drive Rabi frequencies $\Omega/(2\pi) \gg 10^6$ Hz are required in order to get a measurable signal through the cell (the condition to have Doppler-free EIT is $\Omega \gg \Delta\omega_D \sqrt{\gamma_{bc}/\gamma}$). We note that for this range of intensity, the Lorentzian approximation holds, and also that $\partial\chi'/\partial\nu$ is nearly the same for hot and cold gases. Furthermore, for the experimental conditions for the warm gas [98], a reduction of γ_{bc} allows the possibility of reaching much lower group velocities, near 10 m/s. Such a reduction in γ_{bc} is quite possible by increasing laser beam diameter as shown in [18]. A coated cell has properties similar to the properties of a cell with buffer gas, and may provide an opportunity to slow light down to the same values as above for the uncoated cells [23].

On the other hand, cold-atom technology does hold promise for a truly Doppler-free payoff, e.g., nonlinear optical processes involving "sideways" coupling [77] in which the drive and probe lasers are perpendicular. This is not possible in a hot gas. Likewise, EIT experiments in cold gases might be a very interesting tool for studying the properties of and even manipulating the Bose condensate.

Inhomogeneous broadening is usually huge in doped materials. However, in a solid, as in a cold gas, the direction of the drive and probe fields does not change the coupling between the drive and probe lasers. This is because the inhomogeneous broadening in solids does not depend on the velocity of atoms/ions, but rather is determined by strains in the material. As shown below, slow group velocities on the same order of magnitude as gases [192] can be achieved in solids. The atomic density in solids is much larger than in atomic vapors, but coherences in solids and atomic vapors have comparable decay rates. Therefore, solids give much more hope for further reduction of group velocity of light and increase of the refractive index.

E. EARLY SLOW LIGHT EXPERIMENTS INVOLVING ATOMIC COHERENCE

Almost immediately after the discovery of EIT, it was recognized that the transmission of a medium showing EIT is accompanied by steep dispersion [8]. Harris et al. [68] showed that an atomic transition that has been made transparent by means of an additional electromagnetic field exhibits a rapidly varying refractive index with zero group velocity dispersion at the line center. The corresponding delay time, inferred from the results of measurement of medium dispersion in a 10-cm-long cell containing Pb atoms with density 7×10^{15} cm^{-3}, is about 83 ns, corresponding to a group velocity $v_g \simeq c/250$. A similar value, $v_g \simeq c/165$, was later demonstrated by the same group [97].

Xiao et al. [201] have shown that slow group velocity of light can be achieved in a so-called ladder (cascade) configuration of atomic energy levels, unlike previous experiments with Λ-type atoms. Xiao et al. measured the dispersive properties of the $5S_{1/2} \rightarrow 5P_{3/2}$ (D$_2$ line) in atomic Rb in the presence of a strong coupling field applied to the $5P_{3/2} \rightarrow 5D_{5/2}$ transition. The resulting group velocity for the probe beam $v_g = c/13$, was inferred from the measured dispersion curve.

Schmidt et al. [170] measured the index of refraction in the region of an EIT resonance in a cesium vapor cell. They found steep dispersion at low absorption that corresponds to group velocity $v_g = c/3000$.

IV. Ultraslow Light

A. SLOW LIGHT IN A COLD GAS

A Bose–Einstein condensate (BEC) provides an ideal test bed for EIT studies. The high density of atomic vapor with zero atomic motion is an excellent combination. In a landmark experiment, Hau and co-workers [77] demonstrated very small group velocities of 17 m/s. Their experiment was performed with a gas of sodium atoms cooled to well below the BEC transition temperature (which was $T_c = 435$ nK in their setup). The medium consisted of a cloud of atomic sodium containing several million atoms with a peak density of 5×10^{12} cm^{-3}. The coupling and probing lasers propagated at right angles to each other, which is an advantage that can only be realized for motionless atoms.

The setup for atom cooling and trapping has been described in other papers [76]. The experiment used a "candlestick" atomic beam source and the atoms were decelerated with a traditional Zeeman slower and magnetooptical trap (MOT). Roughly 10^{10} atoms were trapped at a temperature of about 1 mK with a density of 6×10^{11} cm^{-3} in the MOT.

Further cooling of the atoms to about 50 μK was accomplished using polarization gradient cooling, and the atoms were optically pumped into the $F = 1$ hyperfine component of the ground state. The atoms were then transferred to a single magnetic sublevel with magnetic filtering. Finally, the cloud was evaporatively cooled to the BEC transition. The trap could then be "softened" by adiabatically adjusting the magnetic fields, leaving 1–2 million atoms in the condensate cooled well below the transition temperature.

To create EIT, a linearly polarized coupling laser was tuned to the transition between the unpopulated hyperfine states $|c\rangle = 3S_{1/2}$, $F = 2$, $M_F = -2$ and $|a\rangle = 3P_{3/2}$, $F = 2$, $M_F = -2$. A left circularly polarized weak probe beam was tuned to the $|b\rangle = 3S_{1/2}$, $F = 1$, $M_F = -1 \rightarrow |a\rangle = 3P_{3/2}$, $F = 2$, $M_F = -2$ transition. This combination produced a "dark" state, in a narrow frequency interval with a width determined by power broadening produced by the strong coupling laser.

EIT allows transmission of the probe laser through the cloud of atoms, which would otherwise be quite opaque because of the the high atomic density. For the parameters of the experiment, below the transition temperature the transmission would be less than e^{-10}.

The group velocity for a propagating electromagnetic pulse is [c.f. Eq. (5)]

$$v_g = \frac{c}{n(v_p) + v_p(dn/dv_p)} \approx \frac{\hbar c}{8\pi v_p} \frac{|\Omega|^2}{\wp_{ab}^2 N}, \quad (27)$$

where $n(v_p)$ is the refractive index at probe frequency, $|\Omega|^2$ is the square of the Rabi frequency for the coupling laser (varies linearly with intensity), $\wp_{ab}$ is the electric dipole matrix element between states $|a\rangle$ and $|b\rangle$, and N is the atomic density. On resonance, the refractive index is unity and is negligible beside the second term in the denominator of Eq. (27).

A series of probe pulses were propagated through the cloud of cold atoms; delays and corresponding cloud sizes were measured for various cloud temperatures. Equation (27) shows that the light speed is inversely proportional to the atom density, which increases with lower temperatures. When the BEC transition temperature was reached, an additional density increase was observed leading to the group velocity decrease. Equation (27) also shows that the group velocity should decrease with lower coupling laser power that was confirmed by the experiment. The minimum group velocity reported in this experiment was 17 m/s, which was achieved for an atom cloud initially prepared as an almost pure Bose–Einstein condensate, where the condensate fraction was ~90%.

B. SLOW LIGHT IN A HOT GAS

To demonstrate ultraslow light effects in a hot vapor, Kash et al. performed an experiment [98] with a thermal ensemble of rubidium atoms. A group delay (T_g) of 0.26 ms for propagation through 2.5-cm-long, optically thick EIT medium was observed. The corresponding group velocity is 100 m/s, much less than the mean thermal speed of the atoms.

The measurements were done for the D_1 resonance line ($\lambda = 795$ nm) of ^{87}Rb (nuclear spin $I = \frac{3}{2}$). The drive laser was tuned to the $5^2 S_{1/2}(F = 2) \rightarrow 5^2 P_{1/2}$ ($F = 2$) transition; a co-propagating probe laser was tuned to the $5^2 S_{1/2}(F = 1) \rightarrow 5^2 P_{1/2}(F = 2)$ transition. Both were external-cavity diode lasers, and they were phase-locked with a frequency offset near the ground-state hyperfine splitting of 6.8 GHz, which was fixed by a tunable microwave-frequency synthesizer. The probe laser power was 5% of the drive laser power. A cell containing isotopically pure ^{87}Rb and 30 torr of Ne buffer gas was heated to about 80°C to obtain atomic density ~10^{12} cm^{-3}. Laser beam diameters were 2 mm, and the ground-state coherence relaxation rate $\gamma_{bc}/(2\pi)$ was found to be about 1 kHz.

The experiment measured the time delay as a function of the power of the drive laser with two methods. First the probe laser was amplitude-modulated by approximately 50% with a sine wave at a frequency that was varied in the range 0.1–10 kHz. The phase delay of the transmitted probe light was measured as a function of this modulation frequency, and the phase delay was found to increase linearly with modulation frequency. The slope of this dependence is the time delay. This technique is very sensitive and can be used even for very small laser powers. To check this method, the time delay of a Gaussian-shaped pulse was measured under conditions of good signal-to-noise ratio and found to be equal to the slope of the modulation.

The experimental results show the inferred *average* group velocity for each measured delay time. As the lasers propagate down the length of the cell, the drive laser power is attenuated. Since the group velocity decreases with drive laser power, the instantaneous velocity is lower toward the output end of the cell than near the input. Hence, the average velocity in the cell was reported.

In the experiment, both the drive beam and the probe beam are transmitted through the cell, and because of nonlinear optical processes additional frequencies are generated by the medium. To isolate the amplitude of the transmitted probe, a small part of the drive is split off before the cell, and frequency shifted down by a small amount (50 MHz). This shifted beam bypasses the cell and is combined on the detector along with the transmitted drive and probe and any generated fields. Because the amplitude of the shifted field is constant, this signal is proportional to the transmitted probe without any contribution from the transmitted drive field.

C. Slow Light and Nonlinear Magnetooptic Rotation

The propagation of nearly resonant light through rubidium vapor contained in a cell with antirelaxation-coated walls was investigated by Budker et al. [23]. Probe light delays up to $T_g \approx 13$ ms were observed, corresponding to an 8 m/s group velocity of light.

The apparatus used for the measurements was largely the same as for study the nonlinear magnetooptic effect (see the discussion of optical magnetometry below). The vapor cell (diameter 10 cm), contained in a four-layer magnetic shield, and a glass Faraday rotator were positioned between crossed polarizers. Diode lasers were used to produce CW light at 795 nm (D_1 transition) or 780 nm (D_2 transition).

The linearly polarized light induces atomic coherence. A current pulse applied to a Faraday rotator coil placed before the cell caused rotation of light polarization at the entrance to the cell. The polarization state after the cell exhibited a transient response, which was measured by a photodetector after the analyzer and its time dependence was investigated.

The optical thickness of the atomic vapor was ≈ 1–2 absorption lengths, which allows investigation of pulse propagation and reshaping phenomena under conditions of EIT. In addition, it was demonstrated that EIT, and consequently the group velocity of light, can be controlled using weak magnetic fields ($B \approx 1$ μG) corresponding to the Larmor frequency $\Omega_L = g_F \mu B \geq \gamma_{bc}$. Here g_F is the ground-state Landé factor and μ is the Bohr magneton. Magnetic field $B > 1\mu$G was used to switch off the wall collision-induced effect contribution to the delay time.

For $F \to F'$ transitions with $F' = F$ or $F - 1$, measuring the time dependence of the output polarization is equivalent to performing an EIT experiment in a Λ system. In traditional EIT, state $|b\rangle$ is populated, and probe light (with frequency ν_p) is tuned to a resonance with the $|a\rangle \to |b\rangle$ transition. Coupling light (ν_c) is applied resonant with the $|c\rangle \to |a\rangle$ transition; states $|c\rangle$ and $|a\rangle$ are both unpopulated in the absence of light.

In the Faraday-type case states $|b\rangle$ and $|c\rangle$ are superpositions, known as dark and bright states, of the (degenerate) ground-state Zeeman sublevels M_F. They are equally populated in the absence of light. When the coupling light is applied, the populations of the ground-state sublevels redistribute due to optical pumping: the bright state depopulates, and the vapor becomes linearly dichroic.

The probe beam was of the same central optical frequency as the coupling light and was produced with the Faraday rotator by slightly rotating the input polarization. As long as the polarization rotation induced by the rotator is small (as is always the case here), the coupling light can be considered constant, while the probe intensity goes as $I_p(t) \sim i^2(t)$, where $i(t)$ is the time-dependent current applied to the Faraday rotator.

To analyze the light propagation dynamics, the probe field E was modulated at the frequency ν_m: $E = E_0 \exp[i\nu_c(n_p z/c - t)] \sin(\nu_m t)$, where z is the propagation direction. A field of this form corresponds to light at the two frequencies $\nu_p^{\pm} = \nu_c \pm \nu_m$. Under certain simplifying assumptions such as a few absorption lengths and zero transverse magnetic fields, the contribution of the narrow feature due to the coupling light to the refractive index and absorption is determined by Eqs. (25) and (26). A probe light pulse, long enough, so that all its Fourier components satisfy $|\nu_p - \nu_c| = \nu_m \ll \gamma_{bc} + |\Omega|^2/\gamma$, propagate without change of shape, where Ω is the total Rabi frequency of the probe and drive fields and γ is the radiative decay rate of the optic transitions.

In a separate experiment, Gaussian current pulses were applied to the Faraday rotator (width 200 ms, maximum polarization rotation 25 mrad) and delayed pulses on the photodetector were observed with $B_z \approx 0$. No delay was observed when a magnetic field of 10 μG was applied or when the laser is tuned off-resonance. To minimize pulse reshaping, the pulse duration has to be long enough thus the pulse delay constitutes a small fraction of the pulse width. It is worth mentioning here that negative time delays were observed at some frequency detunings.

D. SLOW LIGHT IN SOLIDS

It is known that solids can also provide EIT (see in [61, 64, 100], for example). Hemmer et al. [192] demonstrated experimentally that Pr^{3+}-doped Y_2SiO_5 (Pr:YSO) crystal allows one to reduce group velocity up to the order of 45 m/s. The crystal had a thickness of 3 mm in the light propagation direction. Optical transition $^3H_4 \leftarrow\, ^1D_2$ of Pr^{3+} ion with central frequency 605.7 nm was used. An energy-level diagram with crystal-field sublevels is shown in Fig. 9. A re-pump laser is necessary to refill the spectral holes burned by coupling and probe fields.

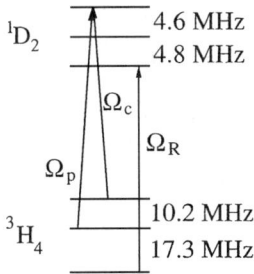

FIG. 9. Scheme of energy levels for Pr^{3+}-doped Y_2SiO_5. Ω_p is the Rabi frequency of the probe laser, Ω_c is the Rabi frequency of the coupling laser, and Ω_R is the Rabi frequency of the repump laser.

V. Storing and Retrieving Quantum Information

The quest for new tools and techniques in quantum information physics has fueled new studies of interest in their own right. Manipulation by single photons always was a tremendous task here. Photons are the fastest, simplest, and very robust carriers of quantum information, but they are difficult to store and process without destruction. All the newest fields in quantum optics, such as quantum cryptography, quantum teleportation, and quantum computation, rely on the assumption that such manipulations are possible.

To achieve the strong coupling between electromagnetic waves that is necessary to manipulate photons, a large nonlinear susceptibility with small losses is necessary. It was suggested that resonant coherent media are very useful here [47, 50, 55, 57, 71, 73, 84, 159, 169, 171, 198, 207]. It was shown that information about single photon can be stored in an atom placed in a high-Q cavity [1a–1e]. Raman passage techniques with time-dependent external control fields can be used to implement a direct transfer of a quantum state to a metastable coherent state of a single atom [1d]. However, experimental realization of these methods is very complicated. It was predicted theoretically and shown experimentally that light can transfer a part of its quantum statistical properties to dense atomic ensemble via usual absorption [59, 60, 108].

Recently, new methods of such transfer and manipulation, coherent and reversible, have been suggested theoretically [45, 46, 126, 127]. These methods are based on the phenomenon of EIT and allow for an ideal transfer of quantum correlations between light fields and metastable states of matter. The methods use trapping quantum states of photons in coherently driven atomic media, in which the group velocity is adiabatically reduced to zero. Another, quite similar technique, which involves synchronized Raman coupling between light and atoms, was proposed in [105]. Possible applications of these techniques are quantum state memories and quantum information processing including teleporting and time reversing state of light.

Several essentially classical proof-of-principle experiments utilizing the above ideas have been conducted [120, 157]. In these experiments, control and signal light pulses propagate in a gas of three-level Λ-type atoms and excite a spatial profile of a long-lived coherence between ground spin states. This coherence profile keeps information about these "writing" pulses after they have left or have been absorbed by the medium. Switching back the same control pulse results in its Raman scattering on the atomic coherence. Ideally, under specific conditions (such as short signal pulse and long control pulse lengths, adiabatic passage, etc.), the scattered pulse can be identical to the signal pulse, i.e., it can possess the same carrying frequency, the same profile and quantum statistics, and propagate in the same direction as the signal pulse.

Another classical proof-of-principle experiment [208] proves that using different reading pulses rather then switching back of the same control pulse provides a wide range of new tools for quantum information storage and processing, such as time reversing, teleportation, color switching, and multiplexing of the signal pulse of light.

In the experiments [208], one pair of control and signal pulses (the "writing" fields) prepare a spin coherence of the lower level. When a third field (a second control pulse, or "reading" field) scatters by this coherence, a new, or "restored," field is generated. This restored field, being a Raman-scattered component of the reading pulse, certainly acquires some properties of the reading field. If the reading field pulse is centered about a frequency other than that of the the writing fields, and propagates in the opposite direction to the writing fields, then the restored field propagates in the same direction as the reading field and indeed has a different frequency from the incident writing fields. The scattering effect persists even if the writing and reading pulses are separated not only in time but also in space. The atomic center of mass motion allows transport of the coherence grating to another point in the atomic cell.

Under ideal conditions (when the signal pulse profile and quantum statistics would be exactly stored by an atomic coherence grating [45, 46, 126, 127]), reading by the backward-propagating pulse would result in scattered back pulse which would represent itself as an exact time-reversed copy of the signal pulse. In other words, the physical process of "restored" pulse propagation in the $-z$ direction would look as if it would be seen recorded on the film and the film would be switched on in the back direction z. At the same time, atomic transverse motion in the direction orthogonal to the direction of propagation of light along with reading by a spatially shifted beam would result in teleportation of light by atoms exactly in the same spirit as teleportation of atoms (i.e., atomic state) by light has been traditionally discussed. Frequency of the "restored" pulse can be easily switched by the frequency shift of the reading pulse. Finally, using a frequency comb of the reading pulses propagating along different spatial channels should result in multiplexing of signal light.

The experiments on usage of atomic coherence for light information storage [120, 157, 208] certainly relates to earlier stimulated photon echo experiments as originally discussed for two-level systems and more recently generalized to multilevel media [26, 62, 79, 164]. However, in the experiments, unlike photon echoes, there is no time delay between reading and generated pulses associated with the spin restoration in an inhomogeneously broadened medium. Due to the Doppler-free configuration of the fields, it is possible to reproduce the signal pulse at any moment of time provided that the spin coherence survives—there is no need for strong π pulses. The reading and writing pulses can in principle be quantum mechanical pulses [45, 46, 126, 127], in contrast to the original photon echo experiments, in which only classical light beams were envisioned.

On the other hand, it is worth noting that combination of the new pulse profile and quantum statistics storage technique with the two-photon echo technique is possible. It requires only one modification of the standard two-photon spin-echo technique. That is, after restoration of the total spin coherence following application of a π microwave pulse, another π microwave pulse has to be applied just before sending the reading pulse. This is necessary since adiabatic passage requires restoration of the population difference between spin levels which was inverted by the first π pulse. The obvious application of such a combination would be (1) realization of quantum information storage in solid materials (where low-frequency transition is typically inhomogeneously broadened, and (2) increase of the storage capacity by the product of the ratio of inhomogeneous to homogeneous linewidths and the ratio of the total cross section of the sample to the diameter of the light beam.

There is also obvious connection of the present research with resonant four-wave mixing experiments in coherent media [58, 63, 67, 69, 72, 80, 87, 119, 136, 185]. But unlike the resonant four-wave mixing experiments, one can spatially and temporally separate the writing fields (which create the coherence grating in the atomic medium) from the reading field.

The restored field is a function of both the writing and reading fields. To be sure that the reading field does not contaminate this information by its own quantum noise, the field should be classical (i.e., having power much larger than the signal fields).

Concerning time reversal, we note that in the case where atomic coherence $\hat{\sigma}_{bc}$ (Fig. 10) is determined primarily by the memory of the writing fields and $\hat{E}_1 \gg \hat{\mathcal{E}}_1$, we have that $\hat{\sigma}_{bc}$ is governed by $-\hat{\mathcal{E}}_1^*/\hat{E}_1$ [157], which gives

$$\left(\frac{\partial}{\partial t} + c\frac{\partial}{\partial z}\right)\hat{E}_2 = -\eta\hat{E}_2 + \frac{\mu\hat{\mathcal{E}}_1^*\hat{\mathcal{E}}_2}{\hat{E}_1} + \hat{\mathcal{F}}_E', \tag{28}$$

where $\hat{\mathcal{F}}_E'$ is the fluctuational force. Equation (28) shows that one can reproduce the time-reversed operator field $\hat{\mathcal{E}}_1^*$ by taking the writing field E_1 and the reading field $\mathcal{E}_2$ to be in classical limit. This allows one to generate time-reversed $\hat{\mathcal{E}}_1$ pulses on demand.

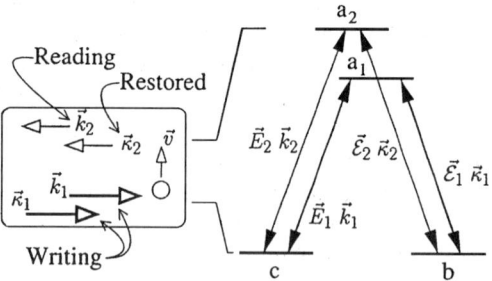

FIG. 10. Scheme of the experiment.

The same reasoning leads to the possibility of "storage of light" in the medium. In reality, according to [157], if writing field $\mathcal{E}_1$ and reading field $\mathcal{E}_2$ are classical and $\hat{E}_1$ is quantum, then it is possible to write the information about the weak pulse $\hat{\mathcal{E}}_1$ on the coherence grating, using adiabatic passage, $\hat{\sigma}_{bc} = -\hat{E}_1/\mathcal{E}_1^*$, and read it by the reading pulse

$$\left(\frac{\partial}{\partial t} + c\frac{\partial}{\partial z}\right)\hat{E}_2 = -\eta\hat{E}_2 + \frac{\mu\hat{E}_1\hat{\mathcal{E}}_2}{\hat{\mathcal{E}}_1} + \hat{\mathcal{F}}_E''. \tag{29}$$

Equation (29) shows that it is possible to reproduce the "direct" operator field $\hat{E}_1$ using the coherent medium.

Another fact of recent work involves "teleportation" of information about light by coherently prepared atoms in the spirit of continuous spatial photon echoes [172]. The atoms transport the coherence grating to another spatial place in the cell. To gain physical insight about spatial information transfer, one can describe the experiment using a beam of atoms moving in the +x direction. The atoms first cross the light beam consisting of the coupling field $\hat{\mathcal{E}}_1$ and the signal field $\hat{E}_1$, and then cross the reading field $\hat{\mathcal{E}}_2$. All light beams propagate in the +z direction, and the writing and reading fields are separated by a distance x_0. Due to the writing fields, the atoms acquire a long-lived coherence

$$\dot{\hat{\sigma}}_{bc} + [\gamma_{bc} + g(|\hat{E}_1|^2 + |\hat{\mathcal{E}}_1|^2)]\hat{\sigma}_{bc} = -g\hat{E}_1\hat{\mathcal{E}}_1^* + \hat{\mathcal{F}}_{bc}, \tag{30}$$

where g is an interaction constant, γ_{bc} is the coherence decay rate, and $\hat{\mathcal{F}}_{bc}$ denotes the quantum noise introduced in the system due to uncertainty of the initial conditions and decoherence processes.

After the light beams are switched off, the atoms freely diffuse in space and the spatial distribution of the atomic coherence $\hat{\sigma}_{bc}(\mathbf{r}, t)$ changes according to (see [15]),

$$D\nabla^2\hat{\sigma}_{bc} = K\hat{\sigma}_{bc} + \left(\frac{\partial}{\partial t} + (\mathbf{v}\nabla)\right)\hat{\sigma}_{bc} \tag{31}$$

where D is the diffusion coefficient, K is a quantity related with the coherence decay due to collisions between the working and and buffer gas atoms, and $\mathbf{v}$ is the velocity of convective flow of the gas.

The interaction of the atoms with the reading field leads to generation of a new, field $\hat{E}_2$. This new field is a function of the writing and reading fields,

$$\left(\frac{\partial}{\partial t} + c\frac{\partial}{\partial z}\right)\hat{E}_2 = -\eta\hat{E}_2 + \mu\hat{\mathcal{E}}_2\hat{\sigma}_{bc} + \hat{\mathcal{F}}_E, \tag{32}$$

where μ is a coupling constant, η is the decay rate of the field, and $\hat{\mathcal{F}}_E$ represents fluctuations due to incoherent interactions.

Equations (30)–(32) clearly show that the field $\hat{E}_2$ contains some quantum information about the signal fields $\hat{E}_1$ and $\hat{\mathcal{E}}_1^*$. The direction of the restored light is determined by a phase-matching condition between the fields $\mathbf{k}_2 = \kappa_2 + \mathbf{k}_1 - \kappa_1$. This phase-matching condition is similar to that in photon echo and four-wave mixing. The information exchange decreases due to the atomic diffusion in the buffer gas as well as velocity-exchange collisions in the experiment.

The amplitude of the restored pulse decreases with increasing time delay between writing and reading pulses which results from the decay of coherence in the medium. The dependence of the relative amplitude of the signal is practically insensitive to the amplitude of either writing or reading pulses and is determined purely by medium properties such as coherence decay rate and atomic motion. It was possible to observe the signal delayed up to 1.5 ms [208], 1 ms [120], and 0.5 ms [157].

As the beams are spatially separated, the recovered pulse amplitude also decreases. For beams of 1.5-mm diameter, the scattering effect has been observed up to 6 mm separation. An example of the effect is shown in Fig. 11.

Raman scattering of the second control pulse on the coherence produced by the writing pulses certainly ensures some correlation between the statistics of the recovered and writing signal light pulses (as is also true for CW four-wave mixing). Thus, the important question to be addressed is to what extent the retrieved pulse inherits the characteristics of the incident signal pulse and to what extent it acquires the properties of the second control pulse under experimentally achievable conditions. Further study of such correlation and its possible application for coherent quantum information storage would be important.

The role of the adiabaticity for the storage of the quantum information was discussed recently [2a, 208], and it was shown that there is no need for adiabatic passage in the storage and retrieval of information in optically thick vapor of

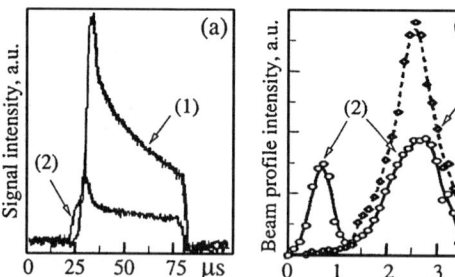

FIG. 11. (a) Restored pulse for spatially overlapped [curve (1)] and separated [curve (2)] writing and reading beams. Zero count on the time scale corresponds to the switch-off time for the writing pulses. (b) Intensity profiles of the spatially overlapped (dashed line) and separated (solid line) writing and reading laser beams. The writing beam is wider than the reading beam.

Lambda-type atoms. This information can be mapped into and retrieved out of long-lived atomic coherence with nearly perfect efficiency by strong writing and reading pulses with steep rising and falling edges for any switching time, if the group velocity of the signal pulse is much less than the speed of light in the vacuum c and the bandwidth of the signal pulse is much less then the width of the two-photon resonance. The maximum loss of the information appears in the case of instantaneous switching of the writing and reading fields compared with adiabatic switching, and is determined by the ratio of the initial group velocity of the signal pulse in the medium and speed of light in the vacuum c, which can be very small.

As is clear from the above discussion, there is no actual stopping of the light in the experiments on long-lived coherence information storage. There is trapping and storage of light's signature. This provides high flexibility in manipulations by stored information using different reading pulses as demonstrated in the experiments. In principle, it is possible to actually stop a pulse of light as opposed to just storing and retrieving information about the pulse, as is discussed below.

VI. The Ultimate Slow Light: Frozen Light

As pointed out above, the limit for slowing of the group velocity on the basis of large temporal dispersion is given by

$$v_g = \frac{8\pi \Omega^2}{3\lambda^2 N \gamma} > 0. \tag{33}$$

This equation is derived under the condition $\Omega^2 \gg \gamma \gamma_{bc}$. It can be shown that the group velocity approaches its minimum $v_{g\,min} = 8\pi \gamma_{bc}/(3\lambda^2 N)$ when $\Omega^2 \approx \gamma \gamma_{bc}$.

It is possible to overcome this limit by using spatial dispersion of the medium, i.e., by using the dependence of the refractive index on the wave number [see Eq. (5)]. In other words, rather than helplessly trying to increase the denominator in Eq. (5), one can fine-tune the numerator to zero [102].

This idea provides the opportunity to "freeze" the light, with $v_g = 0$, and even make group velocity negative so that the propagation direction is opposite to that of the wave vector, so that $v_g < 0$. The simplest example of spatial dispersion is the so-called drift dispersion corresponding to a monovelocity atomic beam or moving sample with velocity v.

Let us denote the group velocity of the light in the frame co-moving with atoms as $\tilde{v}_g$. In this frame, atoms are at rest, and hence there is no spatial dispersion. The Galilean transformation to the laboratory frame, $k = \tilde{k}$, $\nu = \tilde{\nu} - \tilde{k}v$, where v is the atomic velocity, yields the group velocity $v_g = \text{Re}(d\nu/dk) = \tilde{v}_g - v$.

This simple transformation shows "dragging" of light by moving atoms. If the velocity of the atoms with respect to the laboratory frame is just equal and opposite to the velocity of light in the frame where atoms are at rest, then we get "freezing" of light in the laboratory frame. This is just equivalent to the statement that in the frame accompanying the light, the light is not moving. Let us note that drift dispersion results in a very simple modification of the polariton dispersion branch, namely, substracting kv. As a result, the group velocity becomes negative as soon as v exceeds $\tilde{v}$.

This simple result was appreciated and used a long time ago, for example, in discussions on dragging of light by the hypothetical medium ether. More recently, it was also discussed for SIT solitons [138]. Since the lowest group velocity observed with SIT was about 3×10^4 m/s, the freezing of light would require a dense atomic beam propagating with this velocity.

A possibility of light drag via moving coherent medium was also discussed in [4a]. Leonhardt and Piwnicki reviewed the theory of light propagation in moving media [4b]. It is worthwhile to note that in the case of nonrelativistic group velocities and finite length of the sample (or beam), the total time delay of the pulse in the medium would be the same as in the frame where atoms are at rest.

It is remarkable, however, that in the case of EIT, the realization of frozen light does not actually require moving the atomic sample. Provided the group velocity is less than or of the order of the mean thermal speed of the atoms, a pulse of light can be stopped in a stationary cell with a Maxwellian thermal distribututon of atomic velocities (see Fig. 12). Moreover in this case the delay time for pulse passing through the medium would tend to infinity (contrary to the case of the finite atomic beam) in the sense that this pulse would never leave the cell: it would live and die there.

The idea of freezing light in a cell is to tune the frequency of the driving field to resonance with the velocity group of atoms moving into the light with speed equal $\tilde{v}$. If the intensity of the drive is strong enough to provide EIT for the resonant group of atoms, i.e., $\Omega^2 > \gamma_{bc}\gamma$ (see Fig. 13) but at the same time weak enough to avoid interaction with off-resonant atoms, moving with "wrong" velocities (i.e., $\Omega < kv_T\sqrt{\gamma_{bc}/\gamma}$), then it is mainly the atoms in this single velocity group that would support the ultraslow group velocity as a slow EIT polariton. The atoms

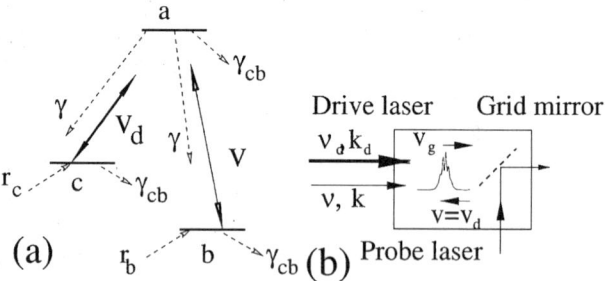

FIG. 12. (a) Three-level atomic Λ-system. (b) Geometry of ultra-slow EIT pulse propagation in the gas of atoms.

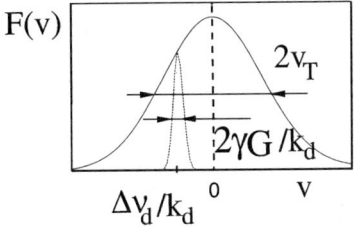

FIG. 13. The velocity distribution of atoms in a cell (solid line). Effective drifting beam (dotted) selected by drive laser.

would act as an effective atomic beam. Under these conditions $v_g = \tilde{v}'_g - v_d$, where $\tilde{v}'_g$ is the group velocity of the light defined by the effective atom beam if those atoms were at rest. In turn, the group velocity in the frame co-moving with this effective beam is $v_g = 8\pi \Omega^2/3\lambda^2 N' \gamma$, where N' is the density of the effective beam (see Fig. 13) which is given by $N' = NF(v_d) \Delta v$. Here Δv is an effective velocity width of the beam defined by the maximal detuning where atoms are still trapped for a given intensity: $k\Delta v = \Omega \sqrt{\gamma/\gamma_{bc}}$. Thus, a simple quadratic equation defining group velocity as a function of driving field detuning is

$$\frac{v_g}{v_T} = x^2 - 2bx + 1, \tag{34}$$

where

$$x = \frac{v_d}{v_T} \qquad b = \frac{3\lambda^3 N \gamma}{16\pi^2 \Omega} \sqrt{\frac{\gamma}{\gamma_{bc}}}.$$

It is clear that the minimum atomic density for freezing light (defined from condition $b = 1$) is $N = 16\pi^2 \gamma_{bc}/(3\lambda^3 \gamma)$. This corresponds to optimal tuning of driving field on resonance with atoms possessing thermal velocity $v = v_T$.

An explicit calculation of the dispersion law $v(k)$ for the EIT polariton in a hot gas in a cell at rest is given in [102]. It results from by an average of the beam susceptibility over a velocity distribution $F(v)$ of atoms in a gas with thermal velocity v_T, $\chi(v, k) = \int_{-\infty}^{+\infty} dv \, F(v) \chi_v(v, k)$. Instead of the Maxwellian thermal distribution we can use a Lorentzian, $F(v) = v_T/[\pi(v_T^2 + v^2)]$, since the far-off-resonant tails are not important. This approximation allows one to obtain simple analytical results. The dispersion relation so obtained is

$$\Delta v = \Delta v_d - v_d \, \Delta k + i\gamma_k - \frac{\Omega^2}{\gamma(1+G)} \left(\frac{\hbar \gamma G \Delta k}{2\pi \mu_{ab}^2 k_d N'} + i \right)^{-1}. \tag{35}$$

This result is shown in Fig. 14. The EIT half-width is $\Delta'_{\text{EIT}} = \gamma_k/\tilde{v}'_g$. For small detuning, $|\Delta k| \ll \Delta k'_{\text{EIT}}$, Eq. (35) yields linear dispersion and a parabolic decay

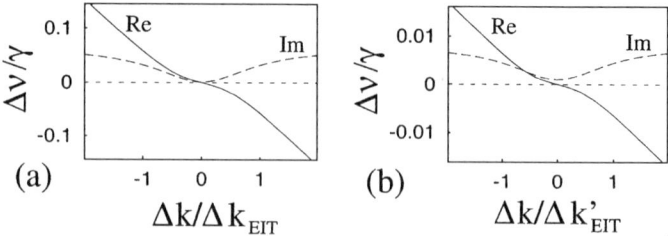

FIG. 14. Dispersion, Re $\Delta \nu = \text{Re}(\nu - \nu_{ab})$, and decay, Im $\Delta \nu$, spectra of the ultraslow EIT polariton according to numerical solution of the dispersion equation for: (a) an atomic beam ($N = 1.1 N_{\text{cr}} \pi F(v_d) \gamma G / k_d$); (b) a stationary cell of hot gas ($N = 1.1 N_{\text{cr}}$); $\Omega = 0.25\gamma$, $v = v_d = v_T$, $k_d v_T = 100\gamma$, $\gamma_{cb} = 0.001\gamma$.

profile, $\Delta \nu = \Delta \omega_d + \Delta k(\tilde{v}_g' - v_d) + i\gamma_{cb} + i\Delta k^2 \tilde{v}'^2 / \gamma_k$. Decay increases twice as much as the EIT minimum value, Im $\Delta \nu = 2\gamma_{cb}$, and at very small detuning $\delta k'_{\text{EIT}} = \sqrt{\gamma_{cb} \gamma_k}/\tilde{v}_g' \ll \Delta k'_{\text{EIT}}$. The group velocity describes pulse kinematics if $d\nu/dk$ has negligible imaginary part, i.e., near the center of the EIT dip, where $|\Delta k| < |\tilde{v}_g - v_d| \gamma_k / \tilde{v}_g'^2$. This inequality does not mean that the pulse cannot be stopped. It just means that when the pulse is frozen, $v_g = \tilde{v}_g' - v_d = 0$, its evolution is governed by the dispersion of absorption.

Figure 14 clearly shows that the ultraslow EIT polariton in a hot gas is similar to that in a monovelocity beam (see Fig. 15), since detuning of driving field picks an effective beam with velocity $v_d = -\Delta \nu_d / k_d$. However, the effective density of atoms supporting the EIT polariton N', and the EIT width $\Delta k'_{\text{EIT}} = \gamma_k / \tilde{v}_g'$ in a hot gas are different because of factors γG and $F(\nu)$. As a result, the group velocity at the EIT resonance, according to Eq. (6), in terms of a critical density is

$$v_g = \frac{\beta N_{\text{cr}}}{N F(v_d)} - v_d \qquad N_{\text{cr}} = \frac{\hbar \Omega}{2\pi^2 \beta \mu_{ab}^2} \sqrt{\frac{\gamma_{cb}}{\gamma}}, \qquad (36)$$

where $\beta = \max[v_d F(v_d)]$. For Lorentzian $F(v_d)$, we have $\beta = 1/2\pi$, and $v_g = (v_d - v_d^{(1)})(v_d - v_d^{(2)}) N_{\text{cr}} / 2N v_T$ is a quadratic polynomial in v_d, i.e., the group velocity is zero for drive detunings $v_d^{(1,2)} = v_T [N/N_{\text{cr}} \pm \sqrt{(N/N_{\text{cr}})^2 - 1}]$ and negative between those roots for densities higher than the critical value, $N > N_{\text{cr}}$, as is shown in Fig. 16. To achieve minimal group velocity, min $v_g = -(v_T N / 2 N_{\text{cr}})[1 - (N_{\text{cr}}/N)^2]$, one has to tune the drive laser frequency to interact with $v_d = v_T N / N_{\text{cr}}$. The condition to freeze or reverse the light ($v_g \leq 0$) means that the group velocity supported by the drifting beam with density $N' = \pi N F(v_d) \gamma G / k_d$ should be equal to or less than the velocity of atoms in the beam, i.e., $\tilde{v}_g' = \tilde{v}_g N'/N \leq v_d$.

If we compare a monovelocity beam with a hot gas at $v_d = v$ and the same N' as the total density N in a beam to provide the same group velocity, $\tilde{v}_g = \tilde{v}_g'$,

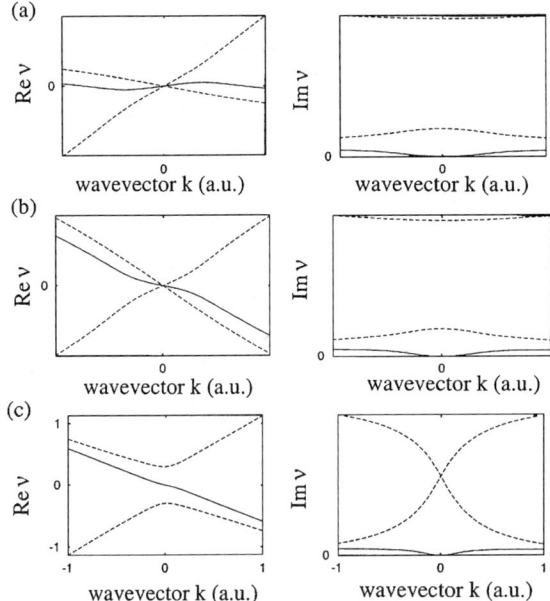

FIG. 15. Real and imaginary parts of the frequency dispersion $\nu = \nu(k)$ for a moving monovelocity beam. The solid curve corresponds to a slow polariton having low relaxation rate. Dashed curves are the other branch of electromagnetic (EM) waves for an atomic beam with high relaxation rates. For different densities of atoms and different velocities of the atomic beam, the dispersion curves are as follows: (a) corresponds to low density, $3\lambda^3 N/16\pi^2 \ll 1$, and positive group velocity, $v_g = \tilde{v}_g - v > 0$; (b) corresponds to low density, $3\lambda^3 N/16\pi^2 \ll 1$, and negative group velocity $v_g = \tilde{v}_g - v < 0$; (c) corresponds to high density, $3\lambda^3 N/16\pi^2 \gg 1$, and negative group velocity $v_g = \tilde{v}_g - v < 0$.

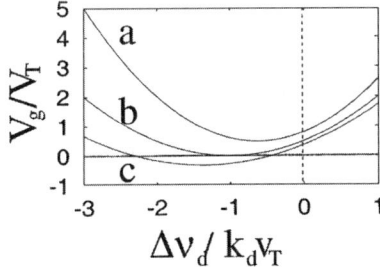

FIG. 16. Ultraslow and negative group velocity of EIT polariton versus detuning of drive laser; $\Omega = 0.25\gamma$, $k_d v_T = 100\gamma$, $\gamma_{cb} = 0.001\gamma$. (a) $N = 0.6 N_{cr}$. (b) $N = N_{cr}$. (c) $N = 1.5 N_{cr}$.

we find that the EIT width and the residual decay in a hot gas are $G \simeq \Omega/\sqrt{\gamma_{bc}\gamma}$ times less than in a beam. To minimize N_{cr} the drive intensity should be as low as possible, to avoid power broadening and to avoid EIT contribution from the atoms with "wrong" (positive) velocities. That is, the drive intensity should be just larger than the threshold of the EIT effect at resonance, $\Omega^2 > \gamma_{cb}\gamma$. For realistic parameters relevant to the experiments with ^{87}Rb vapor [98] the critical density is $N_{cr} \approx 10^{11}$ cm^{-3}. The important conclusion is that the drifting beam provides large enough spatial dispersion $\partial n/\partial k$ [see Eq. (1)] to ensure $v_g \leq 0$. Although the density of drifting atoms is small $N' \ll N$, their resonant contribution dominates. This allows us to make the group velocity zero or even negative.

Absorption or temporal variation of the drive field results in a spatial or temporal dependence of the group velocity in the cell. This allows us to control the input parameters of the pulse in the cell. According to geometric optics, the parameters of the EIT polariton adiabatically follow the local properties of the driven atoms. Figure 17 demonstrates how the ultraslow pulse decelerates up to the point $v_g = 0$, where it becomes frozen.

It is clear that the point z at which light is stopped corresponds to a stable equilibrium. Indeed, if the pulse passes this point, it acquires a negative velocity which brings it back to the same point. Therefore the question arises whether negative group velocity can be realized at all. One possibility is to insert a grid mirror into the cell that has grid stripes of small area, so that atoms can fly freely through the mirror, and small spacing between the grid stripes as compared to the wavelength of light to provide efficient reflection. It atoms were at rest, the light would propagate in the forward direction. However, under appropriate detuning and intensity of the driving field, one should see a frozen or backward pulse.

To observe freezing or backward light one can look, e.g., for scattering, luminescence, delay, or enhanced nonlinear mixing caused by the ultraslow pulse.

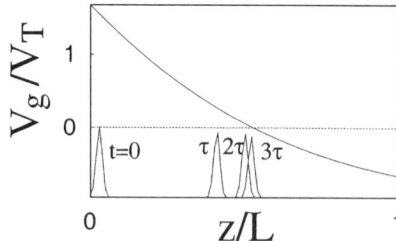

FIG. 17. Kinematics of the deceleration of the ultraslow pulse to the point of freezing ($v_g = 0$) along a cell with decreasing group velocity $v_g(z)$. Positions of pulse are shown at subsequent moments of time $t = m\tau [\tau = 3L/2v_g(0), m = 0, 1, 2, 3]$. $v_g(z)$ is calculated numerically according to decreasing drive intensity found from the wave equation for the same parameters as in Fig. 14, $L = 10$ cm.

VII. Applications

A. MAGNETOMETRY

1. General Concept

The optical detection of magnetic fields is a well-developed technique with applications ranging from geology and medicine [3, 4] to fundamental tests of violations of parity and time-reversal symmetry [22, 75, 81].

Optical magnetometers can be divided in two basic types. In the first type, light absorption at a magnetic resonance is used to detect Zeeman level shifts, while the second type makes use of the associated changes of the index of refraction. The optical pumping magnetometer (OPM) [3, 4] as well as dark-state magnetometers based on absorption measurements [148, 200], belong to the first class. More recently, magnetometers have been based on phase-coherent atomic media [41, 114, 177]. Experiments with picotesla magnetometry with coherent dark states based on the dispersion measurements was reported in [3a]. These and the mean-field laser magnetometer of [21] belong to the second category.

2. Phaseonium Magnetometers

The smallest detectable Zeeman shift (in units of frequency) is determined by the ratio of the noise level of the signal S to its rate of change with respect to frequency,

$$\Delta \nu_{\min} = \frac{S_{\text{noise}}}{|dS/d\nu|}. \tag{37}$$

A fundamental lower limit of S_{noise} results from photon counting errors due to shot noise of the probe electromagnetic wave. The denominator $(dS/d\nu)^{-1}$ is determined by the effective width of the magnetic resonance. The ultimate goal of magnetometer design is to minimize the noise level and the effective width at the same time.

The width of magnetic resonances in optical magnetometers is subject to two types of broadening: resonant power broadening due to the coupling of the optical fields to the probe transition, and a broadening due to ac-Stark shifts resulting from nonresonant couplings to other transitions. As shown in [41, 114, 177], power broadening limits the simultaneous minimization of noise and $(dS/d\nu)^{-1}$ in absorption-based magnetometers. In such devices, increasing the probe laser power reduces the shot noise but also reduces the signal. As a consequence, the sensitivity saturates at rather low power. On the other hand, as shown in [41, 114, 177], this effect can be compensated in a magnetometer that detects *phase shifts* of the probe electromagnetic wave propagating in an optically thick atomic medium under conditions of EIT. Theoretically a complete elimination is possible in a three-level Λ-type system.

3. Magnetometers Based on Nonlinear Faraday Effect

Let us now discuss in detail an EIT magnetometer in resonant nonlinear Faraday configuration [22, 25, 44, 165]. For this we consider the propagation of a strong, linearly polarized light field through an optically dense medium consisting of resonant Λ-type systems (atoms, quantum wells, etc.). For simplicity we ignore optical pumping into lower states other than those included in Λ scheme and assume a closed system. For example, in a resonant $J = 1 \rightarrow J = 0$ transition, optical pumping into the lower $m_J = 0$ state depletes both states $m_J = \pm 1$ in the same way and thus effectively diminishes the optical density but does not affect the signal. Symmetric repumping can be used to maintain the population in the relevant subsystem without affecting the detection scheme.

The two circular components E_- and E_+ of the linear polarized light generate a coherent superposition (dark state) of the states $|b_\pm\rangle \equiv |J = 1, m_J = \pm 1\rangle$. A magnetic field parallel to the propagation axis leads to an antisymmetric level shift of $|b_\pm\rangle$, and thus, by virtue of the large linear dispersion at an EIT resonance, the index of refraction changes by amounts for the two components. This results in the polarization direction being rotated, which is the so-called resonant nonlinear Faraday effect [10, 24, 32, 48, 49, 53, 96, 174]. The difference between this and the linear Faraday effect is the presence of the intensity-dependent dark resonance generated by the action of the strong laser field, as opposed to the usual absorption resonance in the weak-field limit. The rotation of the plane of polarization at the output can be measured by detecting the intensity difference of two linear polarized components $\pm 45°$ rotated with respect to the input polarization.

An aspect of the system which becomes particularly important when strong fields are considered is nonresonant couplings of the two circular components to other levels, which to lowest order give rise to ac-Stark shifts of the states $|b_\pm\rangle$. In a Faraday configuration the ac-Stark shifts of $|b_+\rangle$ and $|b_-\rangle$ are exactly equal and opposite in sign due to symmetry, and thus there is no average effect on the signal and no bias phase shift or rotation. Thus the Faraday magnetometer is not subject to systematic errors associated with ac-Stark shifts. However, ac-Stark shifts cause a coupling between intensity and phase fluctuations which need to be taken into account. For a certain optimum intensity, the fundamental signal-to-noise ratio attains a maximum value which represents the standard quantum limit of the optical magnetometer based on phase-shift measurements. This quantum limit is determined by the dispersion–absorption ratio of the atomic medium and the strength of the intensity–phase noise coupling. The unique property of EIT is to provide a dispersion–absorption ratio which is independent of power broadening and is given by the lifetime of a ground-state coherence. The minimum magnetic level shift corresponding to the quantum limit of EIT magnetometers can thus be orders of magnitude smaller than that of optical pumping devices.

B. Intracavity Electromagnetically Induced Transparency

When an ensemble of coherently prepared atoms is placed inside an optical cavity, the resonator response is drastically modified, resulting in frequency pulling, optical bistability, and a substantial narrowing of spectral features [1, 2, 13, 122, 145, 147, 197]. This effect can be used for frequency-difference stabilization of lasers [122] or other experiments with two-mode light involving broad-band parametric oscillations and phase conjugation [186].

An optical resonator with modified properties due to a highly dispersive medium was considered in [147]. The steep, nonabsorbing dispersion was created with an additional pump field in an atomic beam using the effect of coherent population trapping. The linewidth of such a resonator depends on the slope of the dispersion line, which in turn depends on the atomic density and the intensity of the pump and probe fields. The cavity linewidth in the experiment was reduced by a factor of more than 50 relative to the linewidth of the empty resonator. It was shown that, due to the dispersion of the medium, the resonance frequency was nearly independent of the geometric length of the resonator.

Intracavity EIT results in locking of the beat note of the resonance frequency of a two-photon transition between metastable atomic levels and causes a substantial reduction of quantum and classical noise in the beat signal. The profound effects of intracavity EIT are due to positive large dispersion close to the point of almost vanishing absorption, which can easily exceed the empty cavity dispersion in the case of an optically thick Λ medium.

To illustrate the locking and narrowing mechanism, let us consider following [122] a ring cavity containing a cell of length l with a linear dispersive medium. The medium response is characterized by the real (χ') and imaginary (χ'') parts of the susceptibility, for which we assume $\chi' = \beta(\nu - \nu_0)$ and constant χ'' for frequencies ν that are sufficiently close to some resonance frequency ν_0. β and χ'' are proportional to the atomic density N/V. The cavity response function, i.e., the ratio of circulating to input intensity, is given by

$$S(\nu) = \frac{I_{\text{circ}}}{I_{\text{in}}} = \frac{t^2}{1 + r^2\tilde{\kappa}^2 - 2r\tilde{\kappa}\cos[\Phi(\nu)]}, \tag{38}$$

where t and r are the transmissivity and the reflectivity of the input coupler ($t^2 + r^2 = 1$), $\Phi(\nu) = \nu L/c + \kappa l \chi'/2 \approx \nu L/c + kl\beta(\nu - \nu_0)/2$ is the total phase shift, and $\tilde{\kappa} = \exp(-kl\chi'')$ describes the medium absorption per round trip, L. From the expression for the round-trip phase shift one finds that the resonance frequency of the combined cavity–medium system [$\Phi(\nu_r) = 2\pi m$, m is integer] is governed by a pulling equation:

$$v_r = \frac{1}{1+\eta}v_c + \frac{\eta}{1+\eta}v_0. \tag{39}$$

Here $\eta = (ck/2)(l/L)\beta$ defines a frequency-locking or stabilization coefficient and v_c is the resonance frequency of the empty cavity. Similarly, we get for the width of the cavity resonance Δv the expression

$$\frac{\Delta v}{C} = \frac{1-r\tilde{\kappa}}{\sqrt{\tilde{\kappa}}(1-r)}\frac{1}{1+\eta}, \tag{40}$$

where C is the empty cavity linewidth. The first factor on the r.h.s. of (40) describes an effective with of the cavity–medium system owing to additional losses, and the second one describes the reduction due to the linear dispersion. When EIT regime is established in the intracavity medium, the absorption can be negligible, whereas the dispersion is large, resulting in substantial line narrowing.

To quantify this conclusion we consider the response of a Λ system driven by a strong laser field of Rabi frequency Ω. The corresponding linear susceptibility for a weak probe field of Rabi frequency α in the vicinity of two-photon resonance between probe and drive is

$$\chi' = \xi\frac{\gamma(v-v_0)}{\Omega^2} \qquad \chi'' = \xi\frac{\gamma\gamma_0}{\Omega^2}. \tag{41}$$

Here $\xi = (3/4\pi^2)(N\lambda^2/V)$, γ_0 is the coherence decay rate, and γ is the homogeneous decay rate of both optical transitions of the Λ scheme. In the situation typical for EIT, $\gamma \gg \gamma_0$, and thus the absorption can be small, $\tilde{\kappa} \approx 1$. The ultimate limit of stabilization can be obtained by imposition of the condition that the residual absorption losses in the cell should not exceed the empty-cavity losses. One easily finds that the maximal stabilization coefficient is

$$\eta \leq \frac{C}{2\gamma_0}. \tag{42}$$

For a long-lived ground-state coherence, the ratio C/γ_0 can become large. For example, for a cavity with quality factor $Q = 10^7$ and $v_c = 10^{15}$ s^{-1}, the linewidth is $C = 10^8$ s^{-1}, while for a hot gas of alkali atoms the coherence linewidth is about $\gamma_0 \approx 10^3$ s^{-1}, hence $\eta \simeq 10^5$. In this case the effective resonance frequency of the cavity coincides with v_0 and the cavity width can be reduced by several orders of magnitude, whereas the photon losses are small enough.

The lowest loaded cavity linewidth can be estimated as $\delta v \to 4v_0$, i.e., the linewidth can be orders of magnitude smaller than both the empty-cavity linewidth and the single-atom transparency window (the width of lambda resonance is broader because of power broadening).

The frequency locking influences on the phase-difference noise of two optical modes that are independently oscillating inside the cavity. A three-level medium displaying EIT can be used to lock the beat note of the two-modes to the resonance frequency of two-photon transition ω_{b1b2}. The steady-state solution of the two-mode laser phase equations leads immediately leads to the frequency-pulling equation

$$\nu_1 - \nu_2 = \frac{1}{1+\eta}(\nu_1^c - \nu_2^c) + \frac{\eta}{1+\eta}\frac{\omega_{b1b2}}{2}. \tag{43}$$

For $\eta \gg 1$ the beat-note frequency is locked to ω_{b1b2}.

Atomic coherence can also result in negative atomic dispersion without absorption [187]. This leads to the concept of optical cavities with large buildup but broad-band response (white-light cavities) [147, 199]. For example, strongly driven double-Λ systems can show negative dispersion without absorption, which is needed in order to compensate for the variation of the wavelength with the frequency.

C. Slow Light and Phonons

It follows from common sense that forward stimulated Brillouin scattering (SBS) does not occur because of energy and momentum conservation laws [17, 19, 37, 106, 107, 110, 191], with the strongest scattering typically in the backward direction. Efficient stimulated Brillouin scattering requires phase matching (i.e., momentum conservation) among the incident and scattered light waves and the acoustic wave. Usually, because the frequencies of acoustic waves are small compared with the frequency of light and their wavevectors large compared to optical waves, phase matching can be established only for backward scattering. For forward Brillouin scattering in bulk materials, phase matching is impossible even in principle. Surprisingly, we find that in ultradispersive coherent media that are able to hold an ultraslow group velocity of light, just the opposite can happen: that is, maximum scattering is in the forward direction, while back scattering is forbidden [133].

The SBS changes substantially in confined geometries, for example, in optical fibers. Due to the long length of fibers, nonlinear effects can be effectively developed and accumulated even when the cross section for nonlinear effects is relatively small [5]. One such effect is stimulated Brillouin scattering [85, 189]. SBS in fibers is different from SBS in bulk materials because in the confined geometry of optical fibers, acoustic modes exist that have nonzero cutoff frequencies at zero wave vector [190, 196], and phase-matched forward scattering from these modes occurs [180, 181]. As a result, both stimulated backward Brillouin scattering from acoustic waves with the usual linear dispersion (pure longitudinal waves) and guided acoustic wave Brillouin scattering (forward scattering from modes with nonzero cutoff frequencies) have been observed in optical fibers.

Recently, the use of electromagnetically induced transparency (EIT) as a new way of establishing phase matching for co-propagating optical waves in fibers was suggested [132] and generalized to different types of acoustic modes [160]. This technique provides a possibility of utilizing the ponderomotive nonlinearity of the fiber for new phase modulators, frequency shifters, and sensors on the one hand, and for effective quantum wave mixing, generation of nonclassical states of light, and Quantum Nondemolition Measurements (QND) measurements on the other.

Let us consider propagation of two plane monochromatic waves $E_{p1}e^{-i(\nu_1 t - \mathbf{k}_1 \mathbf{r})}$ and $E_{p2}e^{-i(\nu_2 t - \mathbf{k}_2 \mathbf{r})}$ in a dielectric medium. The resonant interaction of the beat-note intensity of the electromagnetic waves running through the medium with sound vibrations of the medium leads to SBS.

The wave vector of the acoustic wave $\mathbf{q}$ should satisfy Bragg's condition (momentum conservation),

$$\mathbf{k}_2 - \mathbf{k}_1 = \mathbf{q}, \tag{44}$$

and, by energy conservation, the frequency of the acoustic wave ω_b should satisfy

$$\nu_1 - \nu_2 = \omega_b. \tag{45}$$

Condition (44) implies

$$|\mathbf{q}|^2 = |\mathbf{k}_1|^2 + |\mathbf{k}_2|^2 - 2|\mathbf{k}_1||\mathbf{k}_2|\cos\theta$$
$$= (|\mathbf{k}_1| - |\mathbf{k}_2|)^2 + 4|\mathbf{k}_1||\mathbf{k}_2|\sin^2\frac{\theta}{2}. \tag{46}$$

For a dispersionless medium with linear index of refraction n, the absolute values of the wave vectors $|\mathbf{k}_1|$ and $|\mathbf{k}_2|$ almost coincide, i.e., $|\mathbf{k}_2| \approx |\mathbf{k}_1| = \nu n/c$, if $\nu_2 \approx \nu_1 = \nu$, and dispersion relation for the sound wave is

$$|\mathbf{q}| = \frac{\omega_b}{v_s}, \tag{47}$$

where v_s is the speed of sound in the material. Then (46) and (47) imply

$$|\mathbf{q}| = 2\frac{\nu n}{c}\sin\frac{\theta}{2}, \tag{48}$$

$$\omega_b = 2\frac{n v_s}{c}\nu\sin\frac{\theta}{2}. \tag{49}$$

Conditions (48) and (49) show that the forward SBS ($\theta = 0$) vanishes.

In turn, in an ultradispersion medium,

$$|\mathbf{k}_1| - |\mathbf{k}_2| = \frac{v_1 n(v_1)}{c} - \frac{v_2 n(v_2)}{c}$$

$$\approx \frac{v_1 - v_2}{c} \frac{\partial[v n(v)]}{\partial v} = \frac{v_1 - v_2}{v_g}, \quad (50)$$

where v_g is the group velocity of light. When $v_g \approx c/n$, as in a dispersionless medium, one gets the previous results. However, when v_g is small enough, one can find a new expression for the acoustic wave vector.

In particular, the wave vector difference $|\mathbf{k}_1| - |\mathbf{k}_2|$ appearing in Eq. (50) is now nonzero. From (45) and (50) we have $|\mathbf{k}_1| - |\mathbf{k}_2| = \omega_b/v_g$. Taking $|\mathbf{k}_1| = vn/c$, and from (50) $|\mathbf{k}_2| = vn/c - \omega_b/v_g$, we can write Eq. (46) as

$$|\mathbf{q}| = \sqrt{\left(\frac{\omega_b}{v_g}\right)^2 + 4\frac{vn}{c}\left(\frac{vn}{c} - \frac{\omega_b}{v_g}\right)\sin^2\frac{\theta}{2}}, \quad (51)$$

and because of (47), Eq. (51) can be used to obtain the equation for the sound frequency;

$$\omega_b^2 \frac{v_g^2 - v_s^2}{4 v_s^2 v_g^2} - \frac{vn}{c}\left(\frac{vn}{c} - \frac{\omega_b}{v_g}\right)\sin^2\frac{\theta}{2} = 0. \quad (52)$$

Equations (51) and (52) coincide with (48) and (49) for $v_g \gg v_s$.

The dispersion relation (52) shows that there is no scattering if $v_s > v_g$, because both roots of Eq. (52) are nonpositive ($|\mathbf{k}_2| > 0$). The acoustic frequency for backward scattering when $v_g \geq v_s$ is

$$\omega_b^{\text{backward}} = 2\frac{n v_s}{c} v \frac{v_g}{v_g + v_s}, \quad (53)$$

while for the forward scattering,

$$\omega_b^{\text{forward}} \leq \frac{n v_s}{c} v \quad \text{if } v_g = v_s \quad (54)$$

$$\omega_b^{\text{forward}} = 0 \quad \text{if } v_s \neq v_g,$$

which is drastically different from the case of dispersionless medium. For example, when $v_g \to v_s$ the wave vector $|\mathbf{k}_2|$ goes to zero and, therefore, backward scattering is forbidden. In contrast, forward scattering is strongly allowed.

In optical fibers, unlike the bulk material, the confinement plays a significant role. It is instructive to idealize a fiber as a long square bar with side a; the index of refraction of the fiber material is n, and the interaction between acoustic and light waves is provided by electrostriction and the modulation of the refractive index by the acoustic strain. This model facilitates simple calculation of the dispersion relations for acoustic and electromagnetic waves.

Let us consider the acoustic waves which propagate along the fiber and have axial dependence in the form $\exp i(\omega_s t - k_s z)$. The dispersion relation for the acoustic waves depends on whether the fiber is free or fixed, but generally it looks like

$$\omega_s^2 = v_s^2 k_s^2 + \omega_{cs}^2, \qquad (55)$$

where v_s is the phase velocity of acoustic wave, and the cutoff frequency, ω_{cs}, is defined by

$$\omega_{cs} = v_s \sqrt{\left(\frac{\pi l_s}{a}\right)^2 + \left(\frac{\pi m_s}{a}\right)^2}, \qquad (56)$$

l_s and m_s are integers.

The confined waveguide geometry produces two kinds of modes, those with nonzero frequency (cutoff frequency) at zero wave vector, and modes that have zero frequency for zero wave vector ($l_s = m_s = 0$). The pure longitudinal mode that normally exhibits backward stimulated Brillouin scattering in fibers and the lowest transverse mode with ($l_s = m_s = 0$) are examples of the second type. Both types of modes can participate in Brillouin scattering with phase-matching conditions established in a new way using EIT.

In general, the acoustic modes in confined geometries, such as optical fibers, are coupled dilatational and shear modes resulting in more complex higher-order dispersion relations. We would like to underline that the above simple calculations reproduce the main features of dispersion relations for such modes, especially near their cutoff frequencies (see [190, 196]).

The dispersion relation for electromagnetic waves is given by

$$n^2 v^2 = c^2 k^2 + v_c^2. \qquad (57)$$

The cutoff frequency, v_c, is defined by

$$v_c^2 = \left(\frac{\pi l c}{na}\right)^2 + \left(\frac{\pi m c}{na}\right)^2, \qquad (58)$$

where c is the speed of light in the fiber, a is the fiber dimension, and l and m are the electromagnetic mode indices.

In practical fiber waveguides, the optical frequency is well above cutoff [i.e., v_c^2 is small compared to the other terms in Eq. (57)]. Except near the zero-dispersion wavelength, dispersion is dominated by material dispersion (the frequency dependence of n) rather than waveguide dispersion. For a fiber doped with impurities resonant with the light, the index of refraction is changed due to the contribution of dopant ions, as $n_{\text{doped}} = n_{\text{host}} + \chi(\nu)/2n_{\text{host}}$, where χ is the susceptibility due to the dopants. Usually $\chi \ll n_{\text{host}}$, but it can have a steep dispersion in the presence of electromagnetically induced transparency. The main modification to the dispersion is the slope at the photon resonance frequency, whereas the phase velocity remains practically unchanged.

Here we note the main results for a Λ scheme in which one of the allowed transitions is strongly driven with an optical Rabi frequency $\Omega \approx \gamma_{lc} \gamma$, where γ_{lc} is the linewidth of the two-photon Raman transition and γ is the inhomogeneous linewidth of the allowed transition. In this case a light wave near the two-photon resonance condition with the pump experiences a group velocity index $n_g = (2\alpha_0 c/\gamma_{lc})(\Omega^2/\gamma_{lc}\gamma)/(1 + \Omega^2/\gamma_{lc}\gamma)^2$, where α_0 is the linear absorption coefficient of the resonant dopants, and a reduced absorption of $\alpha = \alpha_0/(1 + \Omega^2/\gamma\gamma_{lc})$. These effects exist over a bandwidth given by $\delta = \gamma_{lc}(1 + \Omega^2/\gamma\gamma_{lc})$.

Let us consider the phase-matching condition for forward SBS in the doped fiber [160]. To have effective Brillouin scattering we should meet the resonance frequency condition and the phase-matching condition between the EM waves and the acoustic wave,

$$\nu_1 - \nu_2 = \omega_s \qquad k_1 - k_2 = k_s, \qquad (59)$$

where the subscripts 1 and 2 refer to the incident and scattered fields. Using Eq. (57), after a simple transformation we obtain

$$k_1 - k_2 = k_s = \frac{1}{c}\frac{\partial(\nu n)}{\partial \nu}\omega_s = \frac{\omega_s}{v_g}. \qquad (60)$$

The phase-matching condition can be represented in the form (for the case $k \gg v_c/c$)

$$k_s^2 + \left(\frac{\omega_{cs}}{v_s}\right)^2 = \left(\frac{v_g}{v_s}\right)^2 k_s^2. \qquad (61)$$

Therefore, to meet conditions Eq. (59), the acoustic wave should have a particular wave vector given by

$$k_s = \frac{1}{\sqrt{(v_g/v_s)^2 - 1}}\frac{\omega_{cs}}{v_s}. \qquad (62)$$

This expression is in agreement with the case when $v_g = c/n$, and only acoustic oscillations at the cutoff frequencies are phase-matched with optical fields. Indeed, from Eq. (62) it follows that $k_s = n\omega_{cs}/c \simeq 0$, and this is the case has been investigated in [180, 181].

We see that the possibility to control the group velocity of the light waves leads to interesting consequences. First, as has been found in [132] for pure longitudinal waves ($l = m = 0$, $\omega_{cs} = 0$), phase matching always takes place in the case when $v_g = v_s$ for any frequency of acoustic wave within the bandwidth of the EIT. Furthermore, even in the case $l > 0$, $m > 0$, and $\omega_{cs} > 0$, the phase-matching condition can be modified by controlling v_g. The modification of phase matching via the use of EIT and slow light suggests a new means to control an output of Brillouin fiber laser by controlling steep dispersion via an external optical field. For example, changing the ratio v_s/v_g leads to switching on and off the acoustooptical interactions in the doped fiber.

In addition to the effects discussed above, the reduction of resonant absorption together with the steep dispersion opens the possibility for new kinds of resonant interactions between electromagnetic waves and mechanical degrees of freedom of atoms. For example, it has been shown in [74] that the longitudinal gradient force, acting on a two-level test atom, can be enhanced via spatial compression of an optical pulse moving with ultraslow group velocity in a coherent medium. This enhanced force yields a ballistic atom motion and atom surfing, and a new kind of local ponderomotive light scattering.

It is interesting also to study interaction between a light pulse and the medium where the pulse propagates. It was shown that atomic recoil can change the group velocity of a slow EIT pulse propagating through coherently driven Bose–Einstein condensate [27]. For a sample at rest, the group velocity of the light pulse is the sum of the velocity which one would observe without the mechanical effect and the velocity of the recoiling atoms. This recoil effect may give rise to a lower bound for observable group velocities.

D. From Black Hole to Gyro

Very interesting and exciting application of ultraslow light can be found in physics of moving media. It is well known that a moving dielectric appears to light as an effective gravitational field. And, at low flow velocities the dielectric acts on light in the same way as a magnetic field acts on a charged matter wave [31, 168].

Leonhardt *et al.* have discussed light propagation around a vortex flow [115–117]. They pointed out that for a highly dispersive medium the vortex shows an optical Aharonov–Bohm effect at large distances from the core, and at shorter ranges the vortex may resemble an optical black hole. This "black hole" appears exactly due to slow light. However, the experimental conditions and properties of the refractive media suggested in [115–117] are far beyond the reach of realistic experiments, and the properties of existing materials possessing high dispersions do not resemble the properties of the media considered in [115–117]. Therefore, the proposed phenomenon is rather a *Gedankenexperiment*.

Let us discuss in short the way of reasoning used in [115–117] and consider a moving medium with linear temporal dispersion $dn/dv \neq 0$. At any point of space-time there exists a co-moving frame of reference. The atoms of which the medium consists are at rest relative to this co-moving frame. An electromagnetic wave propagating in the medium can be described by a vector $(v', \mathbf{k}')$ obeying the dispersion relation

$$k'^2 - \left(1 + \frac{\chi(v')}{2}\right) \frac{v'^2}{c^2} = 0, \tag{63}$$

which follows from the Maxwell equation for the field. We assume here that susceptibility $\chi(v')$ does not depend on the field amplitude (linear medium).

Making the transformation to the laboratory frame of reference $(v', \mathbf{k}') \leftrightarrow (v, \mathbf{k})$, we get

$$\mathbf{k}' = \frac{\mathbf{k} - \mathbf{v}v/c^2}{\sqrt{1 - v^2/c^2}} \qquad v' = \frac{v - \mathbf{v}\mathbf{k}}{\sqrt{1 - v^2/c^2}}. \tag{64}$$

The same can be written by introducing a covariant four-wave vector $k_i = (v_0/c, -\mathbf{k})$ and presenting the dispersion relation in the laboratory frame of reference as

$$g^{ij} k_i k_j = 0, \tag{65}$$

where g^{ij} is the contravariant tensor of the moving medium. Now, using the Hamilton–Jacobi method [111] to study light propagation, the covariant wave vector is the negative gradient of the eikonal S, and we get

$$k_i = \partial_i S \quad \text{or} \quad v = -\frac{\partial S}{\partial t} \quad \mathbf{k} = \nabla S. \tag{66}$$

Therefore, geometric optics are well justified to describe the light propagation.

Light rays turn out to be zero-geodesic lines of

$$ds^2 = g_{ij} dx^i dx^j \qquad dx^i = (cdt, d\mathbf{x}), \tag{67}$$

where g^{ij} is the covariant tensor of the moving medium. The moving dielectric appears as a curved space-time and acts as an effective gravitational field on light traveling inside it. It is shown in [117] that in a vortex of an ultradispersive medium this effective gravitational field can be so strong that the vortex resembles a black hole for the light.

Dragging of light by an ultradispersive medium results in high sensitivity to the mechanical motion of devices containing the medium. It was first recognized by Scully [175] that coherence effects can be effectively used for detection of rotation. The use of an extremely high dragging coefficient of a slow light medium allowing one to provide a tool to measure or control motion was estimated in [118].

VIII. Conclusion

We have reviewed recent works concerning propagation of light in an ultradispersive coherent media. The major breakthrough is the reduction of the group velocity of light up to a few meters per second has been achieved in multilevel coherently driven atomic systems. This achievement is based on the phenomenon of electromagnetically induced transparency (EIT), which leads to steep normal dispersion of the medium accompanied by vanishing absorption.

EIT is the transparency of a resonant atomic transition for a weak probe field provided by a strong drive field applied to the adjacent atomic transition under condition of two-photon resonance. Ultraslow light is possible when the two-photon atomic coherence decays much more slowly than the one-photon coherence. Slow decay of the coherence allows one to realize the transparency at sufficiently low intensity of the driving field determined by the expression $\Omega_d^2 \gg \gamma_{bc}\gamma$, where Ω_d is the Rabi frequency of the drive, γ_{bc} is the coherence decay rate, and γ is the polarization decay rate.

The behavior of slow light in three-level media is similar to the behavior of a pulse propagating in any "two-level" dispersive medium, and it can be explained in terms of paritonic modes. It is well known that a pulse of light is spatially compressed entering the dispersive medium. The degree of the spatial squeezing is equal to the ratio c/v_g. From the kinematic point of view, it appears because the front edge of the pulse is moving with slow velocity v_g, while the back edge is moving with the speed of light in a vacuum, c. From the dynamic point of view, pulse reshaping results from light absorption and stimulated reemission by the medium. The lion's share of pluse energy is stored in the two-level medium.

Unlike usual polaritons, almost no energy is stored in the coherent media where ultraslow light is observed. Instead, the coherent media work as a transducer between a slow light pulse and coupling electromagnetic wave. A pulse of a probe field entering the medium experiences stimulated Raman-type scattering into the coupling field on a long-lived coherence grating.

Ultraslow light and atomic coherence have already found many important applications in low-intensity nonlinear optics and metrology. Novel applications in quantum nonlinear optics and quantum information processing seem to be feasible.

The coherence allows one to store information about the probe light and transport it in space. It allows time reversing the light. The coherent medium can be used for multiplexing. Moreover, the coherence allows one to increase coupling between light fields such that it becomes possible to study the interaction between single photons. There are a lot of other applications of the coherence, and this list is growing fast.

IX. References

1. Akulshin, A. M., Celikov, A. A., and Velichansky, V. L. (1991). *Opt. Commun.* **84**, 139.
1a. Parkins, A. S., Marte, M., Zoller, P., and Kimble, H. J. (1993). *Phys. Rev. Lett.* **71**, 3095.
1b. Pelizzari, T., Gardiner, S. A., Cirac, J. I., and Zoller, P. (1995). *Phys. Rev. Lett.* **75**, 3788.
1c. Cirac, J. I., Zoller, P., Mabuchi, H., and Kimble, H. J. (1997). *Phys. Rev. Lett.* **78**, 3221.
1d. Bergmann, K., Theuer, H., and Shore, B. W. (1998). *Rev. Mod. Phys.* **70**, 1003.
1e. Kimble, H. J. (1998). *Physica Scripta* **76**, 127.
2. Akulshin, A. M., and Ohtsu, M. (1994). *Quantum Electron.* **24**, 561.
2a. Matsko, A. B., Rostovtsev, Y. V., Kocharovskaya, O., Zibrov, A. S., and Scully, M. O. "Nonadiabatic Approach to Quantum Optical Information Storage," submitted to *Phys. Rev. A*.
3. Alexandrov, E. B., and Bonch-Bruevich, V. A. (1992). *Opt. Eng.* **31**, 711.
3a. Stahler, M., Knappe, S., Affolderbach, C., Kemp, W., and Wynands, R. (2001). *Europhys. Lett.* **54**, 323.
4. Alexandrov, E. B., Balabas, M. V., Pasgalev, A. S., Vershovskii, A. K., and Yakobson, N. N. (1996). *Laser Physics* **6**, 244.
4a. Artoni, M., Carusotto, I., La Rocca, G. C., and Bassani, F. (2001). *Phys. Rev. Lett.* **86**, 2549.
4b. Leonhardt, U. and Piwnicki, P. (2001). *J. Mod. Opt.* **48**, 977.
5. Agrawal, G. P. (1989). "Nonlinear Fiber Optics." Academic Press, San Diego.
5a. Bennink, R. S., Boyd, R. W., Stroud, C. R., and Wong, V. (2001). *Phys. Rev. A* **63**, 033804.
5b. Park, Q. H., and Boyd, R. W. (2001). *Phys. Rev. Lett.* **86**, 2774.
6. Ankerhold, G., Schiffer, M., Mutschall, D., Scholz, T., and Lange, W. (1993). *Phys. Rev. A* **48**, R4031.
6a. Yuan, S., Man, W. N., Yu, J., and Gao, J. Y. (2001). *Chin. Phys. Lett.* **18**, 364.
7. Arimondo, E., and Orriols, G. (1976). *Nuovo Cimento Lett.* **17**, 333.
8. Arimondo, E. (1996). *In* "Progress in Optics" (E. Wolf, Ed.), Vol. XXXV, p. 257. Elsevier Science, Amsterdam.
9. Asher, I. M., and Scully, M. O. (1971). *Opt. Commun.* **3**, 395.
10. Barkov, L. M., Melikpashaev, D. A., and Zolotarev, M. S. (1989). *Opt. Commun.* **70**, 467.
11. Bayindir, M., and Ozbay, E. (2000). *Phys. Rev. B* **62**, R2247.
12. Bendickson, J. M., Dowling, J. P., and Scalora, M. (1996). *Phys. Rev. E* **53**, 4107.
13. Bentley, C. L., Liu, J. R., and Liao, Y. (2000). *Phys. Rev. A* **61**, 023811.
14. Beterov, I. M., and Chebotaev, V. P. (1969). *Pis'ma Zh. Eksp. Teor. Fiz.* **9**, 216; *Sov. Phys. JETP Lett.* **9**, 127.
15. Bicchi, P., Moi, L., Savino, P., and Zambon, B. (1980). *Nuovo Cimento* **55 B**, 1.
16. Bogomolov, V. N., Gaponenko, S. V., Germanenko, I. N., Kapitnov, A. M., Petrov, E. P., Gaponenko, N. V., Prokofiev, A. V., Ponyanina, A. N., Silvanovich, N. I., and Samoilovich, S. M. (1997). *Phys. Rev. E* **55**, 7619.
17. Boyd, R. W. (1997). "Nonlinear Optics." Academic, San Diego.
18. Brandt, S., Nagel, A., Wynands, R., and Meschede, D. (1997). *Phys. Rev. A* **56**, R1063.

19. Brillouin, L. (1922). *Ann. Phys. (Paris)* **17**, 88.
20. Brillouin, L. (1960). "Wave Propagation and Group Velocity." New York, Academic Press.
21. Bretenaker, F., Lépine, B., Cotteverte, J. C., and Le Floch, A. (1992). *Phys. Rev. Lett.* **69**, 909.
22. Budker, D., Yashchuk, V. V., and Zolotorev, M. (1998). *Phys. Rev. Lett.* **81**, 5788.
23. Budker, D., Kimball, D., Rochester, S., and Yashchuk, V. (1999). *Phys. Rev. Lett.* **83**, 1767.
24. Budker, D., Kimball, D. F., Rochester, S. M., and Yashchuk, V. V. (2000). *Phys. Rev. Lett.* **85**, 2088.
25. Budker, D., Kimball, D. F., Rochester, S. M., Yashchuk, V. V., and Zolotorev, M. (2000). *Phys. Rev. A* **62**, 043403.
26. Carlson, N. W., Babbitt, W. R., Bai, Y. S., and Mossberg, T. W. (1985). *J. Opt. Soc. Am. B* **2**, 908.
27. Carusotto, I., Artoni, M., and La Rocca, G. C. (2000). *Pis'ma Zh. Eksp. Teor. Fiz.* **72**, 420; *JETP Lett.* **72**, 289.
28. Chiao, R. Y. (1994). *Quantum Opt.* **6**, 359.
29. Chen, W., and Mills, D. L. (1987). *Phys. Rev. Lett.* **58**, 160.
30. Cheng, Z., and Kurizki, G. (1996). *Phys. Rev. A* **54**, 3576.
31. Chow, W. W., Gea-Banacloche, J., Pedrotti, L. M., Sanders, V. E., Schleich, W., and Scully, M. O. (1985). *Rev. Mod. Phys.* **57**, 61.
32. Drake, K. H., Lange, W., and Mlynek, J. (1988). *Opt. Commun.* **66**, 315.
33. Dowling, J. P., and Bowden, C. M. (1994). *J. Mod. Opt.* **41**, 345.
34. Dowling, J. P., Scalora, M., Bloemer, M. J., and Bowden, C. M. (1994). *J. Appl. Phys.* **75**, 1896.
35. Eggleton, B. J., deSterke, C. M., and Slusher, R. E. (1997). *J. Opt Soc. Am. B* **14**, 2980.
36. Erdogan, T. (1997). *J. Lightwave Technol.* **15**, 1277.
37. Fain, V. M., and Yashchin, E. G. (1964). *Sov. Phys. JETP* **19**, 474.
38. Feld, M. S., and Javan, A. (1969). *Phys. Rev.* **177**, 540.
39. Fleischhauer, M., Keitel, C. H., Scully, M. O., and Su, C. (1992). *Opt. Commun.* **87**, 109.
40. Fleischhauer, M., Keltel, C. H., Scully, M. O., Su, C., Ulrich, B. T., and Zhu, S.-Y. (1992). *Phys. Rev. A* **46**, 1468.
41. Fleischhauer, M., and Scully, M. O. (1994). *Phys. Rev. A* **49**, 1973.
42. Fleischhauer, M. (1999). *Europhys. Lett.* **45**, 659.
43. Fleischhauer, M., Lukin, M. D., Matsko, A. B., and Scully, M. O. (2000). *Phys. Rev. Lett.* **84**, 3559.
44. Fleischhauer, M., Matsko, A. B., and Scully, M. O. (2000). *Phys. Rev. A* **62**, 013808.
45. Fleischhauer, M., Yelin, S. F., and Lukin, M. D. (2000). *Opt. Commun.* **179**, 395.
46. Fleischhauer, M., and Lukin, M. D. (2000). *Phys. Rev. Lett.* **84**, 5094.
47. Franson, J. D., and Pittman, T. B. (1999). *Phys. Rev. A* **60**, 917.
48. Gawlik, W., Kowalski, J., Neumann, R., and Träger, F. (1974). *Phys. Lett. A* **48**, 283.
49. Gawlik, W. (1994). *In* "Modern Nonlinear Optics," part 3 (M. Evans and S. Kielich, Eds.), Wiley, New York.
50. Gheri, K. M., Alge, W., and Grangier, P. (1999). *Phys. Rev. A* **60**, R2673.
51. Giazatto, A. (1989). *Phys. Rep.* **182**, 365.
52. Gibbs, H. M., and Slusher, R. E. (1970). *Phys. Rev. Lett.* **24**, 638.
53. Giraud-Cotton, S., Kaftandjian, V. P., and Klein, L. (1985). *Phys. Rev. A* **32**, 2211; *Phys. Rev. A* **32**, 2223.
54. Wilson-Gordon, A. D., and Friedmann, H. (1992). *Opt. Commun.* **94**, 238.
55. Grangier, P., Reymond, G., and Schlosser, N. (2000). *Prog. Phys.* **48**, 859.
56. Gray, H. R., Whitley, R. M., and Stroud, C. R. (1978). *Opt. Lett.* **3**, 218.
57. Greentree, A. D., Vaccaro, J. A., de Echaniz, S. R., Durrant, A. V., and Marangos, J. P. (2000). *J. Opt. B* **2**, 252.
58. Hakuta, K., Marmet, L., and Stoicheff, B. P. (1991). *Phys. Rev. Lett.* **66**, 596.
59. Hald, J., Sorensen, J. L., Leich, L., and Polzik, E. S. (1998). *Opt. Express* **2**, 93.

60. Hald, J., Sorensen, J. L., Schori, C., and Polzik, E. S. (1999). *Phys. Rev. Lett.* **83**, 1319.
61. Ham, B. S., Hemmer, P. R., and Shahriar, S. M. (1997). *Opt. Commun.* **144**, 227.
62. Ham, B. S., Shahriar, M. S., Kim, M. K., and Hemmer, P. R. (1997). *Opt. Lett.* **22**, 1849.
63. Ham, B. S., Shahriar, M. S., and Hemmer, P. R. (1997). *Opt. Lett.* **22**, 1138; *Opt. Lett.* **24**, 86.
64. Ham, B. S., Shahriar, S. M., and Hemmer, P. R. (1999). *J. Opt. Soc. Am. B* **16**, 801.
65. Hänch, T., and Toschek, P. (1970). *Z. Phys.* **236**, 213.
66. Happer, W. (1972). *Rev. Mod. Phys.* **44**, 169.
67. Harris, S. E., Field, J. E., and Imamoglu, A. (1990). *Phys. Rev. Lett.* **64**, 1107.
68. Harris, S. E., Field, J. E., and Kasapi, A. (1992). *Phys. Rev. A* **46**, R29.
69. Harris, S. E., Yin, G. Y., Jain, M., Xia, H., and Merriam, A. J. (1997). *Phil. Trans. Roy. Soc.* **355**, 2291.
70. Harris, S. E. (1997). *Phys. Today*, p. 36, June.
71. Harris, S. E., and Yamamoto, Y. (1998). *Phys. Rev, Lett.* **81**, 3611.
72. Harris, S. E., and Sokolov, A. V. (1998). *Phys. Rev. Lett.* **81**, 2894.
73. Harris, S. E., and Hau, L. V. (1999). *Phys. Rev, Lett.* **82**, 4611.
74. Harris, S. E., (2000). *Phys. Rev. Lett.* **85**, 4032.
75. Haroche, S., Gray, J. C., and Grynberg, G. (1991). *Science* **252**, 73.
76. Hau, L. V. (1998). *Phys. Rev. A* **58**, R54.
77. Hau, L. V., Harris, S. E., Dutton, Z., and Behroozi, C. H. (1999). *Nature* **397**, 594.
78. Haus, H. A. (1979). *Rev. Mod. Phys.* **51**, 331.
79. Hemmer, P. R., Cheng, K. Z., Kierstead, J., Shahriar, M. S., and Kim, M. K. (1994). *Opt. Lett.* **19**, 296.
80. Hemmer, P., Katz, D. P., Donoghue, J., Cronin-Golomb, M., Shahriar, M., and Kumar, P. (1995). *Opt. Lett.* **20**, 982.
81. Hinds, E. A. (1988). "Atomic Physics" (L. R. Hunter, Ed.), Vol. 11.
82. Ho, K. M., Chan, C. T., and Soukoulis, C. M. (1990). *Phys. Rev. Lett.* **65**, 3152.
83. Huang, F. (1995). *IEE Proc. Microwaves Antennas and Propagation* **142**, 389.
84. Imamoglu, A., Schmidt, H., Woods, G., and Deutsch, M. (1997). *Phys. Rev. Lett.* **79**, 1467.
85. Ippen, E. P., and Stolen, R. H. (1972). *Appl. Phys. Lett.* **21**, 539.
86. Jackson, J. D. (1975). "Classical Electrodynamics," 2 ed. John Wiley and Sons, New York.
87. Jain, M., Xia, H., Yin, G. Y., Merriam, A. J., and Harris, S. E. (1996). *Phys. Rev. Lett.* **77**, 4326.
88. Javan, A. (1957). *Phys. Rev.* **107**, 1579.
89. John, S. (1987). *Phys. Rev. Lett.* **58**, 2486.
90. John, S., and Rangavajan, R. (1988). *Phys. Rev. B* **38**, 10101.
91. John, S., and Wang, J. (1990). *Phys. Rev. Lett.* **64**, 2418.
92. John, S., and Wang, J. (1991). *Phys. Rev. B* **43**, 12772.
93. John, S., and Akozbek, N. (1993). *Phys. Rev. Lett.* **71**, 1168.
94. John, S., and Quang, T. (1994). *Phys. Rev. A* **50**, 1764.
95. John, S., and Quang, T. (1995). *Phys. Rev. Lett.* **74**, 3419.
96. Kanorsky, S. I., Weis, A., Wurster, J., and Hanch, T. W. (1994). *Phys. Rev. A* **47**, 1220.
97. Kasapi, A., Jain, M., Yin, G. Y., and Harris, S. E. (1995). *Phys. Rev. Lett.* **74**, 2447.
98. Kash, M. M., Sautenkov, V. A., Zibrov, A. S., Hollberg, L., Welch, G. R., Lukin, M. D., Rostovtsev, Y., Fry, E. S., and Scully, M. O. (1999). *Phys. Rev. Lett.* **82**, 5229.
99. Khurgin, J. (2000). *Phys. Rev A* **62**, 013821.
100. Kim, M. K., Ham, B. S., Hemmer, P. R., and Shahriar, S. M. (2000). *J. Mod. Opt.* **47**, 1713.
101. Kocharovskaya, O. (1992). *Phys. Rep.* **219**, 175.
102. Kocharovskaya, O., Rostovtsev, Y., and Scully, M. O. (2001). *Phys. Rev. Lett.* **86**, 628.
103. Kofman, A. G., Kurizki, G., and Sherman, B. (1994). *J. Mod. Opt.* **41**, 345.
104. Kozhekin, A., and Kurizki, G. (1995). *Phys. Rev. Lett.* **74**, 5020.
105. Kozhekin, A. E., Mølmer, K., and Polzik, E. (2000). *Phys. Rev. A* **62**, 033809.

106. Kroll, N. M. (1964). *Bull. Am. Phys. Soc.* **9**, 222.
107. Kroll, N. M. (1965). *J. Appl. Phys.* **36**, 34.
108. Kuzmich, A., Mølmer, K., and Polzik, E. S. (1997). *Phys. Rev. Lett.* **79**, 4782.
109. Kweon, G., and Lawandy, N. M. (1995). *Opt. Commun.* **118**, 388.
110. Landau, L. D., and Lifshitz, E. M. (1960). "Electrodynamics of Continuous Media." Pergamon, Oxford.
111. Landau, L. D., and Lifshitz, E. M. (1976). "Mechanics." Pergamon, Oxford.
112. Lamb, Jr., W. E., and Retherford, R. C. (1951). *Phys. Rev.* **81**, 222.
113. Lamb, Jr., G. L., (1971). *Rev. Mod. Phys.* **43**, 99.
114. Lee, H., Fleischhauer, M., and Scully, M. O. (1998). *Phys. Rev. A* **58**, 2587.
115. Leonhardt, U. and Piwnicki, P. (1999). *Phys. Rev. A* **60**, 4301.
116. Leonhardt, U. (2000). *Phys. Rev. A* **62**, 012111.
117. Leonhardt, U., and Piwnicki, P. (2000). *Phys. Rev. Lett.* **84**, 822.
118. Leonhardt, U., and Piwnicki, P. (2000). *Phys. Rev. A* **62**, 055801.
119. Liang, Q. J., Katsuragawa, M., Le Kien, F., and Hakuta, K. (2000). *Phys. Rev. Lett.* **85**, 2474.
120. Liu, C., Dutton, Z., Behroozi, C. H., and Hau, L. V. (2001). *Nature* **409**, 490.
121. Lukin, M. D., Fleischhauer, M., Zibrov, A. S., Robinson, H. G., Velichansky, V. L., Hollberg, L., and Scully, M. O. (1997). *Phys. Rev. Lett.* **79**, 2959.
122. Lukin, M. D., Fleischhauer, M., Scully, M. O., and Velichansky, V. L. (1998). *Opt. Lett.* **23**, 295.
123. Lukin, M. D., Matsko, A. B., Fleischhauer, M., and Scully, M. O. (1999). *Phys. Rev. Lett.* **82**, 1847.
124. Lukin, M. D., Yelin, S. F., Fleischhauer, M., and Scully, M. O. (1999). *Phys. Rev. A* **60**, 3225.
125. Lukin, M. D., Yelin, S. F., Zibrov, A. S., and Scully, M. O. (1999). *Laser Phys.* **9**, 759.
126. Lukin, M. D., Yelin, S. F., and Fleischhauer, M. (2000). *Phys. Rev. Lett.* **84**, 4232.
127. Lukin, M. D., and Imamoglu, A. (2000). *Phys. Rev. Lett.* **84**, 1419.
128. Maimistov, A. I., Basharov, A. M., Elyutin, S. O., and Sklyarov, Yu. M. (1990). *Phys. Rep.* **191**, 1.
129. Manka, A. S., Dowling, J. P., Bowden, C. M., and Fleischhauer, M. (1994). *Phys. Rev. Lett.* **73**, 1789; (1995) *Phys. Rev. Lett.* **74**, 4965.
130. Marangos, J. P. (1998). *J. Mod. Opt.* **45**, 471.
131. Martorell, J., and Lawandy, N. M. (1990). *Phys. Rev. Lett.* **65**, 1877.
132. Matsko, A. B., Rostovtsev, Y., and Scully, M. O. (2000). *Phys. Rev. Lett.* **84**, 5752.
133. Matsko, A. B., Rostovtsev, Y., Fleischhauer, M., and Scully, M. O. (2001). *Phys. Rev. Lett.* **86**, 2006.
134. Matsko, A. B., Novikova, I., Scully, M. O., and Welch, G. R. "Radiation trapping in coherent media," LANL quant-ph/0101147, to be published.
135. Meade, R. D., Brommer, K. D., Rappe, A. M., and Joannopoulos, J. D. (1991). *Phys. Rev. B* **44**, 13772.
136. Merriam, A. J., Sharpe, S. J., Xia, H., Manuszak, D., Yin, G. Y., and Harris, S. E. (1999). *Opt. Lett.* **24**, 625.
137. McCall, S. L., and Hahn, E. L. (1967). *Phys. Rev. Lett.* **18**, 908.
138. McCall, S. L., and Hahn, E. L. (1969). *Phys. Rev.* **183**, 457.
139. Mills, D. L,. and Trullinger, S. E. (1987). *Phys. Rev. B* **36**, 917.
140. Mills, D. L., and Trullinger, S. E. (1987). *Phys. Rev. B* **36**, 947.
141. Mills, D. L. (1991). "Nonlinear Optics: Basic Concepts." Springer-Verlag, Berlin.
142. Mlynek, J., and Lange, W. (1979). *Opt. Commun.* **30**, 337.
143. Molisch, A. F., and Oehry, B. P. (1998). "Radiation Trapping in Atomic Vapours." Clarendon Press, Oxford.
144. Mompart, J., and Corbalan, R. (2000). *J. Opt. B—Quantum Semiclass. Opt.* **2**, R7.
145. Montes, F. A., and Xiao, M. (2000). *Phys. Rev. A* **62**, 023818.

146. Mossberg, T. W. Private communication.
147. Muller, G., Muller, M., Wicht, A., Rinkleff, R. H., and Danzmann, K. (1997). *Phys. Rev. A* **56**, 2385.
148. Nagel, A., Graf, L., Mariotti, E., Biancalana, V., Meschede, D., and Wynands, R. (1998). *Europhys. Lett.* **44**, 31.
149. Nojima, S. (2000). *Phys. Rev. B* **61**, 9940.
150. Ohtaka, M., Hashimoto, A., Itoh, T., Santo, T., and Yamamoto, A. (1995). *Jpn. J. Appl. Phys.* **34**, 4502.
151. Ozbay, E., Michel, E., Tuttle, G., Biswas, R., Ho, K. M., Bostak, J., and Bloom, D. M. (1994). *Opt. Lett.* **19**, 1155.
152. Ozbay, E., Abeyta, A., Tuttle, G., Tringides, M., Biswas, R., Chan, C. T., Soukoulis, C. M., and Ho, K. M. (1994). *Phys. Rev. B* **50**, 1945.
153. Ozbay, E., Michel, E., Tuttle, G., Biswas, R., Sigolas, M., and Ho, K. M. (1994). *Appl. Phys. Lett.* **64**, 2059.
154. For a review, see articles in "Photonic Band Gaps and Localization" (1993). (C. M. Soukoulis, Ed.), Vol. 38 of NATO Advanced Study Institute, Series B: Physics. Plenum, New York, special issue of *J. Mod. Opt.* **42**, no. 2 (1994).
155. Ecomonou, E. N. (Ed.). (1996). Photonic band gap materials, Vol. 315 of NATO Advanced Study Institute, Series E: Applied Science. Plenum, New York.
156. Peterson, D., and Anderson, L. W. (1991). *Phys. Rev. A* **43**, 4883.
157. Phillips, D. F., Fleischhauer, A., Mair, A., Walsworth, R. L., and Lukin, M. D. (2001). *Phys. Rev. Lett.* **86**, 783.
158. Rathe, U., Fleischhauer, M., Zhu, S.-Y., H"ansch, T. W., and Scully, M. O. (1993). *Phys. Rev. A* **47**, 4994.
159. Rebic, S., Tan, S. M., Parkins, A. S., and Walls, D. F. (1999). *J. Opt. B* **1**, 490.
160. Rostovtsev, Y., Matsko, A. B., Shelby, R. M., and Scully, M. O. "Phase-matching condition between acoustic and optical waves in doped fibers." Submitted.
161. Rupasov, V. I., and Singh, M. (1996). *Phys. Rev. A* **54**, 3614.
162. Russel, P. St. J. (1991). *J. Mod. Opt.* **38**, 1559.
163. Sadeghi, S. M., van Driel, H. M., and Fraser, J. M. (2000). *Phys. Rev. B* **62**, 15386.
164. Samartsev, V. V. (1998). *Laser Phys.* **8**, 1198.
165. Sautenkov, V. A., Lukin, M. D., Bednar, C. J., Novikova, I., Mikhailov, E., Fleischhauer, M., Velichansky, V. L., Welch, G. R., and Scully, M. O. (2000). *Phys. Rev. A.* **62**, 023810.
166. Scalora, M., Dowling, J. P., Bowden, C. M., and Bloemer, M. J. (1994). *Phys. Rev. Lett.* **73**, 1368.
167. Scalora, M., Flynn, R. J., Reinhardt, S. B., Fork, R. L., Bloemer, M. J., Tocci, M. D., Bowden, C. M., Ledbetter, H. S., Bendickson, J. M., Dowling, J. P., and Leavitt, R. P. (1996). *Phys. Rev. E* **54**, R1078.
168. Schleich, W., and Scully, M. O. (1984). *In* "General relativity and modern optics" (G. Grynberg and R. Stora, Eds.), Les Houches, Session XXXVIII. Elsevier, Amsterdam.
169. Schmidt, H., and Imamoglu, A. (1996). *Opt. Lett.* **21**, 1936.
170. Schmidt, O., Wynands, R., Hussein, Z., and Meschede, D. (1996). *Phys. Rev. A* **53**, R27.
171. Schmidt, H., and Imamoglu, A. (1998). *Opt. Lett.* **23**, 1007.
172. Schnurr, C., Stokes, K. D., Welch, G. R., and Thomas, J. E. (1990). *Opt. Lett.* **15**, 1097.
173. Scholz, T., Schiffer, M., Welzel, J., Cysarz, D., and Lange, W. (1996). *Phys. Rev. A* **53**, 2169.
174. Schuller, F., Macpherson, M. J. D., and Stacey, D. N. (1989). *Opt. Commun.* **71**, 61.
175. Scully, M. O. (1987). *Phys. Rev. Lett.* **35**, 452.
176. Scully, M. O. (1991). *Phys. Rev. Lett.* **67**, 1855.
177. Scully, M. O., and Fleischhauer, M. (1992). *Phys. Rev. Lett.* **69**, 1360.
178. Scully, M. O., and Zhu, S.-Y. (1992). *Opt. Commun.* **87**, 134.
179. Scully, M. O. (1992). *Phys. Rep.* **219**, 191.

180. Shelby, R. M., Levenson, M. D., and Bayer, P. W. (1985). *Phys. Rev. Lett.* **54,** 939.
181. Shelby, R. M., Levenson, M. D., and Bayer, P. W. (1985). *Phys. Rev. B* **31,** 5244.
182. Sibilia, C., Nefedov, I. S., Scalora, M., and Bertolotti, M. (1998). *J. Opt. Soc. Am. B* **15,** 1947.
183. Sipe, J. E., Poladian, L., and de Sterke, C. M. (1994). *J. Opt. Soc. Am A* **11,** 1307.
184. Slusher, R. E., and Gibbs, H. M. (1972). *Phys. Rev. A* **5,** 1634.
185. Sokolov, A. V., Walker, D. R., Yavuz, D. D., Yin, G. Y., and Harris, S. E. (2000). *Phys. Rev. Lett.* **85,** 562.
186. Sudarshanam, V. S., Cronin-Golomb, M., Hemmer, P. R., and Shahriar, M. S. (1999). *Opt. Commun.* **160,** 283.
187. Szymanowski, C., Wicht, A., and Danzmann, K. (1997). *J. Mod. Opt.* **44,** 1373.
188. Tarhan, I. I., and Watson, J. H. (1996). *Phys. Rev. Lett.* **76,** 315.
189. Thomas, P. J., Rowell, N. L., van Driel, H. M., and Stegeman, G. I. (1979). *Phys. Rev. B* **19,** 4986.
190. Thurston, R. N. (1992). *J. Sound Vibration* **159,** 441.
191. Chiao, R. Y., Townes, C. H., and Stoicheff, B. P. (1964). *Phys. Rev. Lett.* **12,** 592.
192. Turukhin, A. V., Musser, J. A., Sudarshanam, V. S., Shahriar, M. S., and Hemmer, P. R. LANL e-Print archive quant-ph/0010009.
193. Visser, P. M., and Nienhuis, G. (1997). *Opt. Commun.* **136,** 470.
194. Vlasov, Yu. A. (1997). *Phys. Rev. B* **55,** R13357.
195. Wada, M., Sakoda, K., and Inone, K. (1996). *Phys. Rev. B* **52,** 16297.
196. Waldron, R. A. (1969). *IEEE Trans. Microwave Theory and Techniques,* **MTT-17,** 893.
197. Wang, H., Goorskey, D. J., Burkett, W. H., and Xiao, M. (2000). *Opt. Lett.* **25,** 1732.
198. Werner, M. J., and Imamoglu, A. (1999). *Phys. Rev. A* **60,** 011801.
199. Wicht, A., Danzmann, K., Fleischhauer, M., Scully, M., Muller, G., and Rinkleff, R. H. (1997). *Opt. Commun.* **134,** 431.
200. Wynands, R., and Nagel, A. (1999). *Appl. Phys. B* **68,** 1.
201. Xiao, M., Li, Y., Jin, S., and Gea-Banacloche, J. (1995). *Phys. Rev. Lett.* **74,** 666.
202. Yablonovich, E., (1987). *Phys. Rev. Lett.* **58,** 2059.
203. Yablonovich, E., and Gmitter, T. J. (1989). *Phys. Rev. Lett.* **63,** 1950.
204. Yoo, H. I., and Eberly, J. H. (1985). *Phys. Rep.* **118,** 239.
205. Zhu, Shi-Yao, Chen, Hong, and Huang, Hu (1997). *Phys. Rev. Lett.* **79,** 205.
206. Zibrov, A. S., Lukin, M. D., Hollberg, L. W., Nikonov, D. E., Scully, M. O., Robinson, H. G., and Velichansky, V. L. (1996). *Phys. Rev. Lett* **76,** 3935.
207. Zibrov, A. S., Lukin, M. D., and Scully, M. O. (1999). *Phys. Rev, Lett.* **83,** 4049.
208. Zibrov, A. S., Matsko, A. B., Kocharovskaya, O., Rostovtsev, Y. V., Welch, G. R., and Scully, M. O. "Transporting and time reversing light via atomic Coherence." To be published.
209. Zibrov, A. S., Matsko, A. B., Hollberg, L., and Velichansky, V. L. "Selective reflection in coherent media." To be published.

LONGITUDINAL INTERFEROMETRY WITH ATOMIC BEAMS

S. GUPTA, D. A. KOKOROWSKI, and R. A. RUBENSTEIN[1]
Massachusetts Institute of Technology, Cambridge, Massachusetts 02139

W. W. SMITH
Physics Department, University of Connecticut, Storrs, Connecticut 06269

I. Introduction	243
II. Interference in Ion–Atom Collisions	245
III. Differentially Detuned Separated Oscillatory Fields: DSOF	252
IV. Amplitude Modulation and Rephasing	255
V. Detecting Longitudinal Momentum Coherences	258
VI. Discussion of the Semiclassical Approximation	259
VII. Apparatus and Experimental Techniques	260
VIII. Density Matrix Deconvolution	263
IX. Search for Longitudinal Coherences in an Unmodified Atomic Beam	268
X. Conclusion	272
XI. Acknowledgments	273
XII. References	273

I. Introduction

The study of the wave properties of material particles began with Louis de Broglie's (1924) hypothesis in his Ph.D. thesis which stated that a particle's wavelength is given by $\lambda = h/p$, where h = Planck's constant and p is the magnitude of the particle's momentum. Schrödinger seized on de Broglie's wavelength formula to write the basic equation of wave mechanics, which along with Heisenberg's (equivalent) matrix mechanics, led to the quantum mechanical description of atomic structure in terms of the wave properties of electrons that we know today. de Broglie's hypothesis was quickly verified in the experimental demonstration of the diffraction of electron beams from a crystal lattice by Davisson and Germer (1927a,b) in the United States. About the same time in Scotland, G. P. Thomson and A. Reid presented further evidence of the wave–particle duality by recording on film a Debye–Scherrer powder *electron* diffraction pattern analogous to an X-ray diffraction pattern. Electron matter–wave diffraction studies culminated in the demonstration of an electron interferometer by L. Marton *et al.* (1952, 1954). Beams of atoms and molecules

---
[1] Boston Consulting Group, 4800 Hampden Lane, Bethesda, MD 20814.

were found to exhibit transverse small-angle diffraction effects for particles much heavier than electrons by Estermann and Stern (1930). Soon afterward, neutron diffraction and interferometry developed (Maier-Leibnitz and Springer, 1962), verifying and extending the applicability of de Broglie's formula yet again.

These early investigations into the wave properties of particles have since grown into the field of atom optics. The coherent manipulation of both internal (electronic) and external (momentum) states of atoms has resulted in a plethora of atom optic devices such as diffraction gratings, mirrors, cavities, interferometers, and even active devices such as amplifiers. Key demonstration experiments exploiting atomic beam diffraction in the transverse plane by standing light waves (Gould et al., 1986; Martin et al., 1988) and nanofabricated gratings (Keith et al., 1988) were followed by demonstrations of atom interferometers (Carnal and Mlynek, 1991; Keith et al., 1991; Riehle et al., 1991; Kasevich and Chu, 1991). These have been used to perform measurements of atomic properties as well as precision measurements of fundamental physical quantites.[2]

Most atom interferometers to date have used transverse atom-optical beam splitters, which shift the momentum of the de Broglie wave in a direction *perpendicular* to the direction of propagation. Atom interferometers work by splitting an atomic beam (with an appropriate type of beam splitter) and then recombining the split pieces (with a partially transmitting "mirror" or another beam splitter). The historical predecessors of this basic setup are the Michelson interferometer (for light) and the Ramsey separated oscillatory field (SOF) interferometer (for atoms) (Ramsey, 1950).

The SOF technique, originally invented by Ramsey to improve molecular beam resonance experiments, was extended to four optical fields and reinterpreted as an atom interferometer by Bordé (1989). This generalized Ramsey–Bordé interferometer was demonstrated by Riehle et al. (1991). A further extension to two-photon processes has yielded Raman interferometers (with two-photon momenta kicks) which have been heavily exploited by the groups of Chu and Kasevich to do precision measurements of g (Peters et al., 1999), $\hbar/m$ in cesium (Weiss et al., 1993) (a crucial ingredient for measurements of the fine structure constant α), rotations (Gustavson et al., 1997), and gravity gradients (Snadden et al., 1998). These Raman interferometers, which are fundamentally separated oscillatory fields, are also the interferometers of choice for coherent Bose-Einstein condensates (Kozuma et al., 1999).[3] Another generalization of SOF, experiments described in Sections III and following, introduce a relative detuning between two nearly resonant oscillatory fields to produce an interferometer in longitudinal momentum space.

---

[2]For a review of the techniques and results of modern transverse atom interferometry, see J. Schmiedmayer, et al. (1997). In "Atom Interferometry" (P. Berman, Ed.), Academic Press, San Diego, CA. This article gives many other references.

[3]Typically, Raman processes between the same internal state are used for BECs. The process is then simply Bragg diffraction of de Broglie waves, and the interferometers are called Bragg interferometers.

A fundamental limitation to matter–wave interferometery arises if different matter–wavevectors produce interference of a different phase: the contrast of the total interference pattern will be reduced or completely destroyed. Many experiments circumvent this by using a "white-fringe" geometry, in which the difference in phase $\Delta\phi$ along the two interfering paths is arranged to be independent of velocity.

Alternatively, a velocity-dependent or dispersive phase shift between paths can be exploited to perform novel experiments. In this chapter, we describe two such experimental efforts in which velocity dispersion of the interferometer plays a central role. We first review a series of ion–atom scattering experiments in which novel interference phenomena have been observed that can also be described in terms of velocity-dependent phases. We then turn to new developments in the field of longitudinal matter–wave optics. We describe experiments in which a velocity-dependent phase between interfering paths is used to cancel intrinsic velocity dispersion in an atomic beam, and to deconvolve the beam's longitudinal density matrix. Our understanding of, and ability to manipulate, an atom's longitudinal degree of freedom brings us one step closer to the ultimate engineering goal of atom optics: the ability to coherently control the quantum state of a matter wave.

II. Interference in Ion–Atom Collisions

As early as 1927, Leonard Loeb addressed the question of whether a beam of positive ions could ionize atoms in binary collisions, lamenting that so far, the new quantum mechanics had been unable to solve the problem (Loeb, 1927). Experiments demonstrating ionization were carried out shortly thereafter (see Kessel et al., 1978) and were soon explained by Weizel and Beeck (1932) as resulting from electronic transitions between molecular orbitals of the transitory molecule formed during the collision. This quasi-molecular description of "slow" heavy particle collisions has been refined and extensively verified. It is based on the Born–Oppenheimer (B–O) approximation (Born and Oppenheimer, 1927) as used in molecular spectroscopy: the two-center, adiabatic molecular states can be used to approximate the true molecular orbitals if the velocity of nuclear motion in a diatomic collision is small compared to the Bohr orbital velocity of the electrons in the system. Within the B–O approximation, the wave function of the colliding system approximately factors into the product of the wave functions for the relative translational nuclear motion and for the electronic motion. The same approximation works well for either a bound state (as in molecular spectroscopy) or a continuum collisional state (Kessel et al., 1978). Since total energy is conserved in an inelastic molecular collision, the kinetic energy (or the vibrorotational energy in the case of a bound molecule) must differ along coherently excited pathways of different average electronic potential energy.

Matter–wave interference effects were observed some time ago in heavy-particle atomic collisions. Specific examples abound in the context of ion–atom collisions within the Born–Oppenheimer approximation, as these collisions are intrinsically coherent. In the most studied examples, the interference appears as oscillatory structure in the energy or the angular dependence of the differential scattering cross section. Until recently, for thermal or higher energy heavy-particle atomic collisions, semiclassical scattering theory (Ford and Wheeler, 1959) sufficed to explain the interference structure and generally gave good agreement with experiments. In special cases in which a stationary phase condition exists over a range of impact parameters, interference effects may be seen in the *total* cross section. As specific examples, we mention the Steuckelberg and Rosenthal oscillations in $He^+ + He$ inelastic collisions. These are briefly outlined in this chapter for perspective, as they represent early observations of a kind of longitudinal interferometry within a colliding molecular system. They are reviewed in detail in the Ph.D. thesis of D. A. Clark (1978) and in an earlier review by one of the present authors (Kessel *et al.*, 1978).

By 1963, a number of detailed experiments had been performed on homonuclear ion–atom elastic collisions involving resonant electron transfer (charge transfer) from target to projectile (i.e., $H^+ + H$ or $He^+ + He$). These experiments revealed an oscillatory probability of projectile ion neutralization as a function of collision energy at constant scattering angle (near the forward direction) (Ziemba and Everhart, 1959; Lockwood and Everhart, 1962), as depicted in Fig. 1. Lichten (1963) showed that the period of oscillation in the $He^+ + He$ case could not be explained using adiabatic Born–Oppenheimer quasi-molecular energy-level diagrams.[4] In the related but simpler $H^+ + H$ case, the electron-transfer probability can be calculated from the quantum mechanical interference between two molecular one-electron eigenstates of H_2^+, one state being the attractive or bonding orbital ($\Sigma_g 1s$) and the other being the repulsive or antibonding orbital ($\Sigma_u 1s$). The initial state, with an H^+ projectile and H target, is an equal superposition of these two eigenstates. The final state, consisting of an ionized target and neutralized projectile, is again a superposition of the bonding and antibonding eigenstates but with the opposite relative phase. During the collision, each molecular eigenstate has a different kinetic energy and therefore evolves a different spatial phase. When the colliding particles separate and it becomes appropriate to project onto free-particle eigenstates, the two paths corresponding to different molecular orbitals interfere, producing oscillations in the charge transfer probability. The phase of the interference signal is proportional to the integral of the momentum difference between molecular orbitals, which in turn is a function of the center-of-mass collision energy. Analogous arguments for the case of $He^+ + He$ scattering lead to the oscillations evident in Fig. 1 (Ziemba and Everhart, 1959).

[4]Lichten (1963) suggested using *diabatic* correlation rules and the corresponding diabatic energy levels to explain resonant charge-transfer processes in the kiloelectron-volt energy range.

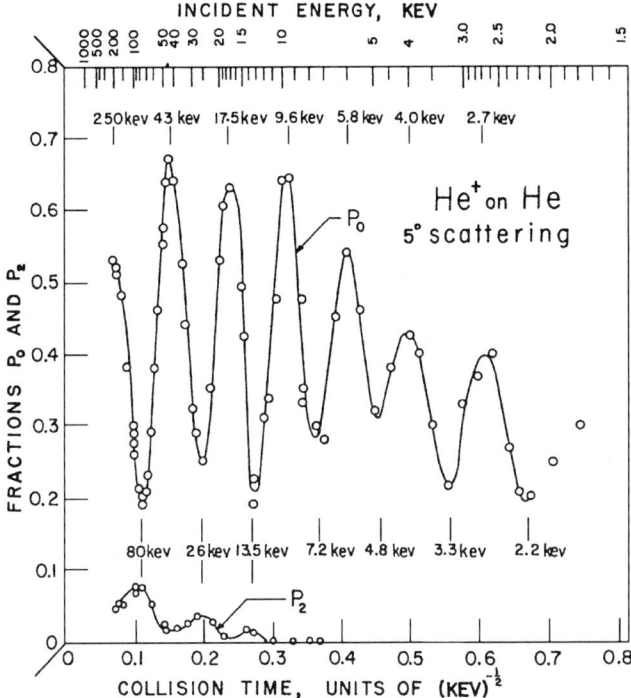

FIG. 1. Scattering of projectiles at 5° from He$^+$ + He collisions are analyzed in terms of the fractions P_0 and P_2 of neutral and doubly ionized helium, respectively. The fractions are plotted both as a function of collision time (lower x axis) and incident energy in kiloelectron volts (upper x axis). Several cycles of interference oscillations are shown. (From Ziemba, F. P., and Everhart, E. *Phys. Rev. Lett.* 1959;2:299.)

These experiments represent an early example of longitudinal interference of matter waves in the context of ion–atom collisions. It should be noted that the charge transfer oscillations in these experiments can be observed only in the *differential* cross section. When the total cross section is measured or calculated by integrating over impact parameter, the oscillations average out to zero.

Similar "Stueckelberg oscillations" (Stueckelberg, 1932) have been observed in inelastic ion–atom collisions as well (Aberth and Lorents, 1966). These result from the localized coupling of two molecular states at an avoided curve crossing of ground- and excited-state Born–Oppenheimer electronic potential curves for the colliding system. In this case, there exist two semiclassical scattering trajectories of different impact parameter that scatter the ions into a common angle (Fig. 2). The localized coupling results from the perturbing effect of kinetic energy terms in the Hamiltonian which are ignored in making the Born–Oppenheimer approximation. Interference can be observed experimentally by measuring the inelastic energy

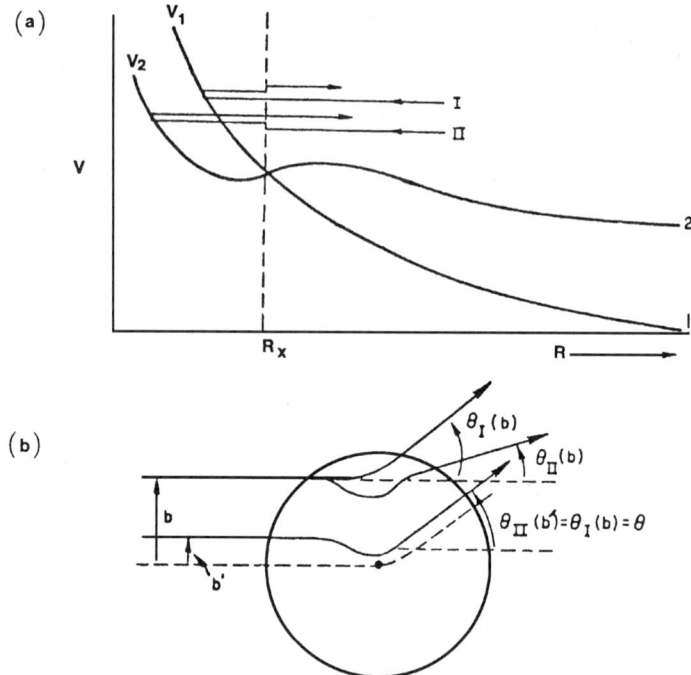

FIG. 2. Explanation of Steuckelberg interference oscillations in the differential inelastic scattering cross section. (a) Two quasi-molecular potential curves V_1 and V_2 cross at internuclear distance R_x, and are mixed at this point by an off-diagonal matrix element due to the motion of the colliding atoms (Landau–Zener model). The two most probable inelastic paths through R_x are indicated as I and II. (b) The scattering paths I and II correspond to two different impact parameters b and b'. Interference occurs when both trajectories I and II lead to the same scattering angle θ.

loss in the collision for a particular scattering angle (i.e., by measuring the energy spectrum of scattered ions as a function of angle).

In inelastic ion–atom collisions, interference structure is sometimes observed as a function of collision energy in the optical emission cross section. In the case of He$^+$ + He collisions in the 100-eV to low-keV range, the optical emission cross sections after collisional excitation of He($1s3s$, $1s3p$, and $1s3d$) show a rich oscillatory structure that is different for each L–S multiplet (Fig. 3). The data were obtained by isolating a particular spectral emission line from the collision region with an interference filter and measuring the intensity of emitted light vs. collision energy. This type of interference structure in the total cross section for excitation of a particular excited state or multiplet has been referred to as *Rosenthal oscillations* (Kessel *et al.*, 1978; Rosenthal and Foley, 1969).

In Fig. 3, excitation of the He($1snl$) excited states by He$^+$ projectiles occurs at the surprisingly low energy of ~30 eV in the center-of-mass frame, only slightly

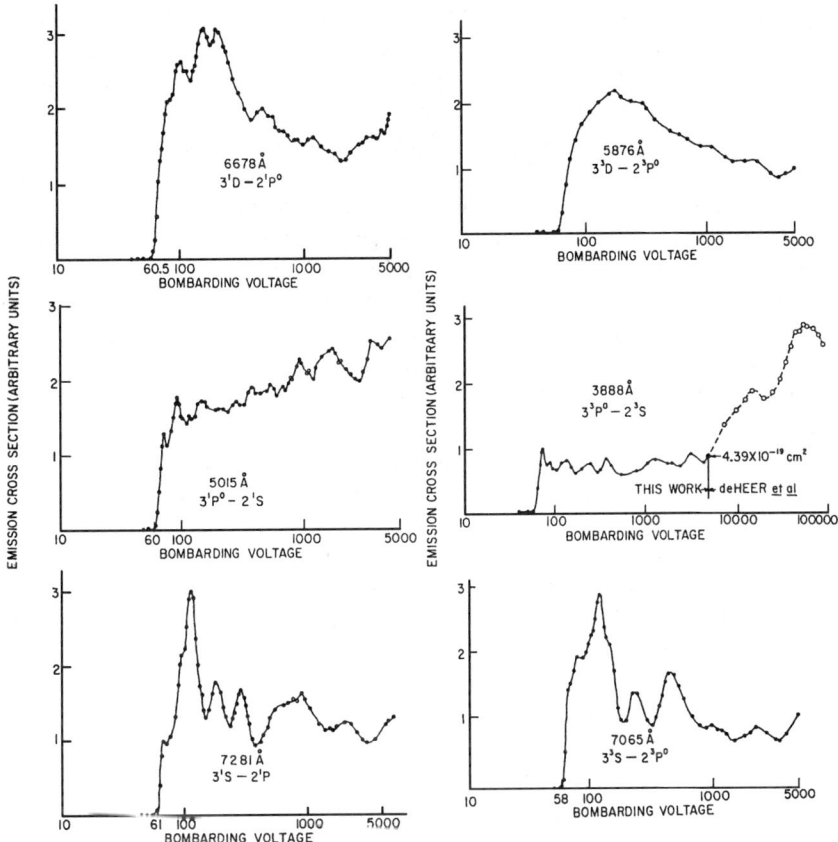

FIG. 3. Optical emission excitation cross sections vs. laboratory ion beam energy for the visible HeI lines originating from the $n = 3$ levels when He gas is bombarded by He^+ ions, under low pressure (approximately single-collision conditions). As estimate of the absolute cross section for the 3^3P state is given. Note the particularly strong interference structure (Rosenthal oscillations) in the lower two curves. (From where details are given.)

above the atomic excitation energy threshold for these states. In many atomic collision systems no inelastic excitation occurs until the center-of-mass collision energy reaches the keV range or higher. An explanation according to the Massey adiabatic criterion and the Landau–Zener (Landau, 1932) curve-crossing model is given by Dworetsky et al. (1967).

When *two* neighboring excited-state Born–Oppenheimer curves are excited coherently at short range ($R_i \simeq 2$–3 Bohr) via avoided (Landau–Zener) crossings with the repulsive ground state $^2\Sigma_g$ curve, the two excited-state amplitudes will evolve differently in time as the collision proceeds because of their different kinetic energies. Rosenthal and Foley (1969) were able to show, via *ab initio* calculations of the Born–Oppenheimer excited-state potential curves for He_2^+, that two of the coherently excited quasimolecular levels were also coupled at *long range* ($R_0 \simeq$ 25 Bohr or more) due to the long-range Stark-effect perturbation of the atom by the incoming ion. These two localized Landau–Zener curve crossings thus form a closed loop of fixed area in the system of potential curves leading to a pair of $n = 3$ excited-state amplitudes which mix at both short and long range. This three-state model, in which a common initial state (the ground state of the ion–atom colliding system) coherently feeds two excited states that are coupled by a closed loop in the energy-level diagram, is shown schematically in Fig. 4. It should be emphasized that the three-state model is only an approximation and that, in general, more than three states may participate in the excitation process.

In the simple three-state picture involving a ground state and two excited states, interference between the two excited states of Fig. 4 gives rise to antiphase oscillations in the energy dependence of the excitation cross sections for this pair of strongly coupled states. Because the difference in momentum of the two excited states is inversely proportional to collision velocity at large internuclear separation, v^*, the oscillations are of approximately constant frequency when plotted vs. $1/v^*$. (Note that the ion beams typically used had a narrow fractional energy spread of 1% or less, which is important for clear observation of these interference effects.) In the Rosenthal–Foley model, the phase evolution between

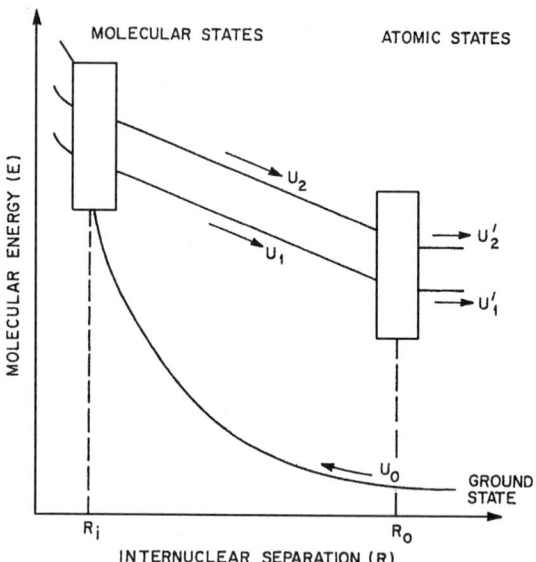

FIG. 4. Three-state Rosenthal–Foley model leading to correlated interference oscillations in the total excitation cross sections for final states U_1' and U_2'. Initial excitation from the ground state is via channel U_0.

the two excited-state amplitudes during the collision is given by

$$\delta\phi = \frac{1}{\hbar}\int_{R1}^{R2}(k_2 - k_1)\,dR \approx \frac{1}{\hbar}\int_{R1}^{R2}(U_2 - U_1)\frac{dR}{v^*}. \qquad (1)$$

Subsequent experiments using the four possible He^3 and He^4 isotopic combinations of projectile and target showed that the phase of the observed oscillations scaled appropriately with relative velocity after excitation (final state velocity) (Dworetsky and Novick, 1969). The interference structure in these isotope experiments *did not* scale as well with the relative velocity before excitation (initial-state velocity), indicating that the interference was produced as a result of collisional excitation.

In certain cases, the Rosenthal oscillations in the optical emission cross section could not be clearly observed as a function of collision energy without looking at the *polarization* of the emitted light from a particular spectral line. Such oscillations were observed by Tolk *et al.* (1976) in polarized optical emission from keV-energy Na^+–Ne collisions and also by Clark and Smith in $He^+ + He$ collisions. Figure 5, from Clark (1978), shows a particularly interesting case of the Rosenthal oscillations in the He_2^+ system, in which polarization selective detection was necessary to clearly bring out strong oscillations as a function of $1/v^*$.

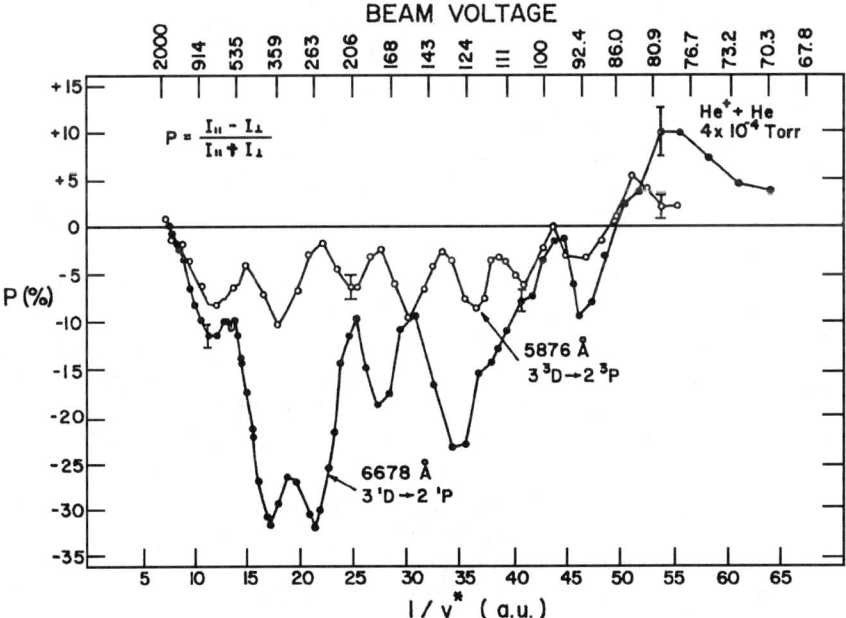

FIG. 5. Linear polarization vs. $1/v^*$ and beam energy for He $3^{3,1}D$ state-selected optical emission resulting from $He^+ + He$ collisions. (From Kessel *et al.*, 1978; Clark, 1978.)

There is a gradual phase slippage of the high-frequency components in the interference pattern for 3^1D vs. 3^3D excitation. This is because the area (or phase) represented by the closed loop in the potential curves leading to excitation of the triplet manifold is slightly different from that of the singlet manifold.

Figure 4, discussed above, brings out the close analogy between interference effects like the Rosenthal oscillations observed in optical emission in ion–atom collisions and the interference observed in atomic-beam radiofrequency (r.f.) resonance using the Ramsey separated oscillatory field (SOF) technique (Ramsey, 1956). In the collision case, the two possible paths evolve with a different spatial phase (due to their different kinetic energy) between the two localized points of strong coupling between the electronic states U_1 and U_2 involved in the collision. In the SOF case, an atom passes through two spatially separated r.f. resonance regions, each tuned to provide strong coupling between two Zeeman or hyperfine levels. Again, because the two interfering paths differ slightly in momentum, a different spatial phase is evolved by each. Interference fringes are observed in both cases by varying the momentum difference between the two interfering paths. In the collision case, the momentum difference is a function of the relative collision velocity. We will see that in the atomic case, the momentum difference can be tuned using the frequency of the applied r.f. radiation.

III. Differentially Detuned Separated Oscillatory Fields: DSOF

We now turn to a discussion of differentially detuned separated oscillatory fields (DSOF), a type of matter–wave interferometry in which interfering paths have different longitudinal momentum. We will describe how the phase shift between paths that results from this momentum difference has been exploited to study the longitudinal coherence structure of atomic beams.

DSOF has its roots in the techniques originally developed by Rabi, Ramsey, and collaborators to study nuclear magnetic resonance in atomic beams—techniques which were subsequently extended to neutron beams as well (Summhammer *et al.*, 1995; Kaiser *et al.*, 1992). Theoretical and experimental work on neutron spin echoes led to the demonstration that oscillating magnetic fields can affect the external longitudinal wave function of a propagating matter wave (Mezei, 1972, 1980; Golub and Gähler, 1987; Golub *et al.*, 1994). In one experiment, small changes in the kinetic energy of neutrons inelastically scattered from a target were measured interferometrically (Golub and Gähler, 1987). Neutron spin resonance has also been combined with transverse neutron interferometry: resonant oscillatory fields of different frequencies were applied to the two spatially separated beams of a single-crystal neutron interferometer (Badurek *et al.*, 1986), resulting in a time-dependent interference signal when the two plane waves differing slightly in total energy were recombined.

In longitudinal matter–wave interferometry, the momentum of the de Broglie wave is increased or decreased slightly in the direction parallel to the propagation. An interesting example of longitudinal Stern–Gerlach interferometry by a group in Paris involved the Zeeman levels of an atomic H(2s) beam and used a *nonresonant* but time-dependent (pulsed), spatially inhomogeneous magnetic field region ("phase object"), between polarizing and analyzing Stern–Gerlach magnets. For each Zeeman level, the total energy is changed by the time-dependent interaction and a net longitudinal momentum transfer can occur. Interference fringes appear with high visibility under state-selective Zeeman-state detection as a function of the amplitude of the pulsed magnetic field "C region" (Chormaic *et al.*, 1994). The purpose of the experiment was to validate the scalar Aharonov–Bohm effect for an atomic beam. An earlier experiment of this group (J. Baudon *et al.*) used a double helical coil between the Zeeman polarizing and analyzing regions to demonstrate the effects of an atomic topological phase (Miniatura *et al.*, 1992).

Our own treatment of DSOF and longitudinal matter–wave optics begins with the standard Hamiltonian for the two-state resonance problem to which a longitudinal kinetic energy term has been added (Smith *et al.*, 1998). Specifically, we consider an atomic plane wave with ground (excited)-state energy $\hbar \omega_g$ ($\hbar \omega_e$) incident upon an oscillatory field of frequency $\omega \neq \omega_e - \omega_g$, where $\hbar \omega_{e,g} = [(\hbar k_{e,g})^2/2m] + U_{e,g}$. In a transition between internal levels there will be an energy surplus/deficit in the amount

$$\hbar[\omega - (\omega_e - \omega_g)] = \hbar\omega - (u_e - u_g). \tag{2}$$

This surplus (deficit) energy $\hbar\delta$ must increase (decrease) the atom's kinetic energy, leading to a corresponding change in the atomic wavevector,

$$\Delta k \equiv k_e - k_g$$
$$= \frac{\sqrt{2m}}{\hbar}[(\hbar\omega_g + \hbar\omega - U_e)^{1/2} - (\hbar\omega_g - U_g)^{1/2}] \simeq \delta/v. \tag{3}$$

The last line applies when the kinetic energy of atoms in the beam is much larger than the Rabi energy $\hbar\omega_R$, the potential energy $U_{e,g}$, and $\hbar\delta$. This assures that the velocities v_e and v_g are approximately equal to v, the average longitudinal velocity of the beam.

Because a resonance region not only changes an atom's internal state but also entangles internal and external degrees of freedom, it can function as a beamsplitter in longitudinal energy/momentum space: an incoming plane wave can emerge in an entangled superposition of ground and excited states, each with a different kinetic energy and hence different wavevectors k_e, k_g.

The maximum attainable momentum difference between the radiatively coupled states is limited by the spatial profile of the oscillatory field $\omega_R(x)$. For large momentum shifts (detunings), the amplitude in the excited state is small, $a_g \simeq 1$, and we can apply first-order perturbation theory to estimate the excited-state amplitude as a function of position,

$$a_e(x) \simeq \frac{-i}{2v} \int_0^x dx' \, \omega_R(x') e^{-i\Delta k x'}. \tag{4}$$

The probability of making a transition is signficant only for $\Delta k < \Delta k_{\max} \approx 1/l$, where l is the characteristic width of the spatial profile $\omega_R(x)$. The corresponding frequency width of the resonance curve, $\delta_{\max} \sim v\Delta k_{\max} \sim v/l$, agrees with the limit derived from the usual finite transit time argument.

The transition-induced momentum shift Δk is not related to or limited by the momentum of free photons of frequency ω, even though the change in total energy is given by $\hbar\omega$. For a r.f. resonance experiment, one can construct regions with l thousands of times smaller than the wavelength of the radiation, λ, yielding a maximum momentum difference, $\hbar \Delta k_{\max} \sim \hbar/l$ thousands of times larger than h/λ, the corresponding photon momentum. Significantly, this large momentum can be transferred with high probability even when the atom undergoes a single cycle of Rabi oscillation (i.e., $\omega_R l/v \simeq \pi$). This large momentum transfer may be ascribed to the interaction of the particle's oscillating dipole moment with the gradients of the oscillating field at the ends of the r.f. region (Hill and Gallagher, 1975).

Combining two r.f. beamsplitters tuned to different frequencies, $\omega_1 = \omega + \delta_1$ and $\omega_2 = \omega + \delta_2$, we can produce an atom whose excited-state wavefunction is a coherent superposition of two different momenta (Fig. 6). We call this arrangement

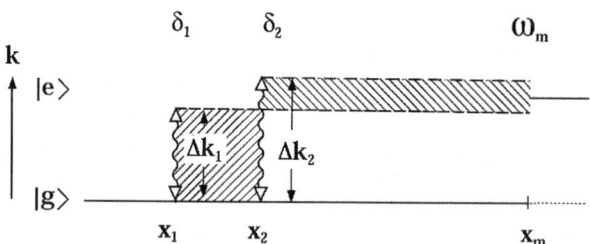

FIG. 6. The DSOF interferometer is formed by oscillatory fields of detuning δ_1, δ_2 applied at x_1, x_2. Ground (excited)-state plane wave components are shown as solid (dashed) lines. The left (right) hatched area denotes the differential phase accumulated by the atoms excited at x_1 and x_2. To detect the superposition of excited states we employ an amplitude modulator of frequency $\omega_m = \delta_2 - \delta_1$ located at position x_m.

of differentially detuned, separated oscillatory fields *DSOF* to emphasize its similarity to the separated oscillatory fields (SOF) method of Ramsey (1950), where $\delta_1 = \delta_2$.

The original SOF configuration and the more general DSOF configuration can be interpreted as interferometers in momentum/internal state space. In each case, an incident particle in state $|g\rangle$ has two paths by which to reach state $|e\rangle$ at the interferometer output, namely, excitation in either the first or second oscillatory field region (see Fig. 6). In SOF, the momentum shift $\hbar \Delta k$ which accompanies the change in internal state is the same for both paths; therefore both are excited into the same final momentum eigenstate with the same total energy. These two paths will exhibit time-independent interference, producing Ramsey fringes in the excited-state probability as a function of detuning. The phase of this interference is just the differential phase, $\phi = \Delta k(x_2 - x_1)$, accrued between the two oscillatory field zones.

A time-dependent interference occurs in the DSOF case because the excited-state portion of the outgoing wavefunction is a coherent superposition of two distinct total energy components, each with a different corresponding longitudinal momentum. The probability of detecting the atom in the excited state at the downstream position x_d is given by

$$P_e \simeq \frac{1}{2}\{1 + \cos[\Delta k_1(x_2 - x_1) - (\Delta k_2 - \Delta k_1)(x_d - x_2) - (\omega_1 - \omega_2)t]\}. \tag{5}$$

The coherent superposition of states with different total energy produces a temporal phase oscillating at the difference frequency $\omega_1 - \omega_2$. Two additional terms represent the spatial phase accumulated between the two field regions $[\Delta k_1(x_2 - x_1)]$, as well as a contribution with opposite sign $[(\Delta k_2 - \Delta k_1)(x_d - x_2)]$ which is the phase accumulated after the second region. When the frequencies of the two oscillatory field regions are equal, $(\Delta k_1 = \Delta k_2 = \delta/v)$, Eq. (5) is consistent with the standard SOF result (Ramsey, 1956).

IV. Amplitude Modulation and Rephasing

The sinusoidal probability distribution as a function of space and time [Eq. (5)] indicates amplitude modulation of the excited state. Amplitude modulation (AM) of a plane wave by definition involves creating a time-varying change in the amplitude of the outgoing wave beyond the point of modulation. This can only result from the coherent superposition of states with more than one longitudinal wavelength.

For a molecular beam, a periodic chopper is the simplest example of an amplitude modulator. The DSOF region discussed in the previous section is a sophisticated amplitude modulator which imparts an additional phase shift depending on the wave vector and internal state of the incoming wave. This property can be employed to create a virtual amplitude modulator located at a preselected position farther along the beam.

Amplitude modulation is difficult to observe in a real molecular beam, because the distribution of velocities in the beam leads to spreading of the initially distinct packets. Typical molecular beams based on supersonic expansions have a roughly 5–10% velocity spread (the spread is even larger for thermal beams), which causes the faster particles from one packet to catch up to the slower ones from the next. In our analysis of amplitude modulation using DSOF, the loss of visibility results from the velocity dependence of the momentum shift $\Delta k \propto 1/v$. If the spread in velocity is large enough that the difference in phase between fast and slow atoms at position x_d is of order π, then after averaging [Eq. (5)] over velocity, there will be no discernible interference fringes.

The amplitude modulation produced by a DSOF region allows velocity dephasing to be compensated, since it includes a velocity-dependent term $[\Delta k(x_2 - x_1) = \delta(x_2 - x_1)/v$ in Eq. (5)]. This term allows the creation of an *initially dephased* momentum distribution which rephases at a later point downstream. Specifically, the spatially dependent phase terms in Eq. (5), which are functions of velocity, produce a total phase

$$\phi(x_d, v) = \frac{(\delta_1 - \delta_2)(x_d - x_2)}{v} + \frac{\delta_1(x_2 - x_1)}{v}. \tag{6}$$

Here $(x_d - x_2)$ is the distance from the second DSOF region to the position at which atom flux will be observed. To make this phase independent of the initial velocity we set it to zero, yielding a rephasing condition,

$$x_{\text{reph}} = x_2 + \frac{(x_2 - x_1)\delta_1}{(\delta_2 - \delta_1)}. \tag{7}$$

This equation tells us that for fixed locations of the oscillatory fields, we can choose detunings which will cause the amplitude modulation to rephase at any desired position x_{reph}.

In a recent experiment (Smith *et al.*, 1998), the difference frequency $(\delta_1 - \delta_2)$ was fixed, and δ_1 was then chosen to satisfy Eq. (7) with x_{reph} set to produce amplitude modulation at downstream position x_m. To detect this modulation, a single oscillatory field tuned to produce transitions between the ground state and some third unobserved state was used to heterodyne the atomic signal at this position to near-dc. The transmission probability of this heterodyne modulator could be adjusted simply by changing the strength of the applied field. The frequency of the heterodyne modulator was set to cancel the time dependence of the rephased amplitude modulation ($\omega_m = \delta_1 - \delta_2$) and produce a time-independent beat signal,

$$\langle P_e(\delta_1, \delta_2, v, x, t)\rangle = \frac{1}{4} + \frac{1}{8}\cos\left[(\delta_1 - \delta_2 + 2\omega_m)\left(\frac{x}{v} - t\right)\right. \\ \left. + \frac{\delta_2 x_2 - \delta_1 x_1 - 2\omega_m x_m}{v} - 2\phi_m\right], \quad (8)$$

where ϕ_m is the phase of the heterodyne modulator relative to the DSOF oscillators. Scanning ϕ_m we observed a periodic modulation of the detected counts, while scanning the global detuning δ_1 of the DSOF yielded "displaced" Ramsey fringes—fringes with the characteristic Ramsey shape, but centered around $\delta_1 = -(\delta_1 - \delta_2)$ $(x_d - x_2)/(x_2 - x_1)$ or, equivalently, around $x_{\text{reph}} = x_m$ (Fig. 7).

At the center of the Ramsey interference pattern, the phases evolved by the particles are independent of their velocity. This configuration constitutes a white-fringe interferometer. It also resembles a "half"-spin echo in that a dephased state is prepared (without first creating an in-phase pulse as in a complete spin echo) that subsequently rephases. Because the relevant phase terms are spatially dependent, this rephasing occurs at a specific point in space, rather than at a specific time as in the usual spin-echo experiment.

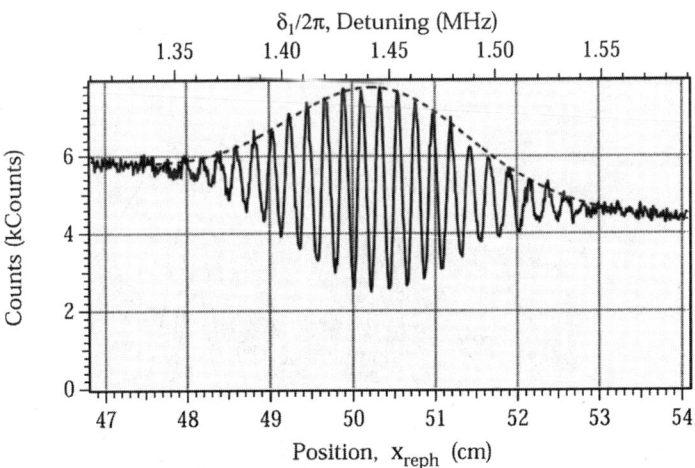

FIG. 7. Rephased fringes for the DSOF with an AM modulator located 50.9 cm downstream of the first DSOF coil and $2\omega_m/2\pi = 500$ kHz. The dashed line is a Gaussian envelope whose contrast is 49% and whose center is at $x_{\text{reph}} = 50.7 \pm 0.1$ cm.

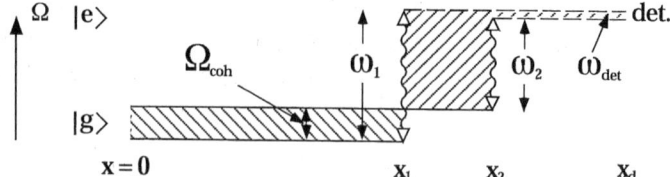

FIG. 8. DSOF interferometer employed to detect amplitude modulation. Two initially coherent ground-state plane-wave components with energies differing by $\hbar\Omega_{coh}$ are driven to the excited state by oscillatory field regions of frequency $\omega_{1,2}$ at positions $x_{1,2}$, respectively. The two incident components thus interfere to produce a time-dependent signal at frequency $\omega_{det} = \Omega_{coh} \pm (\omega_1 - \omega_2)$.

V. Detecting Longitudinal Momentum Coherences

We now discuss how DSOF may be employed as a tool to detect coherent superpositions of longitudinal momentum (coherences) present in the quantum state of an incident matter–wave beam. To do so we consider detecting coherences produced by a suppressed carrier amplitude modulator placed *upstream* of a DSOF interferometer. As before, the finite velocity distribution of the atomic beam causes this coherence to dephase—an intensity detector alone, if placed beyond a short distance downstream of the modulator, cannot detect it. A DSOF interferometer can, however, reverse the effects of this velocity dephasing.

The scheme for rephasing an upstream AM coherence using the DSOF mirrors the heterodyne detection of an AM coherence produced by the DSOF (Fig. 8). We simply arrange DSOF detunings such that x_{reph} is equal to the modulator's

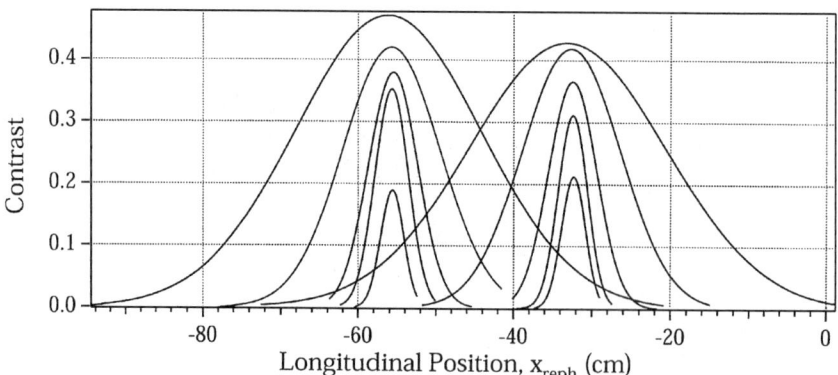

FIG. 9. Envelopes of rephased fringes using AM coils located at 32.8 or 56.1 cm upstream of the first DSOF coil. Modulation frequencies ($2\omega_m$) were 50, 100, 200, 300, and 400 kHz (50 and 400 kHz being the broadest and narrowest curves, respectively).

upstream position. The same expression for excited-state probability downstream of the DSOF [Eq. (8)] applies, except that the sign of ϕ_m is reversed. Scanning x_{reph} yields fringes around x_m (Fig. 9).

Returning to the spin-echo analogy, in experiments with an upstream modulator the AM coil acts like the first $\pi/2$ pulse in the echo, producing a coherence that quickly damps out. The DSOF region can recognize the coherent nature of the ensemble and reveal the location of the modulator, despite the fact that the modulation is dephased. It is also possible to use the DSOF region in a manner directly analogous to the π pulse in a spin echo, by altering the dephased molecules produced by an upstream modulator so that they rephase at the location of a downstream modulator.

VI. Discussion of the Semiclassical Approximation

Much of the quantum treatment of center-of-mass motion given in this chapter has been conducted under the approximation made for Δk in Eq. (3), which ignores terms that are quadratic in $\delta/\omega_{e,g}$. This approximation linearizes the relationship between the detuning of a resonance region, δ, and the resulting momentum shift Δk. This approximation is valid only up to a distance $X_{\max}$ where the error term $\Delta_k^{(1)} \propto (\delta/\omega_{e,g})^2$ does not significantly contribute to the atoms' overall spatial phase: $\Delta_k^{(1)} x_{\max} \ll \pi$. For a modulation frequency (or detunings) of ω_m, our treatment only applies within a downstream distance given by the Talbot length,

$$L_{\text{Talbot}} = \frac{2d^2}{\lambda}, \tag{9}$$

where λ is the atomic de Broglie wavelength and $d = 2\pi v/\omega_m$ is the spatial period of the modulated beam. The Talbot length, a quantity used in transverse optics, is the distance downstream from a periodic structure at which its first near-field image revival occurs (Patorski, 1989). For typical molecular beam experiments ($\omega_m \leq 2\pi \times 10$ MHz), $L_{\text{Talbot}} \gg 10$ m.

In longitudinal atom optics, as in transverse light optics, the Talbot length L_{Talbot} [Eq. (9)] sets the length scale for near-field optical effects (Patorski, 1989), and is useful in explaining how higher order terms in the expansion of Δk alter the wavefunction. At distances much less than L_{Talbot}, the atoms follow ray-like classical trajectories, just like light rays which form a geometric shadow after passing through an opening much larger than their wavelength.

For our current experiments, in which the atoms travel distances $\ll L_{\text{Talbot}}$ and their trajectories may be treated classically, the internal state evolution nevertheless remains entirely quantum mechanical. If the internal state evolution is treated in terms of quantum amplitudes, the result of the near-field approximation is that the total transmission probability for a plane wave propagated through a succession of longitudinal atom optical elements is the product of the transmission probability through each individual element.

For experiments in which all distances are shorter than the Talbot length, neglecting $\Delta_k^{(1)}$ is equivalent to taking a classical position and velocity for the atom while treating its internal evolution quantum mechanically. This is the usual approximation made in the standard treatment of molecular beam resonance (Ramsey, 1956). In this picture, the amplitude and phase of any applied oscillatory field(s) evaluated at the classical position $x = (\hbar k/m)(t - t_i)$, determine whether a particular atom is excited (without any accompanying change in momentum) or whether it remains in the ground state. The equivalence between these two pictures stems from the fact that Eq. (3) forces the additional waves created by any oscillatory fields to have the same group velocity v as the initial wave.

While the quantum and classical center-of-mass treatments lead to identical results for distances much less than the Talbot length, the plane-wave analysis we have adopted in this chapter has several advantages. It extends correctly to the case of slow atoms, where the classical longitudinal velocity picture fails. Perhaps more important is the intuition our analysis provides when considering extensions to the standard techniques of atomic beam resonance. The DSOF rephasing of coherences created by upstream or by downstream amplitude modulators, for example, has an appealingly intuitive explanation in terms of the cancellation of spatial phases.

VII. Apparatus and Experimental Techniques

We have used the principles described above to construct an interferometer for the detection of longitudinal momentum coherences in an atomic beam, and ultimately for the complete reconstruction of the beam's longitudinal density matrix. The basic geometry of our beam machine is shown in Fig. 10. Our atomic beam is formed via supersonic expansion of a noble gas (typically Ar) seeded with Na. The resulting beam is extremely bright $\sim 10^{20}$/(s strad cm^2) and has a typical mean velocity of 1100 m/s and a velocity width of $\sim$3.5% rms. This beam then passes through various state selectors and longitudinal atom optical (LAO) elements comprising the experiment. The atoms are finally detected by a scanning 50 μm Re hot wire on which the Na atoms surface-ionize. The resulting ions are accelerated by electric fields and are incident upon a channel electron multiplier (CEM). The detector frequency response has a roughly Gaussian falloff with a rms width of $\nu_{\text{det}} \sim$ 3 kHz, corresponding to a $\frac{1}{3}$ ms dwell time for an atom on the wire surface.

Because our LAO elements rely on transitions between internal atomic energy levels, we required a method to state select our Na beam. Two Stern–Gerlach magnets, the first placed between the first and second collimation slits and the second at the end of the main chamber, were used to transversely separate the various magnetic sublevels. These SG magnets consisted of two parallel wires

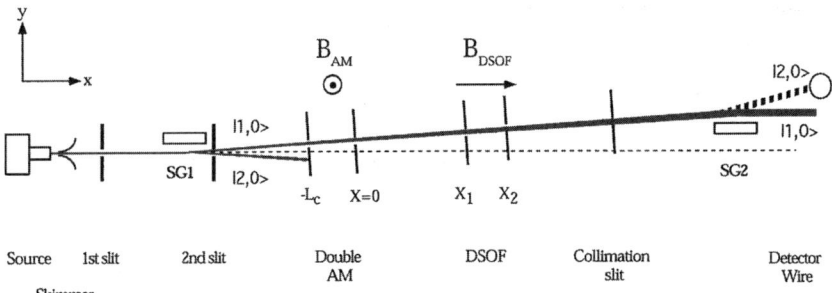

FIG. 10. Top view of beamline employed for LAO and density matrix experiments. The first Stern–Gerlach magnet and collimation slits select the $|F = 1, m_F = 0\rangle$ (solid line) state. The DSOF regions at $x_{1,2}$ drive hyperfine transitions from $|1, 0\rangle \to |2, 0\rangle$, and the $|2, 0\rangle$ state (thick striped line) is selected by a second Stern–Gerlach magnet. The double AM regions drive transitions to the $|1, \pm 1\rangle$ states (not shown), which are not observed by the detector.

($\frac{1}{8}$-in. copper tube) carrying antiparallel currents, which produced $\sim$1000 gauss/cm gradient in the magnetic field magnitude ($\sim$200 gauss) at the position of the atomic beam. This gradient is sufficient to spatially separate the various magnetic sublevels so that they may be detected independently.

The primary component of all the experiments described herein is the DSOF interferometer. The DSOF operates by driving hyperfine transitions between the sodium $|F = 1, m_F = 0\rangle \to |2, 0\rangle$ ($\Delta m = 0$) states, which differ in energy by $\omega_{hf} = 1772$ MHz (Fig. 11). The insensitivity of this transition to external magnetic fields makes it ideal for producing a clean two-state system. The oscillatory field regions (hairpin coils) used to drive transitions in our experiments were constructed of 50 μm copper wire bonded to silicon chips for structural support. The center-to-center spacing of the two wires was about 170 μm, leading to transit time-limited $1/e$ linewidths of $2\pi \times 1.1$ MHz and $2\pi \times 1.5$ MHz for DSOF coils 1 and 2, respectively.

Amplitude modulation, the other major LAO technique that we employ in our experiments, was implemented by driving ($\Delta m = 1$) $|F = 1, m_F = 0\rangle \to |1, \pm 1\rangle$ Zeeman transitions in the $F = 1$ hyperfine manifold (Fig. 11). The coils used to deliver r.f. for amplitude modulation were identical in design to the DSOF coils. We applied time-dependent modulation by mixing this r.f. carrier signal ($\omega_c = 2\pi \times 7.65$ MHz), with a modulation frequency ω_m. For an incident beam $e^{i(k_0 x - \Omega_0 t)} |1, 0\rangle$, this produces a $|1, 0\rangle$ transmission amplitude

$$T(t) = \cos\left[\frac{\omega_R l}{v_0} \sin(\omega_m t)\right] \qquad (10)$$

where $\omega_m \ll \omega_R$, the Rabi frequency of the oscillatory field region, and l is this

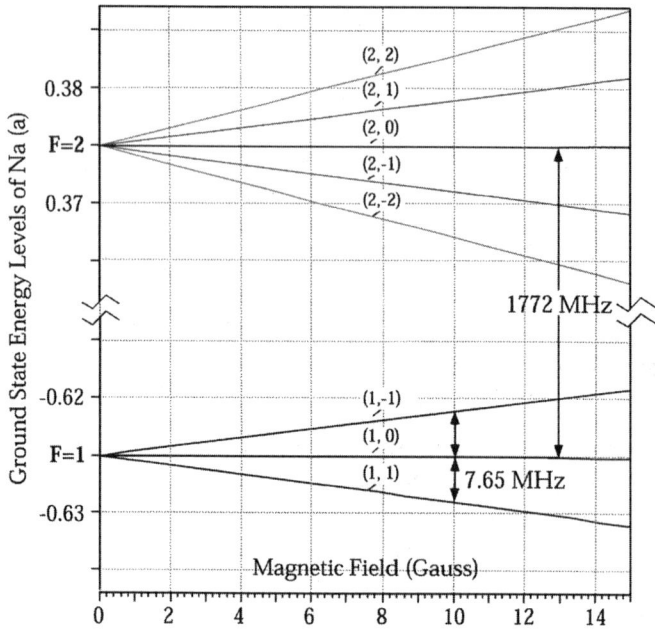

FIG. 11. Level structure of Na $3S_{1/2}$ state, expressed in units of the hyperfine interval $a = 1771.631$ MHz. The $|F = 1, m_F = 0\rangle \rightarrow |2, 0\rangle$ DSOF transition and the $|1, 0\rangle \rightarrow |1, \pm 1\rangle$ transitions are shown.

region's length. The resulting wavefunction produces a wavefunction for atoms in the $|1, 0\rangle$ state is

$$|\Psi_{\text{out}}(x, t)\rangle = \cos\left[\frac{\omega_R l}{v_0} \sin(\omega_m t)\right] |\Psi_{\text{in}}(x, t)\rangle$$

$$= \sum_{-\infty}^{\infty} J_{2n}\left(\frac{\omega_R l}{v_0}\right) e^{i[(k_0 + 2n\frac{\omega_m}{v_0})x - (\Omega_0 + 2n\omega_m)t]}, \quad (11)$$

where J_{2n} is a Bessel function of order $2n$. This indicates that the $|1, 0\rangle$ component contains multiple kinetic energy components spaced at intervals of $2\omega_m$.

A measurement of standard Ramsey fringes is a useful diagnostic, allowing us to adjust the magnetic field in the region between the DSOF coils so that the SOF fringes are centered at the peak of the single-coil resonance lines. In addition, a fit to the SOF fringes (see Fig. 12) yields the contrast of the SOF, which we can optimize by adding fixed attenuators to the input of one or the other coil so that both apply accurate $\pi/2$ pulses.

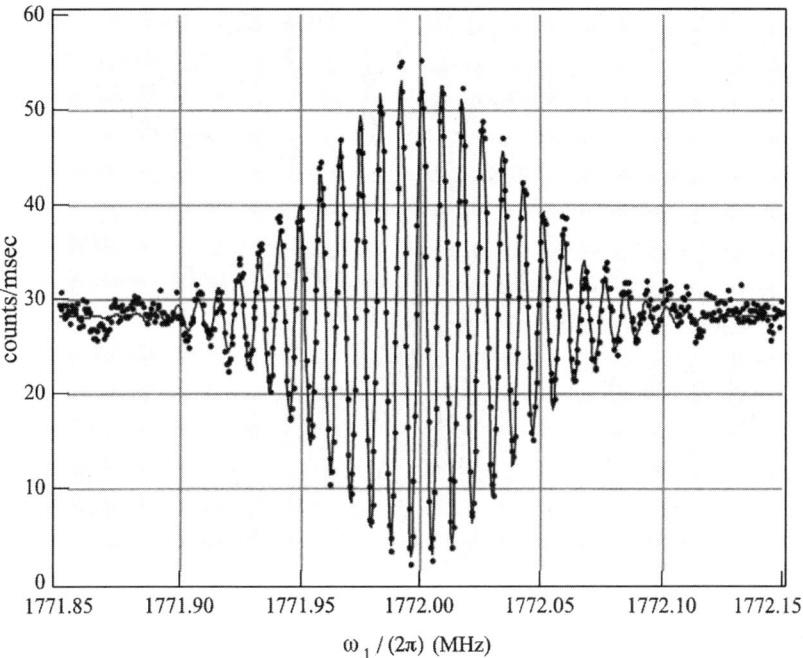

FIG. 12. Ramsey fringes produced by DSOF with same signal driving both coils. Dots are points, and the solid line is a fit, which yields a contrast of 92.7%.

VIII. Density Matrix Deconvolution

The DSOF interferometer is capable of directly measuring both amplitude and phase components of the longitudinal density matrix of an atomic beam. To demonstrate this capability we created a novel longitudinal energy stucture in our supersonic beam using a pair of amplitude modulators (see Fig. 10). We subsequently detected this structure by employing a DSOF configuration downstream. The measured density matrix is in excellent agreement with our theoretical predictions (Rubenstein et al., 1999).

The inherent time dependence of off-diagonal elements necessitates time-dependent techniques for their measurement. In addition, the vacuum dispersion for matter waves,

$$\Omega(k) = \hbar k^2/2m, \qquad (12)$$

reflects the strong dependence of phase velocity on k. Consequently, the finite energy spread of available atomic beams leads to dephasing which washes out any upstream amplitude modulation of atomic probability (Pritchard et al., 1998; Smith et al., 1998). These difficulties are solved by using a DSOF interferometer which operates using two time-dependent interaction regions and is capable of delivering a velocity-dependent phase shift to reverse the vacuum dephasing occurring upstream.

In the longitudinal energy basis, the density matrix ρ undergoes the free evolution,

$$\rho(\Omega, \Omega + \Omega_{\text{coh}}, t) = \overline{\langle \Omega + \Omega_{\text{coh}} | \psi \rangle \langle \psi | \Omega \rangle}$$
$$= \rho_0(\Omega, \Omega + \Omega_{\text{coh}}) e^{i\Omega_{\text{coh}}(t-t_0)}, \quad (13)$$

where ρ_0 is the density matrix at time t_0. The diagonal elements, $\rho(\Omega, \Omega)$, are real and time-independent. They represent the energy distribution of the beam. Off-diagonal elements are in general complex and represent the coherences between different longitudinal energy states. To create off-diagonal elements, in order to demonstrate phase measurement ability, we amplitude-modulated the atomic beam, producing energy coherences at multiples of the modulation frequency.

The configuration of our deconvolution experiment is shown schematically in Fig. 10. The two AM zones are the first LAO elements the sodium beam encounters. This double modulation produces nontrivial off-diagonal structure.

For a perfectly monochromatic beam, the amplitude-modulated $|1, 0\rangle$ state wavefunction downstream of the second AM region is,

$$|\psi(x, t)\rangle = \cos(\theta_1 \sin\{\omega_m[x/v - (t - L_c/v)]\})$$
$$\times \cos\{\theta_2 \sin[\omega_m(x/v - t)]\} e^{i(k_0 x - \Omega_0 t)} |1, 0\rangle, \quad (14)$$

where $\theta_i = (\omega_{Ri} l_i / v)$, ω_{Ri} is the Rabi frequency, l_i is the length of v the ith AM region, ω_m is the modulation frequency, and v is the atomic velocity.

The amplitude modulated wavefunction of Eq. (14) possesses an infinite ladder of energy/momentum sidebands [49]. This leads to an energy-basis density matrix of the form of Eq. (13) with nonzero elements (or coherences) only at $\Omega_{\text{coh}} = 2n\omega_m$ for integer n. In addition to the diagonal, the resulting density matrix then consists of "stripes" of coherences parallel to the diagonal.

The coherence between two momentum components is probed downstream with a DSOF interferometer. If the differential detuning,

$$\delta_{12} = \delta_1 - \delta_2 = \omega_1 - \omega_2, \quad (15)$$

matches the energy difference of the two components, the interference will form a DC signal at the detector. Because the temporal response of the detector is much slower (~3 kHz) than the modulation frequency (~60 kHz), a particular density matrix stripe can be individually detected, while the others average out to zero. Each density matrix stripe forms modified Ramsey fringes at the detector, which can be used to calculate both the amplitude and the phase of the stripe as a function of longitudinal kinetic energy.

We obtain the density matrix from a measurement of $P_e(\delta_{12}, \tilde{\delta}, t)$, the time-dependent probability that an atom will reach the detector in the excited state as a function of δ_{12}, and the "scaled detuning",

$$\tilde{\delta} = \frac{\omega_2 x_2 - \omega_1 x_1}{L} - \omega_{hf}. \tag{16}$$

For the case of our fast atomic beam, the double Fourier transform of $P_e(\delta_{12}, \tilde{\delta}, t)$ with respect to time t and $\tilde{\delta}$ determines the complete translational density matrix,

$$\rho_0(\omega', \omega' + \delta_{12} - \omega_{\text{det}}) = A \int d\tilde{\delta} \tilde{P}_e(\delta_{12}, \tilde{\delta}, \omega_{\text{det}}) e^{-i\tilde{\delta} b/\sqrt{\omega'}}, \tag{17}$$

where the constants A and b are defined by Dhirani et al. (1997) and $\tilde{P}_e$ is the time Fourier transform of P_e. As Eq. (17) shows, $\tilde{P}_e(\delta_{12}, \tilde{\delta}, \omega_{\text{det}})$ contains information about a stripe of the density matrix parallel to, but offset from, the diagonal by an amount $\Omega_{\text{coh}} = \delta_{12} - \omega_{\text{det}}$. Thus a full set of $\tilde{P}_e(\delta_{12}, \tilde{\delta}, \omega_{\text{det}})$ data provides more than enough information to determine the density matrix.

The extremely large kinetic energy of our beam relative to the energy shifts applied in the AM regions and the DSOF ($\Omega_0 \gg \omega_m, \delta_{12}$) permits the use of a semiclassical approximation to obtain a physically intuitive picture of our deconvolution scheme. In this limit, the near-field approximation (Pritchard et al., 1998) to the quantum wave equation leads to solutions in which the atom's external motion follows ray-like classical trajectories, while its internal state evolves quantum mechanically. Thus, the detection probability $P_e(\delta_{12}, \tilde{\delta}, t)$ is the ensemble-averaged product of the double AM transmission probability P_{AM} [absolute value squared of Eq. (14) and a DSOF transmission probability P_{DSOF} (Pritchard et al., 1998)],

$$P_e(\delta_{12}, \tilde{\delta}, t) = \langle P_{\text{AM}} \times P_{\text{DSOF}} \rangle = \left\langle \left\{ \sum_{n=-\infty}^{\infty} [C_n(\theta_1, \theta_2, \Omega) e^{i[2n\omega_m(t-T)]}] \right\} \right.$$
$$\left. \times \frac{1}{2} \left[1 + \cos\left(\frac{\tilde{\delta} L}{v} + \frac{\delta_{12} x_d}{v} - \delta_{12} t\right) \right] \right\rangle. \tag{18}$$

The $\langle\ \rangle$ denotes an average over the initial energy distribution of the atomic beam (approximately Gaussian in velocity space with a mean of 1100 m/s and a rms width of 25 m/s). $T = x_d/v$ is the transit time of an atom from the second AM coil to the detector. The coefficients C_n arise from Bessel function expansions (Arfken, 1985) of the sine and cosine terms in Eq. (14).

We first probed the diagonal of the double AM density matrix by setting $\delta_{12} = 0$, making the DSOF equivalent to a standard separated oscillatory field experiment (Ramsey, 1956). This configuration yields Ramsey fringes (Fig. 13) centered at the hyperfine resonance $\omega_1 = \omega_{hf}$. The Fourier transform of these fringes with respect to $\tilde{\delta}$ gives the energy distribution of the atomic beam as modified by the modulation, i.e., the diagonal of the density matrix (Fig. 14, left).

To obtain the off-diagonal elements of the density matrix, we chose $\delta_{12} \neq 0$ and scanned $\tilde{\delta}$, while maintaining phase coherence between the two DSOF coils. We achieved this by using a single-sideband modulator (Rubenstein, 1999) to generate from ω_1 a phase-coherent signal at $\omega_2 = \omega_1 - \delta_{12}$. We set $\delta_{12} = 2N\omega_m$ (for $N = $ 1, 2, and 3) and observed rephased Ramsey fringes centered around $\omega_1 = \omega_{hf} + [\delta_{12}(x_d - x_2)/L] \sim \omega_{hf} + 2\pi N \times 0.6$ MHz, where the white fringe condition $\tilde{\delta} L + \delta_{12} x_d = 0$ is satisfied at the detector [see Eq. (18)]. A Fourier transform of the fringes for a particular N determined a stripe of the density matrix located $\Omega_{coh} = 2N\omega_m$ away from the diagonal. At large detunings the DSOF regions produce less than the ideal $\pi/2$ pulses assumed in Eq. (18), somewhat reducing the signal/noise. This prevented us from measuring the weak off-diagonal stripes for $N > 3$.

The amplitude of four stripes of the density matrix, calculated from an average of two data sets (a total of about 2 h of data with a mean count rate of 8–10 kcounts/s) is shown in Fig. 14 (left). The multiple peaks for a particular stripe arise from the interference of plane-wave components that accrue relative phase shifts (by multiples of $2\omega_m L_c/v$) while propagating from the first to the second AM region. In the semiclassical picture, these peaks reflect the velocity selection of the two

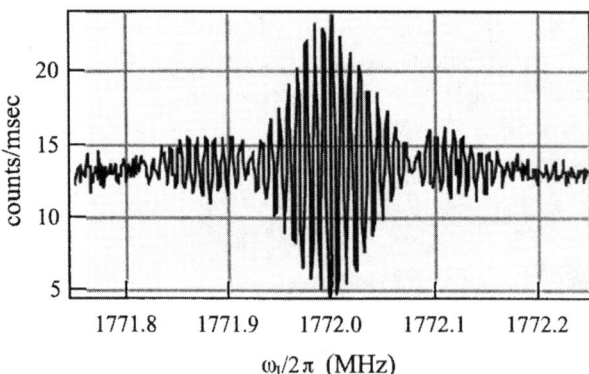

FIG. 13. Ramsey fringes formed by a doubly amplitude modulated atomic beam, obtained by DSOF with $\delta_{12} = 0$. The envelope of the fringes reflects the longitudinal atomic energy distribution produced by the AM regions.

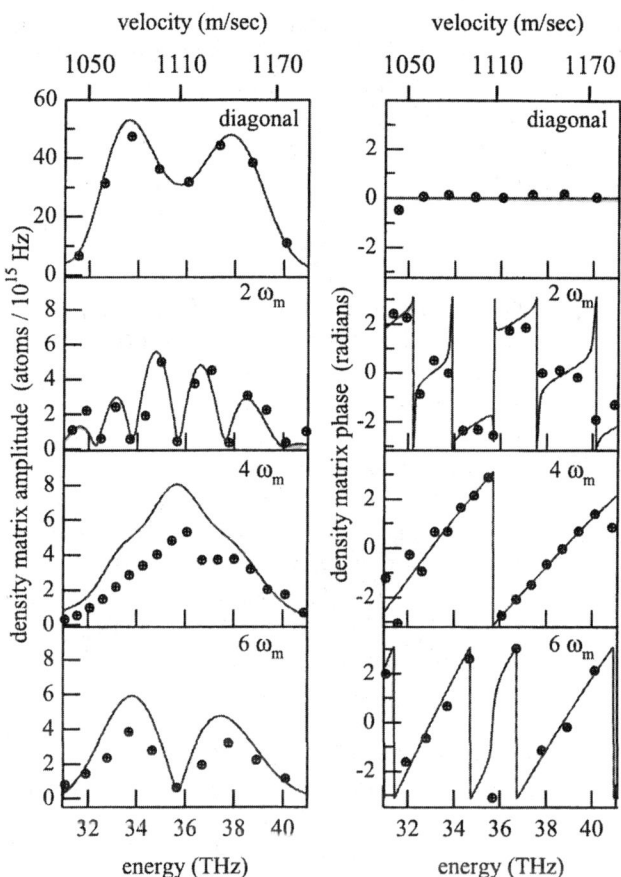

FIG. 14. Measured amplitude (left) and phase (right) of the double AM density matrix with $\omega_m = 2\pi \times 60.9$ kHz. The solid lines represent the theoretical prediction. The amplitudes are normalized to unity for the unmodulated atomic beam. An unimportant overall phase has been subtracted from each set of phase data to compensate for the random phase between sidebands.

AM regions, which together transmit only atoms whose transit times $\tau = L_c/v$ are near-multiples of the modulation period $T_{\text{mod}} = 2\pi/(2N\omega_m)$.

The phase of the density matrix is displayed in Fig. 14 (right). The positive and negative phase regions correspond to velocities in and out of phase with the DSOF heterodyne signal. The overall slopes of the phases reflect the velocity dependence of the transit time between the AM coils, L_c/v. The relative phase between different Ω_{coh} stripes could not be directly determined because the phase of our single sideband mixer was randomized each time the heterodyne frequency δ_{12} was changed.

A theoretical prediction for the density matrix was obtained semiclassically from Eq. (18), and is overlaid on the data in Fig. 14. The excellent fit obtained required slight variations of the Rabi frequencies ω_{R1} and ω_{R2} and the velocity distribution of the beam. These variations were always within the error of independent measurements of these quantitites: $\sim$0.2% for the mean velocity, $\sim$4% for the velocity width, and $\sim$10% for ω_{R1} and ω_{R2}. The large error in $\omega_{R1,2}$ reflects the hysteresis in the AM driver circuit (Rubenstein, 1999).

IX. Search for Longitudinal Coherences in an Unmodified Atomic Beam

Having demonstrated the ability to measure an artificially created longitudinal density matrix, we now turn to the study of the density matrix of an unmodified atomic beam. The coherence structure of matter wave beams has been the subject of great theoretical interest (Golub and Lamoreaux, 1992; Bernstein and Low, 1987; Englert et al., 1994), as well as some controversy [see Kaiser et al., 1983; Klein et al., 1983; Comsa, 1983)]. In a recent experiment, we measured for the first time the longitudinal quantum state of a real atomic beam. In particular, we searched over a wide range of frequencies (from dc to $2\pi \times 100$ kHz) for amplitude modulation produced by a supersonic atom source. This search is capable of revealing the magnitude of any nonzero, off-diagonal density matrix elements (ODEs) in our atomic beam.

While the coherent DSOF density matrix deconvolution scheme described in the previous section could in principle be used to measure off-diagonal elements in our supersonic beam, it is vulnerable to fluctuations in the phase of the incoming atom signal and in the relative phase of the two DSOF coils. Thus, in our search we opted to employ an "incoherent" technique, which measures the power spectrum of the time-dependent atom signal downstream of the DSOF. This method retains the ability to exploit the entire bandwidth of our atom detector, but is insensitive to the phase of the time-dependent oscillations of the detected atom signal. These features allow us to average together multiple power spectra, greatly increasing the sensitivity of our search.

A general problem in searching for time-dependent coherences is that their phase is neither guaranteed to be stable nor known in advance—hence phase-coherent methods of data acquisition [such as are necessary to determine the full density matrix as in Eq. (17)] could average a momentarily coherent signal to zero. We circumvented this by studying the power spectrum of the time-dependent data, $|\tilde{P}_e|^2$, which is not sensitive to the phase. As our data will demonstrate, $|\tilde{P}_e| \approx 0$ for all $(\delta_{12}, \tilde{\delta}, \omega_{\text{det}})$ within our search range. This information alone can rule out the presence of ODEs because as Eq. (17) shows, if $|\tilde{P}_e|$ is equal to zero, then all ODEs $\rho_0(\omega', \omega' + \Omega_{\text{coh}})$ are likewise zero.

We measured $P_e(\delta_{12}, \tilde{\delta}, t)$ on a grid with δ_{12} ranging from 0 to $2\pi \times 100$ kHz in 1-kHz increments.[5] For each δ_{12} we took data at values of $\tilde{\delta}$ between $-2\pi \times 230$ and $2\pi \times 234$ kHz, in $2\pi \times 8$-kHz steps. For each $(\delta_{12}, \tilde{\delta})$ pair, we measured atomic beam flux versus time for approximately 1 s. These data were converted into a quasi-power spectrum $P(\delta_{12}, \tilde{\delta}, \omega_{det})$ which approximates $|\tilde{P}_e|^2$. To increase the sensitivity of our search, we averaged together spectra from four identically obtained data sets. By examining the power spectra, our search scheme employed the full bandwidth of our time-dependent detector to search for ODEs with resolution in Ω_{coh} of better than $2\pi \times 1$ Hz, without requiring us to increment δ_{12} in $2\pi \times 1$-Hz steps.

The ODEs we search for appear as amplitude modulation, which dephases downstream of its origin x_0 due to the finite energy distribution of atoms in the beam. This dephasing is reversed by the DSOF near (Smith et al., 1998; Pritchard et al., 1998),

$$\tilde{\delta}_{reph} \simeq (\Omega_{coh} - \delta_{12})\frac{(x_d - x_0)}{L} - \delta\frac{x_0}{L}, \quad (19)$$

where x_d is the location of the atom detector. This produces a peak in P centered around $\tilde{\delta} = \tilde{\delta}_{reph}$, with a rms width of $2\pi \times 32$ kHz in the variable $\tilde{\delta}$. Figure 15a depicts such a rephasing peak, for 5.3% amplitude modulation produced by a modulated oscillatory field region (Smith et al., 1998; Pritchard et al., 1998) with $\Omega_{coh} = 2\pi \times 50.008$ kHz.

Our search attempts to identify any such rephasing peaks that have amplitude above our statistical noise floor. We take advantage of the $\tilde{\delta}$ width of the rephasing peaks by averaging $P(\delta_{12}, \tilde{\delta}, \omega_{det})$ along $\tilde{\delta}$ with a window size of $2\pi \times 40$ kHz, increasing our signal/noise by about a factor of 2 for amplitude modulation of the atomic beam originating in the neighborhood of the atom source—the most physically likely model for production of ODEs in our beam. We summarize our data by considering only the highest value of the averaged $P(\delta_{12}, \tilde{\delta}, \omega_{det})$ along the $\tilde{\delta}$ axis, $P_{max}(\delta_{12}, \omega_{det})$, as shown in Fig. 15c.

If amplitude modulation at some Ω_{coh} is present, it appears as peaks in the $P_{max}(\delta_{12}, \omega_{det})$ curves for which δ_{12} differs from Ω_{coh} by less than the maximum detector response frequency. Because our detector cannot distinguish between positive and negative frequencies ω_{det}, a particular peak indicates coherence at $\Omega_{coh} = \delta \pm \omega_{det}$. Therefore, for each δ we mapped each $P_{max}(\delta_{12}, \omega_{det})$ to its two possible Ω_{coh} positions. When we averaged together data from five neighboring values of δ_{12} to obtain P_{ave}, real time-dependent signals added together at Ω_{coh}, increasing our signal/noise by a further factor of 2. This averaging process also

---

[5] A simple model for acoustic modulation of our supersonic beam predicts oscillation frequencies in this range.

270   S. Gupta et al.

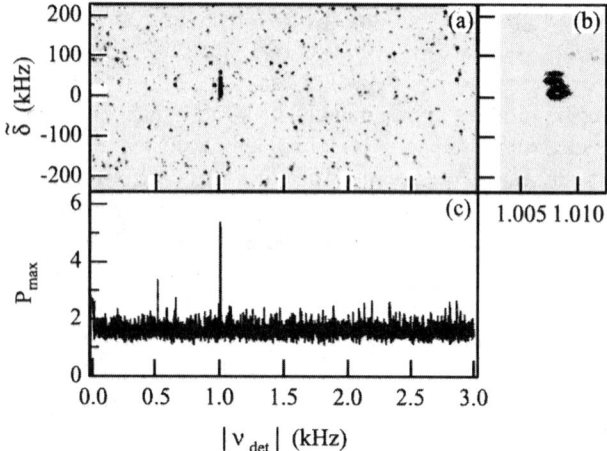

FIG. 15. (a) Quasi-power spectrum $P(\delta_{12}, \tilde{\delta}, \omega_{det})$ with $\delta_{12} = 2\pi \times 49$ kHz, for 5.3% amplitude modulation applied at $\Omega_{coh} = 2\pi \times 50.008$ kHz. The $\tilde{\delta}$ position of the peak reveals the location of the modulation source, while its $|\omega_{det}|$ position at $2\pi \times 1.008$ kHz demonstrates the presence of modulation (and thus of nonzero ODEs) at $\omega_{coh} = \delta_{12} - \omega_{det}$. (b) Close-up of rephasing peak. (c) $P_{max}(\omega_{det})$ for $\delta_{12} = 2\pi \times 49$ kHz showing the expected spike at the frequency corresponding to the applied amplitude modulation.

produced spurious peaks, or images, at other values of Ω_{coh}, but these did not average constructively and hence were of reduced height. Figure 16 shows $P_{ave}(\Omega_{coh})$ for the case of intentionally applied amplitude modulation. The large peak visible at $\Omega_{coh} = 2\pi \times 50.008$ kHz shows the presence of the modulation (and its image at $\Omega_{coh} = 2\pi \times 51.992$) demonstrating the effectiveness and calibrating the sensitivity of our search technique.

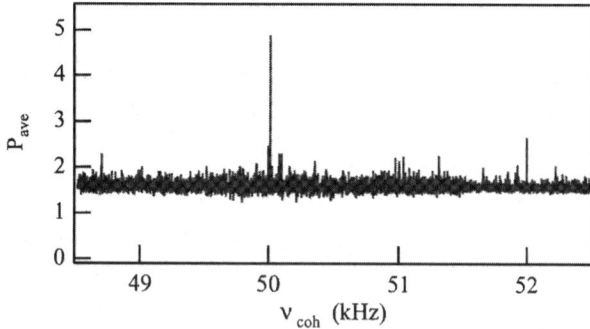

FIG. 16. $P_{ave}(\Omega_{coh})$ reveals the presence of 5.3% amplitude modulation at $\Omega_{coh} = 2\pi \times 50.008$ kHz.

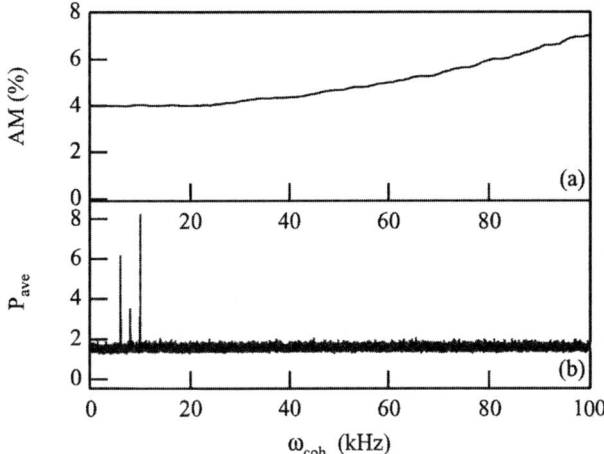

FIG. 17. (a) Amplitude modulation our search could detect with >99% confidence. (b) $P_{ave}(\Omega_{coh})$ for the entire search. Detector sampling rate at $\Omega_{coh} \simeq 2\pi \times 10$ kHz as well as its subsidiary images can be seen.

Figure 17b shows $P_{ave}(\Omega_{coh})$ for the 0 to $2\pi \times 100$-kHz range of our source search. We have removed large time-dependent signals caused by the presence of the DSOF itself (which creates detectable amplitude modulation when $\delta_{12} =$ 1, 2, and 3 kHz) as well as noise at $\omega_{det} = 2\pi \times 60$, 120 Hz and low-frequency noise below $\omega_{det} = 2\pi \times 15$ Hz. Our data sampling rate of about $2\pi \times 10$ kHz created the large spurious peak at this frequency and subsidiary image peaks at $\omega_{coh} \simeq 2\pi \times 6, 8$ kHz.

The smallest amplitude modulation which our search could detect with >99% confidence (the "threshold") was calculated from the mean count rate in the atomic beam and the contrast of the DSOF interferometer, and is shown in Fig. 17a. This threshold rises for large Ω_{coh}, where the frequencies $\omega_{1,2}$ fell outside the linewidths of the oscillating field regions preventing the application of full $\pi/2$ pulses (assumed by Dhirani et al., 1997).

Our data limits the possible size of ODEs to less than those in a 4.0% amplitude modulated[6] atomic beam for $\Omega_{coh} < 2\pi \times 50$ kHz, with the threshold gradually rising to $\sim$7% at $\Omega_{coh} = 2\pi \times 100$ kHz. The search covered a range of 0 to $2\pi \times 100$ kHz in Ω_{coh} with $<2\pi \times 1$-Hz resolution, corresponding to an examination of the density matrix with a resolution of $\sim$3 parts in 10^{14} perpendicular to the diagonal. Our resolution in ω' (determined by our $\tilde{\delta}$ = range) was $\sim 5.1 \times 2\pi \times 10^{11}$ Hz, or about 1.5 parts in 100 parallel to the diagonal of the density matrix.

[6] Amplitude modulation is defined as multiplication of the incoming plane wave by $[1 + m \cos(\omega_m t)]$, where m is the fractional modulation.

X. Conclusion

We have described in this chapter a variety of phenomena in which the longitudinal center-of-mass motion of atoms leads to novel matter–wave effects. In the context of ion–atom scattering, a coherent matter wave analysis, incorporating the interference of multiple center-of-mass paths, leads to an elegant explanation of complex Steuckelberg oscillations in the differential scattering cross section. Similarly, for inelastic ion–atom collisions, a matter–wave analysis can also explain "Rosenthal" oscillations seen in the total optical emission cross section. In this system, the interference results from the coherent interaction of excited-state amplitudes of the colliding "quasi-molecule" with differing electronic energy and longitudinal velocity. Potential applications include cold collision studies and the precision spectroscopy of continuum states using lasers.

The quantum mechanical origin of the interference effects in these colliding systems share a similarity with experiments which employ the DSOF atomic beam interferometer. In particular, both involve interference between paths which have different velocities and therefore evolve a velocity-dependent phase shift. The DSOF interferometer is a flexible tool for creating and probing velocity (momentum) coherences and has been used to address several important issues.

In the first such study, we measured the longitudinal translational density matrix of an atomic beam, the most complete possible description of a quantum system, for a nontrivial state, a doubly amplitude-modulated atomic beam. We determined the density matrix in the longitudinal energy/momentum basis, showing that it possessed theoretically predicted structure both along and off the diagonal. This work was the first experimental demonstration of such a measurement (the subject of several theoretical proposals; Dhirani et al., 1997; Kokorowski and Pritchard, 1997; Raymer, 1997), and it extended previous measurements of the density matrix and Wigner function (Smithey et al., 1993; Dunn et al., 1995; Leibfried et al., 1996; Kurtsiefer et al., 1997) to longitudinally coherent matter waves.

In the second study, we employed a longitudinal atom interferometer to put strict limits (consistent with zero) on the size of possible off-diagonal elements of the longitudinal energy/momentum density matrix of our supersonic Na atomic beam. This ruled out the presence of coherent wavepackets within the range of our search.

In addition to the work we have presented in this chapter, it is interesting to consider some potential future applications of longitudinal atom optics and interferometry. As we discussed in Section III, the DSOF experiments we have described all take place in the ultra-near field. That is, they all probe distances much less than the longitudinal Talbot length L_{Talbot}, which sets the length scale for near-field optical effects (Patorsky, 1989).

At distances much less than L_{Talbot}, the atoms follow ray-like classical trajectories, just like light rays which form a geometric shadow after passing through an opening much larger than their wavelength. To reach the intermediate-field case, where wave optical effects are more evident, we must change either the wavelength of the incident atomic radiation (or, equivalently, the atomic velocity), the periodicity (modulation frequency) of the LAO elements, or both [as demonstrated by Eq. (9)].

To reach the "far-field" case in longitudinal atom optics (LAO), it is necessary to employ beams of cold atoms or Bose–Einstein condensate beams (Mewes *et al.*, 1997; Andrews *et al.*, 1997), which can have narrow energy distributions and large coherence lengths. LAO experiments in these regimes can reach farther into the quantum regime (Mezei, 1980; Pritchard *et al.*, 1999) and provide new tools for studying the longitudinal states of these highly coherent, many-body quantum systems.

Looking to the future, sequences of longitudinal atom optical elements (resonance regions, and amplitude or phase modulators), when combined with novel atom sources such as slow atoms or Bose–Einstein condensates, may allow the creation of useful devices, such as atomic resonators or new longitudinal interferometers. The DSOF technique could be used, for example, to apply relatively large controlled kinetic energy shifts to beams of cold atoms or condensates (Steane *et al.*, 1995).

XI. Acknowledgments

We thank Prof. David E. Pritchard, Ph.D. advisor to co-authors R.R., D.K., and S.G., in whose laboratory at MIT the supersonic beam experiments described were carried out, for many helpful ideas and discussions. Thanks should go to Edward Smith for building much of the DSOF interferometer apparatus, along with Tony Roberts, Jana Lehner, Al-Amin Dhirani, and Herbert Bernstein (Hampshire College) for their contributions to the theory and important parts of the experimental work on the DSOF project. W. Smith thanks David Pritchard and the MIT Physics Department for their hospitality while he was a Visiting Professor during a recent sabbatical from the University of Connecticut. Work at MIT was supported under the following grants: Army Research Office contracts DAAH04-94-G-0170, DAAH04-95-1-0533, DAAG55-97-1-0236, and DAAG55-98-1-0429; Office of Naval Research contracts N000014-96-1-0432, N00014-89-J-1207; Joint Services Electronics Program contracts DAAH04-95-1-0038 and DAAG55-98-1-0080; and National Science Foundation grant PHY-9514795.

XII. References

Aberth, W., and Lorents, D. (1966). *Phys. Rev.* **144**, 109. See also Smith, F. T., Marchi, R. P., Aberth, W., Lorents, D. C., and Heinz, O. (1967). *Phys. Rev.* **161**, 31; Smith, F. T., Fleischmann, H. H., and Young, R. A. (1970). *Phys. Rev. A* **2**, 379.

Andrews, M. R., *et al.* (1997). *Science* **275**, 637.

Arfken, G. (1985). "Mathematical Methods for Physicists," pp. 576–578. Academic Press, San Diego.
Badurek, G., Rauch, H., and Tuppinger, D. (1986). *Phys. Rev. A* **34**, 2600.
Bernstein, H. J., and Low, F. E. (1987). *Phys. Rev. Lett.* **59**, 951.
Bordé, C. (1989). *Phys. Lett. A* **140**, 10.
Born, M., and Oppenheimer, J. R. (1927). *Ann. Phys.* **84**, 457.
Carnal, O., and Mlynek, J. (1991). *Phys. Rev. Lett.* **66**, 2689.
Chormaic, S., *et al.* (1994). *Phys. Rev. Lett.* **72**, 1; see references therein to related interference experiments with neutrons and in the optical domain.
Clark, D. A. (1978). Ph.D. thesis. The University of Connecticut.
Comsa, G. (1983). *Phys. Rev. Lett.* **51**, 1105.
Davisson, C., and Germer, L. H. (1927a). *Nature* **119**, 558.
Davisson, C., and Germer, L. H. (1927b). *Phys. Rev.* **30**, 705.
de Broglie, L. (1924). Ph.D. thesis. L'Université de Paris.
Dhirani, A., *et al.* (1997). *J. Mod. Opt.* **44**, 2583.
Dunn, T. J., Walmsley, I. A., and Mukamel, S. (1995). *Phys. Rev. Lett.* **74**, 884.
Dworetsky, S., and Novick, R. (1969). *Phys. Rev. Lett.* **23**, 1484.
Dworetsky, S., Novick, R., Smith, W. W., and Tolk, N. (1967). *Phys. Rev. Lett.* **18**, 939.
Englert, B.-G., Miniatura, C., and Baudon, J. (1994). *J. de Phys. II* **4**, 2043.
Estermann, I., and Stern, O. (1930). *Z. Phys.* **61**, 95.
Ford, K. W., and Wheeler, J. A. (1959). *Ann. Phys.* **7**, 259.
Golub, R., and Gähler, R. (1987). *Phys. Lett. A* **123**, 43.
Golub, R., Gähler, R., and Keller, T. (1994). *Am. J. Phys.* **62**, 779.
Golub, R., and Lamoreaux, S. K. (1992). *Phys. Lett. A* **162**, 122.
Gould, P. L., Ruff, G. A., and Pritchard, D. E. (1986). *Phys. Rev. Lett.* **56**, 827.
Gustavson, T. L., Bouyer, P., and Kasevich, M. A. (1997). *Phys. Rev. Lett.* **78**, 2046.
Hill, R., and Gallagher, T. (1975). *Phys. Rev. A* **12**, 451.
Kaiser, H., Werner, S. A., and George, E. A. (1983). *Phys. Rev. Lett.* **50**, 560.
Kaiser, H., *et al.* (1992). *Phys. Rev. A* **45**, 31.
Kasevich, M. A., and Chu, S. (1991). *Phys. Rev. Lett.* **67**, 181.
Keith, D. W., Ekstrom, C. R., Turchette, Q. A., and Pritchard, D. E. (1991). *Phys. Rev. Lett.* **66**, 2693.
Keith, D. W., Schattenburg, M. L., Smith, H. I., and Pritchard, D. E. (1988). *Phys. Rev. Lett.* **61**, 1580.
Kessel, Q. C., Pollack, E., and Smith, W. W. (1978). *In* "Collision Spectroscopy," (R. Cooks, Ed.), pp. 147–221. Plenum Press (New York).
Klein, A. G., Opat, G. I., and Hamilton, W. A. (1983). *Phys. Rev. Lett.* **50**, 563.
Kokorowski, D., and Pritchard, D. (1997). *J. Mod. Opt.* **44**, 2575.
Kozuma, M., *et al.* (1999). *Phys. Rev. Lett.* **82**, 871.
Kurtsiefer, C., Pfau, T., and Mlynek, J. (1997). *Nature* **386**, 150.
Landau, L. (1932). *Phys. Z. Sowjetunion* **2**, 46; Zener, C. (1932). *Proc. R. Soc. A* **137**, 696; see also Landau, L. D., and Lifshitz, E. M. (1958) "Quantum Mechanics," pp. 304–312. Addison-Wesley, (Reading, MA).
Leibfried, D., *et al.* (1996). *Phys. Rev. Lett.* **77**, 4281.
Lichten, W. (1963). *Phys. Rev.* **131**, 229.
Lockwood, G. J., and Everhart, E. (1962). *Phys. Rev.* **125**, 567.
Loeb, L. B. (1927). *Science* **66**, 627.
Maier-Leibnitz, H., and Springer, T. (1962). *Z. Phys.* **167**, 386.
Martin, P. J., Oldaker, B. G., Miklich, A. H., and Pritchard, D. E. (1988). *Phys. Rev. Lett.* **60**, 515.
Marton, L. (1952). *Phys. Rev.* **85**, 1057.
Marton, L., Simpson, J. A., and Suddeth, J. A. (1954). *Rev. Sci. Instrum.* **25**, 1099.
Mewes, M.-O., *et al.* (1997). *Phys. Rev. Lett.* **78**, 582.
Mezei, F. (1972). *Z. Phys.* **255**, 146.

Mezei, F. (1980). *In* "Imaging Processes and Coherence in Physics: Proceedings of a Workshop Held at the Centre de Physique, Les Houches, France, March 1979" (M. Schlenker, *et al.*, Eds.), pp. 282–295. Springer-Verlag, Berlin.
Miniatura, C., *et al.* (1992). *Phys. Rev. Lett.* **69,** 261, where other references are given, including related experiments with neutrons.
Patorski, K. (1989). *In* "Progress in optics, XXVII." (E. Wolf, Ed.), pp. 1–108. North-Holland, Amsterdam.
Peters, A., Chung, K. Y., and Chu, S. (1999). *Nature* **400,** 849.
Pritchard, D. E., *et al.* (1999). *Phys. Rev. A* **59,** 4641.
Riehle, F., *et al.* (1991). *Phys. Rev. Lett.* **67,** 177.
Ramsey, N. (1950). *Phys. Rev.* **78,** 695.
Ramsey, N. F. (1956). "Molecular Beams." Oxford University Press, Oxford, .
Raymer, M. G. (1997). *J. Mod. Opt.* **44,** 2565.
Rosenthal, H., and Foley, H. (1969). *Phys. Rev. Lett.* **23,** 1480; Rosenthal, H. (1971). *Phys. Rev. A* **4,** 1030.
Rubenstein, R. A. (1999). Ph.D. thesis. Massachusetts Institute of Technology,
Rubenstein, R. A., *et al.* (1999). *Phys. Rev. Lett.* **83,** 2285.
Smith, E. T., *et al.* (1998). *Phys. Rev. Lett.* **81,** 1996.
Smithey, D. T., Beck, M., Raymer, M. G., and Faridani, A. (1993). *Phys. Rev. Lett.* **70,** 1244.
Snadden, M. J., *et al.* (1998). *Phys. Rev. Lett.* **81,** 971.
Steane, A., Szriftgiser, P., Desbiolles, P., and Dalibard, J. (1995). *Phys. Rev. Lett.* **74,** 4972.
Stueckelberg, E. C. G. (1932). *Helv. Phys. Acta* **5,** 369.
Summhammer, J., *et al.* (1995). *Phys. Rev. Lett.* **75,** 3206.
Tolk, N. H., Tully, J. C., White, C. W., Kraus, J., Monge, A. A., Simms, D. L., Robbins, M. F., Neff, S. H., and Lichten, W. (1976). *Phys. Rev. A* **13,** 969. See also Tolk, N. H., *et al.,* (1973). *Phys. Rev. Lett.* **31** 671; Bobashev, S. V. (1970) *JETP Lett.* **11,** 260.
Weiss, D. W., Young, B. C., and Chu, S. (1993). *Phys. Rev. Lett.* **70,** 2706.
Weizel, W., and Beeck, O. (1932). *Z. Phys.* **76,** 250.
Ziemba, F. P., and Everhart, E. (1959). *Phys. Rev. Lett.* **2,** 299.

Index

A

Adiabatic condition, stimulated Raman adiabatic passage (STIRAP), 90–92
Adiabatic control technique, quantum control, 3–4
Adiabatic elimination, three-state system, 78–79
Adiabatic evolution
  class of change, 66–67
  tripod linkage, 137–141
Adiabatic following, two-state system, 72–73
Adiabatic momentum transfer
  atomic beam splitters, 130–131
  atomic interferometers, 131–133
  atomic mirrors, 127–130
  coherent manipulation of laser-cooled and trapped atoms, 133–134
  coherent matter-wave manipulation, 127–133
  experimental results showing deflection of beam of ^{20}Ne* atoms, 129
  experimental setup for Ne* atomic mirror, 128
  interference fringes for atomic interferometer, 132
  measurement of weak magnetic fields with Larmor velocity filter, 134–135
  STIRAP-based atomic mirrors, 130
  STIRAP for atomic interferometry, 129–130
Adiabatic passage by light-induced potentials (APLIP)
  application of STIRAP, 161
  mechanism for transfer of wave packet, 162
Adiabatic passage phase, stimulated Raman adiabatic passage (STIRAP), 87

Adiabatic states
  stimulated Raman adiabatic passage (STIRAP), 83–85
  two-state system, 72–73
Adiabatons
  propagation, 169
  propagation phenomenon, 167–168
Amplitude modulation (AM)
  definition, 255
  differentially detuned separated oscillatory fields (DSOF) interferometer detecting, 258
  DSOF region, 256
  longitudinal atom optical (LAO) technique, 261–262
  molecular beams, 256
  Ramsey interference pattern, 257
  rephasing conditions, 256
  smallest detection, 271
Atomic beam splitters, coherent matter-wave manipulation, 130–131
Atomic coherence, early slow light experiments involving, 209
Atomic interferometers
  coherent matter-wave manipulation, 131–133
  interference fringes, 132
Atomic mirrors
  coherent matter-wave manipulation, 127–130
  deflection of beam of ^{20}Ne* atoms, 129
  experimental setup, 128
Atom interferometers, transverse atom-optical beam splitters, 244
Atom optics, wave properties of particles, 244
Atoms, laser cooled and trapped, coherent manipulation, 133–134
Autler-Townes phase
  pump-induced, 88
  Stokes-induced, 87

INDEX

Autler-Townes splitting
  experimental demonstration in neon, 82
  three-state system, 79–81
Avoided crossings
  exciting neighboring curves, 249–250
  two-state system, 74

B

Beam geometry, sensitivity to stimulated Raman adiabatic passage (STIRAP), 100–101
Black hole, slow light, 234–235
Born-Oppenheimer approximation, slow heavy particle collisions, 245
Branched-chain excitation
  branched linkage patterns, 135
  experimental demonstration, 141–143
  tripod linkage, 136–141
  See also Tripod linkage
Brumer-Shapiro two-pathway, coherent control, 30

C

Calcium
  accessing problem of inversion, 37–40
  energy level diagram, 38–40
  optimization of double ionization, 38
  potential energy diagram, 39
Chemical reactions
  application of stimulated Raman adiabatic passage (STIRAP), 156
  control, 1–2
Chirped-pulse amplification (CPA), femtosecond pulse shaping, 27–28
Coherence transfer
  between matter and light, 170–172
  light pulse storage in ^{87}Rb cell, 172
Coherent excitation
  goal of theory, 58
  partial coherence: density matrix, 68–69
  principles, 67–68
  three states, 76–81
  two-state system, 69–75
Coherent manipulation
  coherent population transfer, 64–67
  laser state selection, 59–62
Coherent matter-wave manipulation
  atomic beam splitters, 130–131

atomic interferometers, 131–133
atomic mirrors, 127–130
Coherent population transfer
  Rabi oscillations, 65
  resonant coherent excitation, 65
  three-state systems, 66–67
Cold gas, slow light, 209–210
Cold molecules, photoassociative SITRAP as source, 161–162
Complete population return, two-state system, 73
Continuous-wave (CW) lasers
  experimental demonstrations, 103–107
  metastable neon atoms, 104–107
  sodium dimers, 103–104
Continuum
  intermediary for population transfer, 144
  laser-induced structure, 145–150
  population transfer, 150–154
  population transfer via, of intermediate states, 143–145
  tripod coupling, 155–156
Control knobs, quantum control, 4–5
CpFe(CO$_2$)Cl
  correlation diagram, 37
  many-parameter quantum control, 35
  reduction of parameter space, 36
  selective dissociation and complexity analysis with, 34–37

D

Dark-state polaritons
  coherence transfer between matter and light, 170–172
  stopping and reaccelerating, 171
Decoupling, complete, light and matter, 163–165
Definitions
  amplitude modulation, 255
  population trapping, 149
  stimulated Raman adiabatic passage (STIRAP), 82–83
Degenerate states
  adiabatic path connecting initial and final states, 116–117
  avoiding one-photon resonances, 117
  examining evolution of energies of adiabatic states, 113–114

INDEX 279

level scheme in neon experiment, 111–112
population transfer in Ne* versus magnetic field strength, 112–113, 115
stimulated Raman adiabatic passage (STIRAP), 111–117
Zeeman splittings, 114–115
Delay, sensitivity to, STIRAP, 93–94
Density matrix deconvolution
amplitude modulated wavefunction, 264
amplitude of four stripes of density matrix, 266–267
averaging process, 269–270
coherence between two momentum components, 264–265
configuration of experiment, 264
detection and transmission probabilities, 265
free evolution, 264
inherent time dependence of off-diagonal elements, 263–264
longitudinal interferometry, 263–268
obtaining off-diagonal elements, 266
ODEs dephasing, 269
off-diagonal density matrix elements (ODEs), 268
phase of density matrix, 267
rephasing peaks, 269
scaled detuning, 265
semiclassical approximation, 265
theoretical prediction, 268
unmodified atomic beam, 268–271
Differentially detuned separated oscillatory fields (DSOF)
configuration, 255
detecting amplitude modulation, 258
detecting longitudinal momentum coherences, 258–259
envelopes of rephased fringes, 258
formation, 254
future, 273
longitudinal matter-wave interferometry, 253
matter-wave interferometry, 252
maximum attainable momentum difference, 254
measuring both amplitude and phase components, 263
operation, 261

rephased fringes, 257
resonance region functioning as beam splitter, 253
roots, 252
time-dependent interference, 255
transition-induced momentum shift, 254
treatment, 253

E

Electromagnetically induced transparency (EIT)
complete decoupling of light and matter, 163–165
elimination of absorption: slow light, 165–167
intracavity, 226–229
phase matching co-propagating optical waves, 229
phenomenon, 163
pump-induced phase, 88
slow light propagation, 166
Stokes-induced phase, 87
ultraslow EIT polariton, 221–222
Electron diffraction pattern, Debye-Scherrer powder, 243
Energy velocity, light, 192
Equations, STIRAP, 82–83
Excitation *See* Branched-chain excitation

F

Fabry-Perot etalon
scheme, 200
slow light, 199–200
Fano profile, laser-induced continuum structure, 146–148
Fano-type suppression, incoherent ionization, 153–154
Faraday effect, nonlinear, magnetometers, 225–226
$Fe(CO)_5$
comparing many-parameter and one-parameter control, 32
evaluation of photoproduct mass spectra, 31
many-parameter control using different fitness functions, 34
many-parameter quantum control, 30–34

$Fe(CO)_5$ (Contd.)
  photodissociation channels, 31
  pulse duration, 16–17
  pump-probe delay, 18–19
  quantum control, 15, 16
Femtosecond pulse shaping
  adaptive, 23–29
  chirped-pulse amplification (CPA) systems, 27–28
  evolutionary algorithm, 25
  experimental examples, 26–29
  frequency-domain, 24
  learning loop setup, 28–29
  measured spectrum before and after optimization, 29
  technology, 23–26
  titanium:sapphire oscillator laser pulses, 26–27
  ultrashort laser pulses, 5
Femtosecond quantum control
  control parameter, 7–8
  control type, 8
  detection, 8
  experiment, 8
  experimental techniques, 6–12
  femtosecond laser system, 7, 9
  future directions, 47–48
  gas-phase experiments, 10–11
  laser diagnostics, 10
  liquid phase experiments, 11–12
  pulses, 10
  schematic of experimental setup, 7
  setup, 8, 10–12
  summary, 46–47
  time-of-flight (TOF) spectrometer, 11
  See also Quantum control
Front velocity, light, 192
Frozen light
  absorption or temporal variation of drive field, 223
  dispersion and decay spectra, 222
  geometry of ultraslow EIT pulse propagation in gas, 220
  idea of freezing light in cell, 220–221
  kinematics of deceleration of ultraslow pulse to freezing point, 223
  observing freezing or backward light, 223
  three-level atomic Λ sytem, 220
  ultimate slow, 219–223
  ultraslow and negative group velocity of EIT polariton versus detuning of drive laser, 222
  ultraslow EIT polariton, 221–222
  velocity distribution of atoms in cell, 220

G

Gas
  slow light in cold, 209–210
  slow light in hot, 210–211
Gas phase
  accessing problem of inversion with atomic calcium, 37–40
  Brumer-Shapiro two-pathway coherent control, 30
  comparing pulse shapes with one-parameter control schemes, 32–33
  control by linear chirp, 21–23
  control by phase-sensitive excitation, 19–21
  control by pulse duration, 16–17
  control by pulse energy, 13–16
  control by pump-probe delay, 17–19
  many-parameter quantum control, 29–40
  one-parameter quantum control, 12–23
  selective dissociation and complexity analysis with $CpFe(CO)_2Cl$, 34–37
  testing technology with $Fe(CO)_5$, 30–34
Gedankenexperiment, black hole to gyro, 234–235
Group velocity
  contributions, 194
  EIT polariton, 222
  interference of two monochromatic waves, 193
  kinematics, 193–196
  light, 192
  propagating electromagnetic pulse, 210
Gyro, black hole to, 234–235

H

Helium
  energy-level diagram in SCRAP demonstration, 160
  excitation by He+ projectiles, 248–249
  laser-induced continuum structure, 148
  Rosenthal oscillations, 251

scattering of projectiles in ion-atom collision, 247
Heterostructures, slow light, 199–201
High refractive index, slow light and, 196–199
Hot gas, slow light, 210–211
Hyper-Raman STIRAP, 156–157

I

Incoherent excitation, two-state system, 62–63
Incoherent ionization
  Fano-type suppression, 153–154
  suppressing, 152–154
Incoherent population transfer
  incoherent excitation of two-state systems, 62–63
  optical pumping, 63–64
  radiative rate equations, 62
  stimulated emission pumping (SEP), 64
Interference
  ion-atom collisions, 245–252
  quantum mechanical origin of effects, 272
Intermediate state population, STIRAP, 88–89
Intermediate states, population transfer via continuum, 143–145
Intermediate states, multiple, STIRAP, 101–102
Intracavity electromagnetically induced transparency (EIT)
  atomic coherence, 228–229
  frequency of locking, 228
  locking and narrowing mechanism, 227
  slow light, 226–229
Inversion, accessing problem with atomic calcium, 37–40
Ion-atom collisions
  analogy between interference effects and atomic-beam radiofrequency resonance, 252
  antiphase oscillations, 250
  avoided Landau-Zener crossings, 249–250
  Born-Oppenheimer approximation, 245
  excitation of He($1snl$) excited states by He$^+$ projectiles, 248–249

experiments on homonuclear, 246
inelastic, 248
interference in, 245–252
longitudinal interference of matter waves, 247
matter-wave interference effects, 246
optical emission excitation cross sections vs. laboratory ion beam energy for He, 249
Rosenthal-Foley model, 250–251
Rosenthal oscillations, 248
Rosenthal oscillations in He$_2^+$ system, 251
scattering of projectiles in He$^+$ + He collisions, 247
Stueckelberg oscillations, 247–248
Ionization, optimization of calcium double, 38

K

Kinematics, group velocity, 193–196
Kinetic control, chemical reactions, 2

L

Λ linkage, three-state system, 76
Λ systems
  slow light in, 204–209
  transmission and dispersion spectra of coherently prepared vapor, 207
Ladder, three-state system, 76
Landau-Zener crossings, exciting neighboring curves, 249–250
Larmor velocity filter
  measurement of weak magnetic fields, 134–135
  variation of flux of deflected Ne* atoms, 134
Laser beam geometries, STIRAP, 101
Laser-cooled atoms, coherent manipulation, 133–134
Laser-induced continuum structure
  coherent and incoherent channels, 146
  dark and bright states, 149–150
  equations, 145–146
  Fano parameter, 147–148
  Fano profile, 146–148
  observation in helium, 148

Laser-induced continuum (*Contd.*)
   population transfer, 145–150
   population trapping, 149–150
Laser state selection
   development of Raman laser, 60–61
   development of stimulated Raman adiabatic passage (STIRAP), 61–62
   early days, 59–62
   population depletion by optical pumping, 59
   profile of pump and Stokes laser intensities, 61
   stimulated emission pumping (SEP), 60
   Stokes laser beam, 61–62
Laser technology, photochemistry, 2
Learning loop setup, femtosecond pulse shaping, 28–29
Light
   potential control tool, 2
   propagation, 191–192
   wave velocities, 191–192
   See also Frozen light; Slow light; Ultraslow ligh–4
Light and matter
   coherence transfer between, 170–172
   complete decoupling, 163–165
Linear chirp, control by, 21–23
Liquid phase
   correlation diagram of emission versus second-harmonic generation (SHG) efficiency, 44
   experiment, 41–43
   Husimi transforms, 45
   many-parameter quantum control, 40–45
   maximization of molecular emission and SHG, 43–45
   optimization of $[Ru(dpb)_3]^{2+}$ emission in methanol, 42
   ruthenium(II) polypyridyl family, 42–43
   structure of $[Ru(dpb)_3]^{2+}$, 42
Longitudinal atom optical (LAO)
   amplitude modulation, 261–262
   far-field case, 273
Longitudinal interferometry
   amplitude modulation and rephasing, 255–257
   apparatus and experimental techniques, 260–262
   density matrix deconvolution, 263–268
   detecting longitudinal momentum coherences, 258–259
   differentially detuned separated oscillatory fields (DSOF), 252–255
   ion-atom collisions, 245–252
   search for coherences in unmodified atomic beam, 268–271
   semiclassical approximation, 259–260
   Talbot length, 259–260
Losses from intermediate state, sensitivity to, STIRAP, 97, 99–100

M

Magnetometry
   concept, 224
   magnetometers based on nonlinear Faraday effect, 225–226
   phaseonium magnetometers, 224
Many-atom systems, information processing, 177–179
Matched pulses, propagation phenomenon, 168–170
Matter and light
   coherence transfer between, 170–172
   complete decoupling, 163–165
Matter-wave interferometry, limitation, 245
Matter-wave manipulation, coherent, 127–133
Mechanism, STIRAP, 85–86
Michelson interferometer, preceding atom interferometers, 244
Molecular beam resonance, separated oscillatory field (SOF), 244
Momentum transfer, basis of atom optics, 126–127
Multistate stimulated Raman adiabatic passage (STIRAP)
   adiabatic population transfer in four- and five-state systems, 123, 124
   chainwise excitation, 117
   dressed-state picture, 124–125, 126
   extension of STIRAP to multistate, 118
   final-state population in four- and five-state system, 125, 126
   Hamiltonian of five-state chain, 119

off-resonance case, 121–124
optimization, 124–125, 126
Rabi frequencies of five-state example, 119, 120
resonantly driven chains with even number of states, 120–121
resonantly driven chains with odd number of states, 118–120
rotation wave approximation (RWA) Hamiltonian, 118

N

Neon, experimental demonstration of Autler-Townes splitting, 82
Neon atoms, metastable
continuous-wave lasers, 104–107
demonstration of tripod scheme, 141–143
laser-induced fluorescence from final and intermediate state, 105
momentum distribution, 143
transfer efficiency versus displacement, 106
Nitrous oxide molecules
linkage scheme, 108
pulsed lasers, 108, 109
No-crossing case, two-state system, 73
Nonadiabatic transitions, STIRAP, 92–93
Nonlinear magnetooptic rotation, slow light and, 211–213
Null eigenvalue, STIRAP, 88–89

O

Off-diagonal density matrix elements (ODEs)
averaging process, 269–270
dephasing, 269
limiting size, 271
rephasing, 269
search, 268
Off-resonance stimulated Raman scattering (ORSRS), 59
Optical pumping
incoherent population transfer, 63–64
population depletion through, 59
Optimal control theory, quantum control, 4–5

Overtone pumping (OTP), laser state selection, 59

P

Phaseonium magnetometers, slow light, 224
Phase-sensitive excitation, control by, 19–21
Phase velocity, light, 192
Phonons
acoustic frequency for backward and forward scattering, 231
acoustic waves propagating along fiber, 231–232
cutoff frequency, 232
dispersionless medium, 230
forward stimulated Brillouin scattering (SBS), 229
phase-matching condition for forward SBS, 233
possibility of controlling group velocity of light waves, 233–234
resonant interactions, 234
slow light, 229–234
ultradispersion medium, 230
Photoassociative STIRAP, source for cold molecules, 161–162
Photochemical routes
problem of exerting control, 2–3
product formation, 2
Photochemistry, laser technology, 2
Photodissociation, $Fe(CO)_5$, 31
Population transfer
continuum, 150–154
continuum of intermediate states, 143–145
dressed-state picture, 124–125, 126
Fano profile, 146–148
Fano-type suppression of incoherent ionization, 153–154
ideal case, 150–151
incoherent schemes, 62–64
laser-induced continuum structure (LICS), 145–150
LICS coherent and incoherent channels, 146
LICS equations, 145–146
off-resonance case, 121–124

Population transfer (*Contd.*)
  optimization of multistate-STIRAP, 124–125, 126
  population trapping: dark and bright states, 149–150
  resonantly driven chains with even number of states, 120–121
  resonantly driven chains with odd number of states, 118–120
  satisfying trapping condition, 151
  STIRAP-like, in multistate chains, 117–125
  suppressing incoherent ionization, 152–154
  target-state population versus peak ionization rate, 154
  tripod coupling via continuum, 155–156
Population trapping
  dark and bright states, 149–150
  definition, 149
Probability loss, two-state system, 71
Propagation phenomena
  adiabatons, 167–168
  coherence transfer between matter and light, 170–172
  complete decoupling of light and matter, 163–165
  electromagnetically induced transparency (EIT), 163–167
  elimination of absorption: slow light, 165–167
  light pulse storage in ^{87}Rb cell, 172
  matched pulses, 168–170
  propagation of adiabatons, 169
  pulse matching, 169
  slow light propagation, 166
  stopping and reaccelerating dark-state polariton, 171
Pulsed lasers
  experimental demonstrations, 107–111
  general considerations, 107–108
  nitrous oxide molecules, 108
  rubidium atoms in ladder system, 111
  sulfur dioxide molecules, 109–111
Pulse duration, control by, 16–17
Pulse energy, control by, 13–16
Pulse matching, propagation phenomenon, 168–170

Pulse sequences
  stimulated Raman adiabatic passage (STIRAP), 89–90
  three-state system, 77–78
Pulse width, sensitivity to, STIRAP, 94
Pump-induced Autler-Townes phase, STIRAP, 88
Pump-induced EIT phase, STIRAP, 88
Pump-probe delay, control by, 17–19

Q

Quantum bits, information processing, 175
Quantum control
  accessing problem of inversion with atomic calcium, 37–40
  adiabatic control technique, 3–4
  basis of one-parameter, 4
  categories, 3–4
  experimental examples, 5–6
  $Fe(CO)_5$, 15, 16, 30–34
  future directions, 47–48
  historical development, 1–6
  initial development, 3
  linear chirp, 21–23
  many-parameter in gas phase, 29–40
  many-parameter in liquid phase, 40–45
  one-parameter in gas phase, 12–23
  optimal control theory, 4–5
  optimization process, 5
  phase-sensitive excitation, 19–21
  potential energy diagram of Na_3 at 618 nm, 13
  pulse duration, 16–17
  pulse energy, 13–16
  pump-probe delay, 17–19
  pump-probe transients of Na_3^+ and power spectral density, 14
  selective dissociation and complexity analysis with $CpFe(CO_2)Cl$, 34–37
  spectral-domain technique, 3
  time-domain technique, 3
  See also Femtosecond quantum control
Quantum electrodynamics, single-atom cavity, 173–174
Quantum information
  classical proof-of-principle experiments, 214

experiments on usage of atomic coherence for light information storage, 215
possibility of storage of light in medium, 216
reading by backward-propagating pulse, 215
resonant four-wave mixing experiments, 216
role of adiabaticity for storage, 218
scheme of experiment, 216
storing and retrieving, 213–219
teleportation of information, 217
time reversal, 216
two-photon cell technique, 215–216
writing and reading beams, 218
Quantum logic gates, cavity QED, 175–177
Quantum networking, information processing, 177

R

Rabi cycling, two-state system, 70–71
Rabi frequency, sensitivity to, STIRAP, 94
Rabi oscillations, coherent population transfer, 65
Raman interferometers, extension to two-photon processes, 244
Raman laser, development, 60–61
Ramsey fringes, measurement, 262
Ramsey separated oscillatory field (SOF), preceding atom interferometers, 244
Rapid adiabatic passage, two-state system, 73–75
Resonant coherent excitation, population transfer, 65
Retrieving. See Quantum information
Rosenthal-Foley model, phase evolution between two excited-state amplitudes, 250–251
Rosenthal oscillations
analogy with atomic-beam radiofrequency resonance, 252
helium system, 251
interference structure, 248
Rotating wave approximation (RWA)
Hamiltonian for multistate STIRAP, 118
Hamiltonian for tripod scheme, 136–137
STIRAP beyond RWA, 102–103

three-state, Hamiltonian, 76–77
two-state system, 69–70
Rubidium atoms
STIRAP in ladder system, 111
transmission and dispersion spectra, 207

S

Second-harmonic generation, maximizing, 46
Self-induced transparency (SIT), slow light, 202–204
Semiclassical approximation, longitudinal interferometry, 259–260
Sensitivity to interaction parameters
beam geometry, stimulated Raman adiabatic passage (STIRAP), 100–101
losses from intermediate state, STIRAP, 97, 99–100
multiple intermediate states, 101–102
Rabi frequency and pulse width, STIRAP, 94
single-photon detuning, STIRAP, 94–95
STIRAP, 93–103
STIRAP beyond rotating wave approximation (RWA), 102–103
STIRAP to delay, 93–94
two-photon detuning, STIRAP, 95–97
Separated oscillatory fields (SOF)
preceding atom interferometers, 244
See also Differentially detuned separated oscillatory fields (DSOF)
Signal velocity, light, 192
Single-atom cavity quantum electrodynamics, application of STIRAP, 173–174
Single-photon detuning, sensitivity to, STIRAP, 94–95
Slow light
1D photonic band gap crystal, 200
applications, 167, 224–235
black hole to gyro, 234–235
cold gas, 209–210
dispersion of band-gap material, 200
dispersion of EM waves in superstructure, 202
early experiments involving atomic coherence, 209
elimination of absorption, 165–167
heterostructures, 199–201

Slow light (*Contd.*)
  high refractive index and, 196–199
  hot gas, 210–211
  intracavity electromagnetically induced transparency (EIT), 226–229
  Λ systems, 204–209
  magnetometry, 224–226
  nonlinear magnetooptical rotation, 211–213
  phonons, 229–234
  propagation, 166
  scheme of Fabry-Perot etalon, 200
  selective reflection of probe beam without and with driving field, 199
  self-induced transparency (SIT), 202
  solids, 213
  transmission and dispersion spectra of ^{87}Rb vapor, 207
  two-level systems, 202–204
  See also Frozen light; Ultraslow light
Sodium dimer, Na$_2$
  continuous-wave lasers, 103–104
  linear chirp, 21–23
  phase-sensitive excitation, 20–21
  potential energy diagram, 18, 20
  quantum control by pump-probe delay, 17–18
  quantum-mechanical calculations, 22
Sodium triatomic, Na$_3$
  potential energy diagram, 13
  pump-probe transients and power spectral density, 14
  quantum control, 13–15
Solids, slow light in, 213
Spectral-domain technique, quantum control, 3
Stark-chirped rapid adiabatic passage (SCRAP)
  alternative to STIRAP, 157–161
  energy-level diagram of helium atom, 160
  experimental demonstration, 160–161
  illustration, 158
  theory, 158–160
State vectors, techniques to control, 58
Stimulated Brillouin scattering (SBS)
  forward, 229
  phase-matching condition for forward, 233

Stimulated emission pumping (SEP)
  incoherent population transfer, 64
  laser state selection, 60
Stimulated Raman adiabatic passage (STIRAP)
  adiabatic passage of light-induced potentials (APLIP), 161
  applications in quantum optics and quantum information, 173–179
  avoiding decoherence, 173
  coherent control of atomic and molecular processes, 57–58
  control of chemical reactions, 156
  development, 61–62
  experimental demonstration of Stark-chirped rapid adiabatic passage (SCRAP), 160–161
  extensions and applications, 156–162
  first evidence of vacuum stimulated Raman scattering into cavity mode via STIRAP, 175
  hyper-Raman STIRAP, 156–157
  many-atom systems, 177–179
  photoassociative STIRAP as source for cold molecules, 161–162
  population transfer in multistate chains, 117–126
  quantum bits (qubits), 175
  quantum control, 3–4
  quantum logic gates based on cavity QED, 175–177
  quantum networking, 177
  SCRAP, 157–161
  simplified energy level diagram of helium in SCRAP demonstration, 160
  single-atom cavity quantum electrodynamics, 173–174
  summary, 179–180
  theory of SCRAP, 158–160
  tripod version, 136–141
  See also Multistate stimulated Raman adiabatic passage (STIRAP)
Stimulated Raman adiabatic passage (STIRAP), three-state
  adiabatic condition: local and global criteria, 90–92
  adiabatic passage phase, 87
  adiabatic states, 83–85
  basic equations and definitions, 82–83
  basic properties, 81–93

beyond rotating wave approximation
    (RWA), 102–103
degenerate or nearly degenerate states,
    111–117
experimental demonstrations with
    continuous-wave lasers, 103–107
experimental demonstrations with
    pulsed lasers, 107–111
experimentally measured transfer
    efficiency, 98
five-stage description, 86–88
importance of null eigenvalue, 88–89
intermediate state population, 88–89
intuitive versus counterintuitive pulse
    sequences, 89–90
laser beam geometries, 101
metastable neon atoms, 104–107
multiple intermediate states, 101–102
nitrous oxide molecules, 108, 109
nonadiabatic transitions, 92–93
numerically calculated populations vs.
    pulse delay, 91
numerically calculated transfer
    efficiency, 98
pump-induced Autler-Townes phase, 88
pump-induced EIT phase, 88
rubidium atoms in ladder system, 111
sensitivity to beam geometry, 100–101
sensitivity to delay, 93–94
sensitivity to losses from intermediate
    state, 97, 99–100
sensitivity to Rabi frequency and pulse
    width, 94
sensitivity to single-photon detuning,
    94–95
sensitivity to two-photon detuning,
    95–97
sodium dimers, 103–104
STIRAP mechanism, 85–86
Stokes-induced Autler-Townes phase, 87
Stokes-induced EIT phase, 87
sulfur dioxide molecules, 109–111
time dependencies of frequencies and
    populations, 86
vector picture, 87
Stokes-induced Autler-Townes phase,
    STIRAP, 87
Stokes-induced EIT phase, STIRAP, 87
Stokes radiation, generation in cavity and
    external source, 61–62
Storing. *See* Quantum information
Stueckelberg oscillations, ion-atom
    collisions, 247–248
Sulfur dioxide
    pulsed lasers, 109–111
    STIRAP vs. two-photon detuning, 100
    STIRAP vs. pulse delay, 100
Suppression, incoherent ionization,
    152–154

T

Talbot length, longitudinal atom optics,
    259–260
Tannor-Kosloff-Rice scheme, quantum
    control, 17–18
Teleportation, quantum information, 217
Thermodynamic control, chemical
    reactions, 1–2
Three-state system
    adiabatic elimination, 78–79
    Autler-Townes splitting, 79–81
    coherent population transfer, 66–67
    linkages for three-state coupling, 76
    pulse sequences, 77–78
    rotating wave approximation (RWA)
        Hamiltonian, 76–77
    time evolution of energies with
        two-photon detuning, 96
Time-domain technique, quantum
    control, 3
Titanium:sapphire oscillator laser,
    femtosecond pulse shaping, 26–27, 28
Transfer efficiency, STIRAP, 98
Transition probabilities, two-state
    system, 75
Tripod coupling
    continuum, 155–156
    Fano parameter, 155
    Hamiltonian describing, 155
    trapping conditions, 155
Tripod linkage
    adiabatic evolution, 137–141
    concept, 136
    experimental demonstration, 141–143
    rotating wave approximation (RWA)
        Hamiltonian, 136–137
    various possible orderings for pump,
        Stokes, and control laser pulses,
        139

Two-level system, slow light in, 202–204
Two-photon detuning, sensitivity to STIRAP, 95–97
  time evolution of energies in three-state system with, 96
Two-state system
  adiabatic states and adiabatic following, 72–73
  adiabatic transfer state, 73
  avoided crossing, 74
  coherent excitation, 69–75
  complete population return, 73
  estimating transition probabilities, 75
  incoherent excitation, 62–63
  no-crossing case, 73
  probability loss, 71
  Rabi cycling, 70–71
  rapid adiabatic passage, 73–75
  rotating wave approximation (RWA), 69–70
  time evolution of energies and populations, 74

U

Ultraslow light
  slow light and nonlinear magnetooptic rotation, 211–213
  slow light in cold gas, 209–210
  slow light in hot gas, 210–211
  slow light in solids, 213
  See also Frozen light

V

Vector picture, STIRAP, 87
Vee linkage, three-state system, 76
Velocities, wave, light, 191–192

W

Wave-particle duality, history, 243
Wave properties
  historical study, 243
  particles, 244
Wave velocities, light, 191–192
White-fringe geometry, independent of velocity, 245

Contents of Volumes in This Serial

Volume 1

Molecular Orbital Theory of the Spin Properties of Conjugated Molecules, *G. G. Hall and A. T. Amos*

Electron Affinities of Atoms and Molecules, *B. L. Moiseiwitsch*

Atomic Rearrangement Collisions, *B. H. Bransden*

The Production of Rotational and Vibrational Transitions in Encounters between Molecules, *K. Takayanagi*

The Study of Intermolecular Potentials with Molecular Beams at Thermal Energies, *H. Pauly and J. P. Toennies*

High-Intensity and High-Energy Molecular Beams, *J. B. Anderson, R. P. Andres, and J. B. Fen*

Volume 2

The Calculation of van der Waals Interactions, *A. Dalgarno and W. D. Davison*

Thermal Diffusion in Gases, *E. A. Mason, R. J. Munn, and Francis J. Smith*

Spectroscopy in the Vacuum Ultraviolet, *W. R. S. Garton*

The Measurement of the Photoionization Cross Sections of the Atomic Gases, *James A. R. Samson*

The Theory of Electron–Atom Collisions, *R. Peterkop and V. Veldre*

Experimental Studies of Excitation in Collisions between Atomic and Ionic Systems, *F. J. de Heer*

Mass Spectrometry of Free Radicals, *S. N. Foner*

Volume 3

The Quantal Calculation of Photoionization Cross Sections, *A. L. Stewart*

Radiofrequency Spectroscopy of Stored Ions I: Storage, *H. G. Dehmelt*

Optical Pumping Methods in Atomic Spectroscopy, *B. Budick*

Energy Transfer in Organic Molecular Crystals: A Survey of Experiments, *H. C. Wolf*

Atomic and Molecular Scattering from Solid Surfaces, *Robert E. Stickney*

Quantum, Mechanics in Gas Crystal-Surface van der Waals Scattering, *E. Chanoch Beder*

Reactive Collisions between Gas and Surface Atoms, *Henry Wise and Bernard J. Wood*

Volume 4

H. S. W. Massey—A Sixtieth Birthday Tribute, *E. H. S. Burhop*

Electronic Eigenenergies of the Hydrogen Molecular Ion, *D. R. Bates and R. H. G. Reid*

Applications of Quantum Theory to the Viscosity of Dilute Gases, *R. A. Buckingham and E. Gal*

Positrons and Positronium in Gases, *P. A. Fraser*

Classical Theory of Atomic Scattering, *A. Burgess and I. C. Percival*

Born Expansions, *A. R. Holt and B. L. Moiselwitsch*

Resonances in Electron Scattering by Atoms and Molecules, *P. G. Burke*

Relativistic Inner Shell Ionizations, *C. B. O. Mohr*

Recent Measurements on Charge Transfer, *J. B. Hasted*

Measurements of Electron Excitation Functions, *D. W. O. Heddle and R. G. W. Keesing*

Some New Experimental Methods in Collision Physics, *R. F. Stebbings*

Atomic Collision Processes in Gaseous Nebulae, *M. J. Seaton*

Collisions in the Ionosphere, *A. Dalgarno*

The Direct Study of Ionization in Space, *R. L. F. Boyd*

Volume 5

Flowing Afterglow Measurements of Ion-Neutral Reactions, *E. E. Ferguson, F. C. Fehsenfeld, and A. L. Schmeltekopf*

Experiments with Merging Beams, *Roy H. Neynaber*

Radiofrequency Spectroscopy of Stored Ions II: Spectroscopy, *H. G. Dehmelt*

The Spectra of Molecular Solids, *O. Schnepp*

The Meaning of Collision Broadening of Spectral Lines: The Classical Oscillator Analog, *A. Ben-Reuven*

The Calculation of Atomic Transition Probabilities, *R. J. S. Crossley*

Tables of One- and Two-Particle Coefficients of Fractional Parentage for Configurations $s^{\lambda}s^{\prime u}p^{q}$, *C. D. H. Chisholm, A. Dalgarno, and F. R. Innes*

Relativistic Z-Dependent Corrections to Atomic Energy Levels, *Holly Thomis Doyle*

Volume 6

Dissociative Recombination, *J. N. Bardsley and M. A. Biondi*

Analysis of the Velocity Field in Plasmas from the Doppler Broadening of Spectral Emission Lines, *A. S. Kaufman*

The Rotational Excitation of Molecules by Slow Electrons, *Kazuo Takayanagi and Yukikazu Itikawa*

The Diffusion of Atoms and Molecules, *E. A. Mason and T. R. Marrero*

Theory and Application of Sturmian Functions, *Manuel Rotenberg*

Use of Classical mechanics in the Treatment of Collisions between Massive Systems, *D. R. Bates and A. E. Kingston*

Volume 7

Physics of the Hydrogen Master, *C. Audoin, J. P. Schermann, and P. Grivet*

Molecular Wave Functions: Calculations and Use in Atomic and Molecular Processes, *J. C. Browne*

Localized Molecular Orbitals, *Harel Weinstein, Ruben Pauncz, and Maurice Cohen*

General Theory of Spin-Coupled Wave Functions for Atoms and Molecules, *J. Gerratt*

Diabatic States of Molecules—Quasi-Stationary Electronic States, *Thomas F. O'Malley*

Selection Rules within Atomic Shells, *B. R. Judd*

Green's Function Technique in Atomic and Molecular Physics, *Gy. Csanak, H. S. Taylor, and Robert Yaris*

A Review of Pseudo-Potentials with Emphasis on Their Application to Liquid Metals, *Nathan Wiser and A. J. Greenfield*

Volume 8

Interstellar Molecules: Their Formation and Destruction, *D. McNally*

Monte Carlo Trajectory Calculations of Atomic and Molecular Excitation in Thermal Systems, *James C. Keck*

Nonrelativistic Off-Shell Two-Body Coulomb Amplitudes, *Joseph C. Y. Chen and Augustine C. Chen*

Photoionization with Molecular Beams, *R. B. Cairns, Halstead Harrison, and R. I. Schoen*

The Auger Effect, *E. H. S. Burhop and W. N. Asaad*

Volume 9

Correlation in Excited States of Atoms, *A. W. Weiss*
The Calculation of Electron–Atom Excitation Cross Sections, *M. R. H. Rudge*
Collision-Induced Transitions between Rotational Levels, *Takeshi Oka*
The Differential Cross Section of Low-Energy Electron–Atom Collisions, *D. Andrick*
Molecular Beam Electric Resonance Spectroscopy, *Jens C. Zorn and Thomas C. English*
Atomic and Molecular Processes in the Martian Atmosphere, *Michael B. McElroy*

Volume 10

Relativistic Effects in the Many-Electron Atom, *Lloyd Armstrong, Jr. and Serge Feneuille*
The First Born Approximation, *K. L. Bell and A. E. Kingston*
Photoelectron Spectroscopy, *W. C. Price*
Dye Lasers in Atomic Spectroscopy, *W. Lange, J. Luther, and A. Steudel*
Recent Progress in the Classification of the Spectra of Highly Ionized Atoms, *B. C. Fawcett*
A Review of Jovian Ionospheric Chemistry, *Wesley T. Huntress, Jr.*

Volume 11

The Theory of Collisions between Charged Particles and Highly Excited Atoms, *I. C. Percival and D. Richards*
Electron Impact Excitation of Positive Ions, *M. J. Seaton*
The R-Matrix Theory of Atomic Process, *P. G. Burke and W. D. Robb*
Role of Energy in Reactive Molecular Scattering: An Information-Theoretic Approach, *R. B. Bernstein and R. D. Levine*
Inner Shell Ionization by Incident Nuclei, *Johannes M. Hansteen*
Stark Broadening, *Hans R. Griem*
Chemiluminescence in Gases, *M. F. Golde and B. A. Thrush*

Volume 12

Nonadiabatic Transitions between Ionic and Covalent States, *R. K. Janev*
Recent Progress in the Theory of Atomic Isotope Shift, *J. Bauche and R.-J. Champeau*
Topics on Multiphoton Processes in Atoms, *P. Lambropoulos*
Optical Pumping of Molecules, *M. Broyer, G. Goudedard, J. C. Lehmann, and J. Vigué*
Highly Ionized Ions, *Ivan A. Sellin*
Time-of-Flight Scattering Spectroscopy, *Wilhelm Raith*
Ion Chemistry in the D Region, *George C. Reid*

Volume 13

Atomic and Molecular Polarizabilities—A Review of Recent Advances, *Thomas M. Miller and Benjamin Bederson*
Study of Collisions by Laser Spectroscopy, *Paul R. Berman*
Collision Experiments with Laser-Excited Atoms in Crossed Beams, *I. V. Hertel and W. Stoll*
Scattering Studies of Rotational and Vibrational Excitation of Molecules, *Manfred Faubel and J. Peter Toennies*
Low-Energy Electron Scattering by Complex Atoms: Theory and Calculations, *R. K. Nesbet*
Microwave Transitions of Interstellar Atoms and Molecules, *W. B. Somerville*

Volume 14

Resonances in Electron Atom and Molecule Scattering, *D. E. Golden*

The Accurate Calculation of Atomic Properties by Numerical Methods, *Brian C. Webster, Michael J. Jamieson, and Ronald F. Stewart*

(e, 2e) Collisions, *Erich Weigold and Ian E. McCarthy*

Forbidden Transitions in One- and Two-Electron Atoms, *Richard Marrus and Peter J. Mohr*

Semiclassical Effects in Heavy-Particle Collisions, *M. S. Child*

Atomic Physics Tests of the Basic Concepts in Quantum Mechanics, *Francis M. Pipkin*

Quasi-Molecular Interference Effects in Ion–Atom Collisions, *S. V. Bobashev*

Rydberg Atoms, *S. A. Edelstein and T. F. Gallagher*

UV and X-Ray Spectroscopy in Astrophysics, *A. K. Dupree*

Volume 15

Negative Ions, *H. S. W. Massey*

Atomic Physics from Atmospheric and Astrophysical Studies, *A. Dalgarno*

Collisions of Highly Excited Atoms, *R. F. Stebbings*

Theoretical Aspects of Positron Collisions in Gases, *J. W. Humberston*

Experimental Aspects of Positron Collisions in Gases, *T. C. Griffith*

Reactive Scattering: Recent Advances in Theory and Experiment, *Richard B. Bernstein*

Ion–Atom Charge Transfer Collisions at Low Energies, *J. B. Hasted*

Aspects of Recombination, *D. R. Bates*

The Theory of Fast Heavy Particle Collisions, *B. H. Bransden*

Atomic Collision Processes in Controlled Thermonuclear Fusion Research, *H. B. Gilbody*

Inner-Shell Ionization, *E. H. S. Burhop*

Excitation of Atoms by Electron Impact, *D. W. O. Heddle*

Coherence and Correlation in Atomic Collisions, *H. Kleinpoppen*

Theory of Low Energy Electron-Molecule Collisions, *P. G. Burke*

Volume 16

Atomic Hartree–Fock Theory, *M. Cohen and R. P. McEachran*

Experiments and Model Calculations to Determine Interatomic Potentials, *R. Düren*

Sources of Polarized Electrons, *R. J. Celotta and D. T. Pierce*

Theory of Atomic Processes in Strong Resonant Electromagnetic Fields, *S. Swain*

Spectroscopy of Laser-Produced Plasmas, *M. H. Key and R. J. Hutcheon*

Relativistic Effects in Atomic Collisions Theory, *B. L. Moiseiwitsch*

Parity Nonconservation in Atoms: Status of Theory and Experiment, *E. N. Fortson and L. Wilets*

Volume 17

Collective Effects in Photoionization of Atoms, *M. Ya. Amusia*

Nonadiabatic Charge Transfer, *D. S. F. Crothers*

Atomic Rydberg States, *Serge Feneuille and Pierre Jacquinot*

Superfluorescence, *M. F. H. Schuurmans, Q. H. F. Vrehen, D. Polder, and H. M. Gibbs*

Applications of Resonance Ionization Spectroscopy in Atomic and Molecular Physics, *M. G. Payne, C. H. Chen, G. S. Hurst, and G. W. Foltz*

Inner-Shell Vacancy Production in Ion–Atom Collisions, *C. D. Lin and Patrick Richard*

Atomic Processes in the Sun, *P. L. Dufton and A. E. Kingston*

Volume 18

Theory of Electron–Atom Scattering in a Radiation Field, *Leonard Rosenberg*

Positron–Gas Scattering Experiments, *Talbert S. Stein and Walter E. Kauppila*

Nonresonant Multiphoton Ionization of Atoms, *J. Morellec, D. Normand, and G. Petite*

Classical and Semiclassical Methods in Inelastic Heavy-Particle Collisions, *A. S. Dickinson and D. Richards*

Recent Computational Developments in the Use of Complex Scaling in Resonance Phenomena, *B. R. Junker*

Direct Excitation in Atomic Collisions: Studies of Quasi-One-Electron Systems, *N. Anderson and S. E. Nielsen*

Model Potentials in Atomic Structure, *A. Hibbert*

Recent Developments in the Theory of Electron Scattering by Highly Polar Molecules, *D. W. Norcross and L. A. Collins*

Quantum Electrodynamic Effects in Few-Electron Atomic Systems, *G. W. F. Drake*

Volume 19

Electron Capture in Collisions of Hydrogen Atoms with Fully Stripped Ions, *B. H. Bransden and R. K. Janev*

Interactions of Simple Ion–Atom Systems, *J. T. Park*

High-Resolution Spectroscopy of Stored Ions, *D. J. Wineland, Wayne M. Itano, and R. S. Van Dyck, Jr.*

Spin-Dependent Phenomena in Inelastic Electron–Atom Collisions, *K. Blum and H. Kleinpoppen*

The Reduced Potential Curve Method for Diatomic Molecules and Its Applications, *F. Jenč*

The Vibrational Excitation of Molecules by Electron Impact, *D. G. Thompson*

Vibrational and Rotational Excitation in Molecular Collisions, *Manfred Faubel*

Spin Polarization of Atomic and Molecular Photoelectrons, *N. A. Cherepkov*

Volume 20

Ion–Ion Recombination in an Ambient Gas, *D. R. Bates*

Atomic Charges within Molecules, *G. G. Hall*

Experimental Studies on Cluster Ions, *T. D. Mark and A. W. Castleman, Jr.*

Nuclear Reaction Effects on Atomic Inner-Shell Ionization, *W. E. Meyerhof and J.-F. Chemin*

Numerical Calculations on Electron-Impact Ionization, *Christopher Bottcher*

Electron and Ion Mobilities, *Gordon R. Freeman and David A. Armstrong*

On the Problem of Extreme UV and X-Ray Lasers, *I. I. Sobel'man and A. V. Vinogradov*

Radiative Properties of Rydberg States in Resonant Cavities, *S. Haroche and J. M. Ralmond*

Rydberg Atoms: High-Resolution Spectroscopy and Radiation Interaction—Rydberg Molecules, *J. A. C. Gallas, G. Leuchs, H. Walther, and H. Figger*

Volume 21

Subnatural Linewidths in Atomic Spectroscopy, *Dennis P. O'Brien, Pierre Meystre, and Herbert Walther*

Molecular Applications of Quantum Defect Theory, *Chris H. Greene and Ch. Jungen*

Theory of Dielectronic Recombination, *Yukap Hahn*

Recent Developments in Semiclassical Floquet Theories for Intense-Field Multiphoton Processes, *Shih-I Chu*

Scattering in Strong Magnetic Fields, *M. R. C. McDowell and M. Zarcone*

Pressure Ionization, Resonances, and the Continuity of Bound and Free States, *R. M. More*

Volume 22

Positronium—Its Formation and Interaction with Simple Systems, *J. W. Humberston*

Experimental Aspects of Positron and Positronium Physics, *T. C. Griffith*

Doubly Excited States, Including New Classification Schemes, *C. D. Lin*

Measurements of Charge Transfer and Ionization in Collisions Involving Hydrogen Atoms, *H. B. Gilbody*

Electron–Ion and Ion–Ion Collisions with Intersecting Beams, *K. Dolder and B. Pearl*

Electron Capture by Simple Ions, *Edward Pollack and Yukap Hahn*

Relativistic Heavy-Ion–Atom Collisions, *R. Anholt and Harvey Gould*

Continued-Fraction Methods in Atomic Physics, *S. Swain*

Volume 23

Vacuum Ultraviolet Laser Spectroscopy of Small Molecules, *C. R. Vidal*

Foundations of the Relativistic Theory of Atomic and Molecular Structure, *Ian P. Grant and Harry M. Quiney*

Point-Charge Models for Molecules Derived from Least-Squares Fitting of the Electric Potential, *D. E. Williams and Ji-Min Yan*

Transition Arrays in the Spectra of Ionized Atoms, *J. Bauche, C. Bauche-Arnoult, and M. Klapisch*

Photoionization and Collisional Ionization of Excited Atoms Using Synchroton and Laser Radiation, *F. J. Wuilleumier, D. L. Ederer, and J. L. Picqué*

Volume 24

The Selected Ion Flow Tube (SIDT): Studies of Ion–Neutral Reactions, *D. Smith and N. G. Adams*

Near-Threshold Electron–Molecule Scattering, *Michael A. Morrison*

Angular Correlation in Multiphoton Ionization of Atoms, *S. J. Smith and G. Leuchs*

Optical Pumping and Spin Exchange in Gas Cells, *R. J. Knize, Z. Wu, and W. Happer*

Correlations in Electron–Atom Scattering, *A. Crowe*

Volume 25

Alexander Dalgarno: Life and Personality, *David R. Bates and George A. Victor*

Alexander Dalgarno: Contributions to Atomic and Molecular Physics, *Neal Lane*

Alexander Dalgarno: Contributions to Aeronomy, *Michael B. McElroy*

Alexander Dalgarno: Contributions to Astrophysics, *David A. Williams*

Dipole Polarizability Measurements, *Thomas M. Miller and Benjamin Bederson*

Flow Tube Studies of Ion–Molecule Reactions, *Eldon Ferguson*

Differential Scattering in He—He and He$^+$—He Collisions at KeV Energies, *R. F. Stebbings*

Atomic Excitation in Dense Plasmas, *Jon C. Weisheit*

Pressure Broadening and Laser-Induced Spectral Line Shapes, *Kenneth M. Sando and Shih-I Chu*

Model-Potential Methods, *G. Laughlin and G. A. Victor*

Z-Expansion Methods, *M. Cohen*

Schwinger Variational Methods, *Deborah Kay Watson*

Fine-Structure Transitions in Proton-Ion Collisions, *R. H. G. Reid*

Electron Impact Excitation, *R. J. W. Henry and A. E. Kingston*

Recent Advances in the Numerical Calculation of Ionization Amplitudes, *Christopher Bottcher*

The Numerical Solution of the Equations of Molecular Scattering, *A. C. Allison*

High Energy Charge Transfer, *B. H. Bransden and D. P. Dewangan*

Relativistic Random-Phase Approximation,
  W. R. Johnson
Relativistic Sturmian and Finite Basis Set
  Methods in Atomic Physics, G. W. F. Drake
  and S. P. Goldman
Dissociation Dynamics of Polyatomic
  Molecules, T. Uzer
Photodissociation Processes in Diatomic
  Molecules of Astrophysical Interest,
  Kate P. Kirby and Ewine F. van Dishoeck
The Abundances and Excitation of
  Interstellar Molecules, John H. Black

Volume 26

Comparisons of Positrons and Electron
  Scattering by Gases, Walter E. Kauppila
  and Talbert S. Stein
Electron Capture at Relativistic Energies,
  B. L. Moiseiwitsch
The Low-Energy, Heavy Particle
  Collisions—A Close-Coupling Treatment,
  Mineo Kimura and Neal F. Lane
Vibronic Phenomena in Collisions of Atomic
  and Molecular Species, V. Sidis
Associative Ionization: Experiments,
  Potentials, and Dynamics, John Weiner,
  Françoise Masnou-Sweeuws, and
  Annick Giusti-Suzor
On the β Decay of ^{187}Re: An Interface of
  Atomic and Nuclear Physics and
  Cosmochronology, Zonghau Chen,
  Leonard Rosenberg, and Larry Spruch
Progress in Low Pressure Mercury-Rare Gas
  Discharge Research, J. Maya and
  R. Lagushenko

Volume 27

Negative Ions: Structure and Spectra,
  David R. Bates
Electron Polarization Phenomena in
  Electron–Atom Collisions,
  Joachim Kessler
Electron–Atom Scattering, I. E. McCarthy
  and E. Weigold

Electron–Atom Ionization, I. E. McCarthy
  and E. Weigold
Role of Autoionizing States in Multiphoton
  Ionization of Complex Atoms, V. I. Lengyel
  and M. I. Haysak
Multiphoton Ionization of Atomic Hydrogen
  Using Perturbation Theory, E. Karule

Volume 28

The Theory of Fast Ion–Atom Collisions,
  J. S. Briggs and J. H. Macek
Some Recent Developments in the
  Fundamental Theory of Light,
  Peter W. Milonni and Surendra Singh
Squeezed States of the Radiation Field,
  Khalid Zaheer and M. Suhail Zubairy
Cavity Quantum Electrodynamics,
  E. A. Hinds

Volume 29

Studies of Electron Excitation of Rare-Gas
  Atoms into and out of Metastable Levels
  Using Optical and Laser Techniques,
  Chun C. Lin and L. W. Anderson
Cross Sections for Direct Multiphoton
  Ionionization of Atoms, M. V. Ammosov,
  N. B. Delone, M. Yu. Ivanov, I. I. Bondar,
  and A. V. Masalov
Collision-Induced Coherences in Optical
  Physics, G. S. Agarwal
Muon-Catalyzed Fusion, Johann Rafelski and
  Helga E. Rafelski
Cooperative Effects in Atomic Physics,
  J. P. Connerade
Multiple Electron Excitation, Ionization, and
  Transfer in High-Velocity Atomic and
  Molecular Collisions, J. H. McGuire

Volume 30

Differential Cross Sections for Excitation of
  Helium Atoms and Helium-Like Ions by
  Electron Impact, Shinobu Nakazaki

Cross-Section Measurements for Electron Impact on Excited Atomic Species, *S. Trajmar and J. C. Nickel*

The Dissociative Ionization of Simple, Molecules by Fast Ions, *Colin J. Latimer*

Theory of Collisions between Laser Cooled Atoms, *P. S. Julienne, A. M. Smith, and K. Burnett*

Light-Induced Drift, *E. R. Eliel*

Continuum Distorted Wave Methods in Ion-Atom Collisions, *Derrick S. F. Crothers and Louis J. Dubé*

Volume 31

Energies and Asymptotic Analysis for Helium Rydberg States, *G. W. F. Drake*

Spectroscopy of Trapped Ions, *R. C. Thompson*

Phase Transitions of Stored Laser-Cooled Ions, *H. Walther*

Selection of Electronic States in Atomic Beams with Lasers, *Jacques Baudon, Rudolf Düren, and Jacques Robert*

Atomic Physics and Non-Maxwellian Plasmas, *Michèle Lamoureux*

Volume 32

Photoionization of Atomic Oxygen and Atomic Nitrogen, *K. L. Bell and A. E. Kingston*

Positronium Formation by Positron Impact on Atoms at Intermediate Energies, *B. H. Bransden and C. J. Noble*

Electron–Atom Scattering Theory and Calculations, *P. G. Burke*

Terrestrial and Extraterrestrial H_3^+, *Alexander Dalgarno*

Indirect Ionization of Positive Atomic Ions, *K. Dolder*

Quantum Defect Theory and Analysis of High-Precision Helium Term Energies, *G. W. F. Drake*

Electron–Ion and Ion–Ion Recombination Processes, *M. R. Flannery*

Studies of State-Selective Electron Capture in Atomic Hydrogen by Translational Energy Spectroscopy, *H. B. Gilbody*

Relativistic Electronic Structure of Atoms and Molecules, *I. P. Grant*

The Chemistry of Stellar Environments, *D. A. Howe, J. M. C. Rawlings, and D. A. Williams*

Positron and Positronium Scattering at Low Energies, *J. W. Humberston*

How Perfect are Complete Atomic Collision Experiments?, *H. Kleinpoppen and H. Handy*

Adiabatic Expansions and Nonadiabatic Effects, *R. McCarroll and D. S. F. Crothers*

Electron Capture to the Continuum, *B. L. Moiseiwitsch*

How Opaque Is a Star? *M. J. Seaton*

Studies of Electron Attachment at Thermal Energies Using the Flowing Afterglow–Langmuir Technique, *David Smith and Patrik Španěl*

Exact and Approximate Rate Equations in Atom-Field Interactions, *S. Swain*

Atoms in Cavities and Traps, *H. Walther*

Some Recent Advances in Electron-Impact Excitation of $n=3$ States of Atomic Hydrogen and Helium, *J. F. Williams and J. B. Wang*

Volume 33

Principles and Methods for Measurement of Electron Impact Excitation Cross Sections for Atoms and Molecules by Optical Techniques, *A. R. Filippelli, Chun C. Lin, L. W. Andersen, and J. W. McConkey*

Benchmark Measurements of Cross Sections for Electron Collisions: Analysis of Scattered Electrons, *S. Trajmar and J. W. McConkey*

Benchmark Measurements of Cross Sections for Electron Collisions: Electron Swarm Methods, *R. W. Crompton*

Some Benchmark Measurements of Cross Sections for Collisions of Simple Heavy Particles, *H. B. Gilbody*

The Role of Theory in the Evaluation and Interpretation of Cross-Section Data, *Barry I. Schneider*

Analytic Representation of Cross-Section Data, *Mitio Inokuti, Mineo Kimura, M. A. Dillon, Isao Shimamura*

Electron Collisions with N_2, O_2 and O: What We Do and Do Not Know, *Yukikazu Itikawa*

Need for Cross Sections in Fusion Plasma Research, *Hugh P. Summers*

Need for Cross Sections in Plasma Chemistry, *M. Capitelli, R. Celiberto, and M. Cacciatore*

Guide for Users of Data Resources, *Jean W. Gallagher*

Guide to Bibliographies, Books, Reviews, and Compendia of Data on Atomic Collisions, *E. W. McDaniel and E. J. Mansky*

Volume 34

Atom Interferometry, *C. S. Adams, O. Carnal, and J. Mlynek*

Optical Tests of Quantum Mechanics, *R. Y. Chiao, P. G. Kwiat, and A. M. Steinberg*

Classical and Quantum Chaos in Atomic Systems, *Dominique Delande and Andreas Buchleitner*

Measurements of Collisions between Laser-Cooled Atoms, *Thad Walker and Paul Feng*

The Measurement and Analysis of Electric Fields in Glow Discharge Plasmas, *J. E. Lawler and D. A. Doughty*

Polarization and Orientation Phenomena in Photoionization of Molecules, *N. A. Cherepkov*

Role of Two-Center Electron–Electron Interaction in Projectile Electron Excitation and Loss, *E. C. Montenegro, W. E. Meyerhof, and J. H. McGuire*

Indirect Processes in Electron Impact Ionization of Positive Ions, *D. L. Moores and K. J. Reed*

Dissociative Recombination: Crossing and Tunneling Modes, *David R. Bates*

Volume 35

Laser Manipulation of Atoms, *K. Sengstock and W. Ertmer*

Advances in Ultracold Collisions: Experiment and Theory, *J. Weiner*

Ionization Dynamics in Strong Laser Fields, *L. F. DiMauro and P. Agostini*

Infrared Spectroscopy of Size Selected Molecular Clusters, *U. Buck*

Femtosecond Spectroscopy of Molecules and Clusters, *T. Baumer and G. Gerber*

Calculation of Electron Scattering on Hydrogenic Targets, *I. Bray and A. T. Stelbovics*

Relativistic Calculations of Transition Amplitudes in the Helium Isoelectronic Sequence, *W. R. Johnson, D. R. Plante, and J. Sapirstein*

Rotational Energy Transfer in Small Polyatomic Molecules, *H. O. Everitt and F. C. De Lucia*

Volume 36

Complete Experiments in Electron–Atom Collisions, *Nils Overgaard Andersen, and Klaus Bartschat*

Stimulated Rayleigh Resonances and Recoil-Induced Effects, *J.-Y. Courtois and G. Grynberg*

Precision Laser Spectroscopy Using Acousto-Optic Modulators, *W. A. van Wijngaarden*

Highly Parallel Computational Techniques for Electron–Molecule Collisions, *Carl Winstead and Vincent McKoy*

Quantum Field Theory of Atoms and Photons, *Maciej Lewenstein and Li You*

Volume 37

Evanescent Light-Wave Atom Mirrors, Resonators, Waveguides, and Traps, *Jonathan P. Dowling and Julio Gea-Banacloche*

Optical Lattices, *P. S. Jessen and I. H. Deutsch*

Channeling Heavy Ions through Crystalline Lattices, *Herbert F. Krause and Sheldon Datz*

Evaporative Cooling of Trapped Atoms, *Wolfgang Ketterle and N. J. van Druten*

Nonclassical States of Motion in Ion Traps, *J. I. Cirac, A. S. Parkins, R. Blatt, and P. Zoller*

The Physics of Highly-Charged Heavy Ions Revealed by Storage/Cooler Rings, *P. H. Mokler and Th. Stöhlker*

Volume 38

Electronic Wavepackets, *Robert R. Jones and L. D. Noordam*

Chiral Effects in Electron Scattering by Molecules, *K. Blum and D. G. Thompson*

Optical and Magneto-Optical Spectroscopy of Point Defects in Condensed Helium, *Serguei I. Kanorsky and Antoine Weis*

Rydberg Ionization: From Field to Photon, *G. M. Lankhuijzen and L. D. Noordam*

Studies of Negative Ions in Storage Rings, *L. H. Andersen, T. Andersen, and P. Hvelplund*

Single-Molecule Spectroscopy and Quantum Optics in Solids, *W. E. Moerner, R. M. Dickson, and D. J. Norris*

Volume 39

Author and Subject Cumulative Index Volumes 1–38

Author Index

Subject Index

Appendix: Tables of Contents of Volumes 1–38 and Supplements

Volume 40

Electric Dipole Moments of Leptons, *Eugene D. Commins*

High-Precision Calculations for the Ground and Excited States of the Lithium Atom, *Frederick W. King*

Storage Ring Laser Spectroscopy, *Thomas U. Kühl*

Laser Cooling of Solids, *Carl E. Mungan and Timothy R. Gosnell*

Optical Pattern Formation, *L. A. Lugiato, M. Brambilla, and A. Gatti*

Volume 41

Two-Photon Entanglement and Quantum Reality, *Yanhua Shih*

Quantum Chaos with Cold Atoms, *Mark G. Raizen*

Study of the Spatial and Temporal Coherence of High-Order Harmonics, *Pascal Salières, Ann L'Huiller Philippe Antoine, and Maciej Lewenstein*

Atom Optics in Quantized Light Fields, *Matthias Freyburger, Alois M. Herkommer, Daniel S. Krähmer, Erwin Mayr, and Wolfgang P. Schleich*

Atom Waveguides, *Victor I. Balykin*

Atomic Matter Wave Amplification by Optical Pumping, *Ulf Janicke and Martin Wilkens*

Volume 42

Fundamental Tests of Quantum Mechanics, *Edward S. Fry and Thomas Walther*

Wave-Particle Duality in an Atom Interferometer, *Stephan Dürr and Gerhard Rempe*

Atom Holography, *Fujio Shimizu*

Optical Dipole Traps for Neutral Atoms, *Rudolf Grimm, Matthias Weidemüller, and Yurii B. Ovchinnikov*

Formation of Cold ($T \leqslant 1K$) Molecules, *J. T. Bahns, P. L. Gould, and W. C. Stwalley*

High-Intensity Laser-Atom Physics, *C. J. Joachain, M. Dorr, and N. J. Kylstra*

Coherent Control of Atomic, Molecular, and Electronic Processes, *Moshe Shapiro and Paul Brumer*

Resonant Nonlinear Optics in Phase Coherent Media, *M. D. Lukin, P. Hemmer, and M. O. Scully*

The Characterization of Liquid and Solid Surfaces with Metastable Helium Atoms, *H. Morgner*

Quantum Communication with Entangled Photons, *Harald Weinfurter*

Volume 43

Plasma Processing of Materials and Atomic, Molecular, and Optical Physics: An Introduction, *Hiroshi Tanaka and Mitio Inokuti*

The Boltzmann Equation and Transport Coefficients of Electrons in Weakly Ionized Plasmas, *R. Winkler*

Electron Collision Data for Plasma Chemistry Modeling, *W. L. Morgan*

Electron–Molecule Collisions in Low-Temperature Plasmas. The role of Theory, *Carl Winstead and Vincent McKoy*

Electron Impact Ionization of Organic Silicon Compounds, *Ralf Basner, Kurt Becker, Hans Deutsch, and Martin Schmidt*

Kinetic Energy Dependence of Ion–Molecule Reactions Related to Plasma Chemistry, *P. B. Armentrout*

Physicochemical Aspects of Atomic and Molecular Processes in Reactive Plasmas, *Yoshihiko Hatano*

Ion–Molecule Reactions, *Werner Lindinger, Armin Hansel, and Zdenek Herman*

Uses of High-Sensitivity White-Light Absorption Spectroscopy in Chemical Vapor Deposition and Plasma Processing, *L. W. Anderson, A. N. Goyette, and J. E. Lawler*

Fundamental Processes of Plasma–Surface Interactions, *Rainer Hippler*

Recent Applications of Gaseous Discharges: Dusty Plasmas and Upward-Directed Lightning, *Ara Chutjian*

Opportunities and Challenges for Atomic, Molecular, and Optical Physics in Plasma Chemistry, *Kurl Becker, Hans Deutsch, and Mitio Inokuti*

Volume 44

Mechanisms of Electron Transport in Electrical Discharges and Electron Collision Cross Sections, *Hiroshi Tanaka and Osamu Sueoka*

Theoretical Consideration of Plasma-Processing Processes, *Mineo Kimura*

Electron Collision Data for Plasma-Processing Gases, *Loucas G. Christophorou and James K. Olthoff*

Radical Measurements in Plasma Processing, *Toshio Goto*

Radio-Frequency Plasma Modeling for Low-Temperature Processing, *Toshiaki Makabe*

Electron Interactions with Excited Atoms and Molecules, *Loucas G. Christophorou and James K. Olthoff*

Volume 45

Comparing the Antiproton and Proton, and Opening the Way to Cold Antihydrogen, *G. Gabrielse*

Medical Imaging with Laser-Polarized Noble Gases, *Timothy Chupp and Scott Swanson*

Polarization and Coherence Analysis of the Optical Two-Photon Radiation from the Metastable $2^2S_{1/2}$ State of Atomic Hydrogen, *Alan J. Duncan, Hans Kleinpoppen, and Marlan O. Scully*

Laser Spectroscopy of Small Molecules, *W. Demtröder, M. Keil, and H. Wenz*

Coulomb Explosion Imaging of Molecules, *Z. Vager*

Volume 46

Femtosecond Quantum Control, *T. Brixner, N. H. Damrauer and G. Gerber*

Coherent Manipulation of Atoms and Molecules by Sequential Laser Pulses, *N. V. Vitanov, M. Fleischhauer, B. W. Shore, and K. Bergmann*

Slow, Ultraslow, Stored, and Frozen Light, *Andrey B. Matsko, Olga Kocharovskaya, Yuri Rostovtsev, George R. Welch, Alexander S. Zibrov, and Marlan O. Scully*

Longitudinal Interferometry with Atomic Beams, *S. Gupta, D. A. Kokorowski, R. A. Rubenstein, and W. W. Smith*

ISBN 0-12-003846-3